Operator Theory: Advances and Applications
Vol. 155

Advances in Pseudo-Differential Operators

Ryuichi Ashino
Paolo Boggiatto
M. W. Wong
Editors

Springer Basel AG

Editors:

Ryuichi Ashino
Division of Mathematical Sciences
Osaka Kyoiku University
Kashiwara, Osaka 582-8582
Japan
e-mail: ashino@cc.osaka-kyoiku.ac.jp

M.W. Wong
Department of Mathematics and Statistics
York University
4700 Keele Street
Toronto, Ontario M3J 1P3
Canada
e-mail: mwwong@pascal.math.yorku.ca

Paolo Boggiatto
Dipartimento di Matematica
Università di Torino
Via Carlo Alberto, 10
10123 Torino
Italy
e-mail: boggiatto@dm.unito.it

2000 Mathematics Subject Classification: Primary 35J30, 35J55, 35Q40, 35S05, 35S50, 42B35, 42C40, 44A35, 46E35, 47B10, 47G30, 65T60; Secondary 35A27, 35B65, 35Q51, 46B50, 46F12, 47B37, 58J32, 93B30

A CIP catalogue record for this book is available from the Library of Congress, Washington D.C., USA

Bibliographic information published by Die Deutsche Bibliothek
Die Deutsche Bibliothek lists this publication in the Deutsche Nationalbibliografie; detailed bibliographic data is available in the Internet at <http://dnb.ddb.de>.

ISBN 978-3-7643-7140-1 ISBN 978-3-0348-7840-1 (eBook)
DOI 10.1007/978-3-0348-7840-1

Originally published by Birkhäuser Verlag, Basel - Boston - Berlin in 2004

Printed on acid-free paper produced from chlorine-free pulp. TCF ∞
Cover design: Heinz Hiltbrunner, Basel

ISBN 978-3-7643-7140-1

Contents

Preface

The Fourth Congress of the International Society for Analysis, its Applications and Computation (ISAAC) was held at York University from August 11, 2003 to August 16, 2003. It was supported by the Academic Initiative Fund of the Faculty of Arts, NSERC grants from some members of the Department of Mathematics and Statistics and the Office of the Vice-President Academic of York University. In spite of two SARS outbreaks in Toronto in 2003, the ISAAC Congress was held as scheduled and was well attended by mathematicians from all over the world. There were nine plenary lectures and seventeen special sessions representing most major themes in analysis. Among these were two plenary lectures and a special session on pseudo-differential operators organized by Ryuichi Ashino of Osaka Kyoiku University, Paolo Boggiatto of Universitè di Torino and M. W. Wong of York University.

In the summer of 2003, M. W. Wong had the idea of putting together the lectures on pseudo-differential operators in a volume to be published in a series that advocates operator theory and its applications. In early August of 2003, when Israel Gohberg of Tel Aviv University was consulted about the possibility of publishing a volume entitled "Advances in Pseudo-Differential Operators" in his series "Operator Theory: Advances and Applications", he replied immediately endorsing the proposal enthusiastically.

In this volume there are thirteen articles representing recent advances in pseudo-differential operators. Chapters 1 and 2 originate, respectively, from the plenary lectures of Luigi Rodino and B.-W. Schulze. In Chapters 3–5, pseudo-differential operators in the context of partial differential equations are studied. It is well-known that pseudo-differential operators arise in quantum mechanics, and as such, they are known as Weyl transforms. Weyl transforms are represented by Chapters 6 and 7. In recent years, pseudo-differential operators are used in signal analysis as filters or localization operators. Localization operators, which are also known as Wick operators in quantization, form the subject matter of Chapters 1, 6, 8 and 9. Modulation spaces, which form a critical mass in Chapters 8–10, are defined in terms of the short-time Fourier transform and are used in the analysis of pseudo-differential operators and their variants. Also important are the connections of pseudo-differential operators with wavelets, and these are dealt with in the last three chapters.

Operator Theory:
Advances and Applications, Vol. 155, 1–17

Microlocal Analysis and Applications

E. Cordero, F. Nicola and L. Rodino

Abstract. We first give a short survey on the methods of Microlocal Analysis. In particular we recall some basic facts concerning the theory of pseudo-differential operators. We then present two applications. We first discuss lower bounds for operators with multiple characteristics. Then we give a new formula for the composition of Wick operators.

Mathematics Subject Classification (2000). Primary 35S05; Secondary 47G30.

Keywords. Pseudo-differential operators, Wick operators, Weyl calculus, lower bounds, Wigner distributions, short-time Fourier transforms.

1. Introduction

The aim of this paper is to introduce non-experts to methods and results of Microlocal Analysis. Rather than trying to give a complete historical survey, we shall limit ourselves in Section 2 to a presentation of some general ideas and a few classical references. Sections 3 and 4 are then devoted to lower bounds for pseudo-differential operators and compositions of Wick operators, respectively. These topics are representative of the new applications of the microlocal techniques. The problem on lower bounds is quite a classical problem that has recently received new momentum. In Section 3, after recalling a recent result of Nicola and Rodino [40] and related results on lower bounds for pseudo-differential operators with multiple characteristics, we give a proof of an estimate for symbols with limited regularity using arguments in [40] and Tataru [49]. In Section 4, we deal with Wick pseudo-differential operators (time-frequency localization operators). We prove here a new formula for composition, which can be regarded as a symmetric version of the formula in Cordero and Rodino [14]. We observe that there exists a deep connection between lower bounds and Wick operators. In fact, non-negative symbols produce non-negative Wick operators, and this fact can be used to recapture lower bounds. See, for example, Boggiatto, Buzano and Rodino [5] and the above mentioned paper [49]. More precisely, using the fact that the difference between the Weyl quantization and the Wick quantization is an operator of order 1 lower, we get lower bounds whenever this error is negligible.

Let us emphasize, in conclusion, that Microlocal Analysis has now grown up into an independent part of Mathematical Analysis and there are applications all around Mathematics and Physics. By limiting to the above mentioned topics, we cannot claim to cover all new trends. But, following our personal taste, we hope to give a flavour of current activities in the field.

2. Pseudo-Differential Operators

As our starting point of Microlocal Analysis we may consider Fourier's classical treatise (1822). We shall refer in the following to the Fourier transform, defined as

$$(\mathcal{F}u)(\xi) = \hat{u}(\xi) = \int e^{-ix\xi} u(x)\, dx. \tag{1}$$

We are using here standard notation for $x \in \mathbb{R}^n$, $\xi \in \mathbb{R}^n$, namely $x\xi = \sum_{j=1}^n x_j \xi_j$, $dx = dx_1 \dots dx_n$, etc. The Fourier transform $\hat{u}(\xi)$ is well defined in $L^\infty(\mathbb{R}^n)$ if $u \in L^1(\mathbb{R}^n)$, the integral being taken in the Lebesgue sense. According to the Theory of Distributions of Schwartz [46], we may extend $\mathcal{F}$ to the distribution spaces $\mathcal{S}'(\mathbb{R}^n)$ and $\mathcal{E}'(X)$, $X \subset \mathbb{R}^n$.

Classical applications of the Fourier techniques concern the heat and wave equations and other linear equations with constant coefficients. Starting from 1950-60, in the framework of distributions and Sobolev spaces, some technical tools were introduced, based on the Fourier transform and mainly addressed to the study of variable-coefficient, linear and nonlinear partial differential equations. Somewhat improperly, we may say that such techniques, in their whole, constitute the modern Microlocal Analysis. Let us list: pseudo-differential operators (1965) as proposed by Hörmander [26], Kohn and Nirenberg [33], wave front sets and Fourier integral operators (1971), see Hörmander [28], paradifferential operators (1981), see Bony [8], etc.

In view of the applications of the next two sections, let us in the following address attention to pseudo-differential operators. They are, in a broad sense, linear operators P defined by expressions of the type

$$Pu(x) = (2\pi)^{-n} \int e^{ix\xi} p(x,\xi) \hat{u}(\xi)\, d\xi. \tag{2}$$

The function $p(x,\xi)$, with $x \in \mathbb{R}^n$ (or $x \in X \subset \mathbb{R}^n$) and $\xi \in \mathbb{R}^n$, satisfying suitable estimates, is called the symbol of P. The operator P is also denoted by $p(x,D)$. A linear partial differential operator

$$P = \sum_{|\alpha| \le m} c_\alpha(x) D^\alpha \tag{3}$$

can be regarded as a pseudo-differential operator with symbol

$$p(x,\xi) = \sum_{|\alpha|\le m} c_\alpha(x)\xi^\alpha. \tag{4}$$

In (3), (4) we are using the notation $D^\alpha = D_{x_1}^{\alpha_1}\dots D_{x_n}^{\alpha_n}$, with $\alpha = (\alpha_1,\dots,\alpha_n)$, $D_{x_j} = -i\partial_{x_j}$, $|\alpha| = \alpha_1+\dots+\alpha_n$, $\xi^\alpha = \xi_1^{\alpha_1}\dots\xi_n^{\alpha_n}$. In particular, if $p(x,\xi)=1$ then $P = p(x,D)$ is the identity operator.

In the presentation of [26],[33] the symbol $p(x,\xi)$ was assumed to be a smooth function, admitting for $\xi\to\infty$ the asymptotic expansion

$$p(x,\xi) \sim \sum_{j=0}^{\infty} p_{m-j}(x,\xi) \tag{5}$$

where $p_{m-j}(x,\xi)$ is positively homogeneous, with respect to ξ, of order $m-j$. The symbols satisfying (5) are called "classical" symbols, and the relative pseudodifferential operators "classical pseudo-differential operators". Note that in (5) we allow m to be any real number. Investigation was then extended by Hörmander [27] to symbols $p(x,\xi)$ of type ρ,δ, $0\le\delta,\rho\le 1$, i.e.

$$|D_x^\alpha D_\xi^\beta p(x,\xi)| \le C_{\alpha\beta}(1+|\xi|)^{m-\rho|\beta|+\delta|\alpha|}. \tag{6}$$

Take note that the symbols in (5) satisfy (6) with $\delta=0$, $\rho=1$; we use the notation S^m for the class of the symbols in this case.

As proved later (1972) by Calderón and Vaillancourt [11], if $0\le\delta\le\rho\le 1$, $\delta<1$, then for every $s\in\mathbb{R}$

$$p(x,D): H^s \to H^{s-m} \text{ continuously}, \tag{7}$$

where H^s are the standard Sobolev spaces, whose norm is defined for $u\in C_0^\infty(\mathbb{R}^n)$ by

$$\|u\|_s^2 = \int (1+|\xi|^2)^s |\hat{u}(\xi)|^2\, d\xi. \tag{8}$$

Subsequently, several other generalizations and variants of symbol classes appeared, see Beals and Fefferman [3], Beals [2], Hörmander [30]. In the next Section 3 we shall refer to $\mathrm{OPS}^m(X)$ as the space of pseudo-differential operators with symbols satisfying (5) in any $K\subset\subset X$, where X is an open subset of $\mathbb{R}^n$. In Section 4 we shall instead use the classes G_ρ^m of Shubin [47], defined globally in $\mathbb{R}^n\times\mathbb{R}^n$ by imposing for $0<\rho\le 1$:

$$|D_x^\alpha D_\xi^\beta p(x,\xi)| \le C_{\alpha\beta}(1+|x|+|\xi|)^{m-\rho(|\alpha|+|\beta|)}. \tag{9}$$

A basic step in the pseudo-differential calculus is the study of the composition of two operators of the form (2), say P and Q with symbols p and q, respectively. When p and q do not depend on the variables x, then the symbol of the composition is exactly the product $p(\xi)q(\xi)$. This is not true in general when p and q depend on x. Nevertheless, under suitable assumptions on $p(x,\xi)$ and $q(x,\xi)$, in particular if (5) or (9) is satisfied, the symbol of the composition PQ can be written as

$$p(x,\xi)q(x,\xi) + r(x,\xi), \tag{10}$$

where the remainder $r(x,\xi)$ has "lower order" with respect to the product of the symbols. More precisely, we have for $r(x,\xi)$ the asymptotic expansion

$$r(x,\xi) \sim \sum_{\alpha\neq 0} (\alpha!)^{-1}\partial_\xi^\alpha p(x,\xi) D_x^\alpha q(x,\xi). \tag{11}$$

The role of (10), (11) is fundamental in the theory of pseudo-differential operators, since they permit to turn algebraic properties of the symbols into differential properties of the equations. A basic application is the construction of parametrices for elliptic operators, that is, operators with principal symbols $p_m(x,\xi) \neq 0$ for $\xi \neq 0$. In short: if $P = p(x,D)$ is elliptic then we can define $q_{-m}(x,\xi) = 1/p_m(x,\xi)$. In view of (10) the symbol of $p(x,D)q_{-m}(x,D)$ is given by

$$p(x,\xi)q_{-m}(x,\xi) + r(x,\xi) = 1 + \text{lower order terms}.$$

At this moment, taking advantage of (11), we may easily construct by recurrence q_{-m-1}, $q_{-m-2}, \dots$ and consider $q \sim \sum_{j=0}^{\infty} q_{-m-j}$, such that $p(x,D)q(x,D) = q(x,D)p(x,D) =$ identity, modulo operators with rapidly decreasing symbols. The pseudo-differential operator $Q = q(x,D)$ is said to be a parametrix of P. As a corollary, we recapture results on the interior regularity of the solutions of the elliptic partial differential equations, in the framework of distributions and Sobolev spaces.

We can now give motivation to the reader for the adjective "microlocal". Roughly, local analysis means analysis in a small neighborhood V of a fixed point x_0. Given a function, or a distribution u, our concern is then limited to the restriction of u to V; consequently, given a partial differential or pseudo-differential operator acting on u, with symbol $p(x,\xi)$, we are allowed to cut off $p(x,\xi)$ with respect to x near x_0. Microlocal Analysis means that we apply the same arguments to $\hat{u}(\xi)$ and $p(x,\xi)$, in a neighborhood Γ of a point ξ_0. Since we want to perform localization in x_0 and ξ_0 at the same time, we have to take into account the Heisenberg principle: essentially, V and Γ cannot be simultaneously small. The standard choice, that works in the distributional and Sobolev settings, is to take then as Γ a small conic neighborhood of ξ_0. This agrees with the homogeneity with respect to ξ required in the definition of the classical pseudo-differential operators.

The microlocal point of view turns out to be very useful in the study of non-elliptic equations. Namely, given $P \in \mathrm{OPS}^m(X)$, we define the characteristic manifold

$$\Sigma = \{(x,\xi) \in T^*X \setminus 0 : \ p_m(x,\xi) = 0\}. \tag{12}$$

We say that P is micro-elliptic at a point (x_0,ξ_0), $\xi_0 \neq 0$, if $(x_0,\xi_0) \notin \Sigma$. When studying the singularities of the solutions of the equation $Pu = f \in C^\infty(X)$, the micro-elliptic points have no influence, since cutting-off in a neighborhood $V \times \Gamma$ we may apply the preceding construction of parametrices and conclude micro-regularity. So what actually matters is the microlocal study of P in a neighborhood of Σ. To interpret these ideas, we need of course a precise language, which is given by wave-front sets, Fourier integral operators, etc. As references in this connection, let us limit ourselves to the monumental work of Hörmander [31], although there exist at present other excellent books addressed to non-experts, giving a quick introduction to the calculus.

To prepare for the applications in the next sections, we finally discuss the problem of the quantization, namely the way to associate an operator with a symbol $p(x,\xi)$. First observe that we may re-write (2) as

$$Pu(x) = (2\pi)^{-n} \int e^{i(x-y)\xi} p(x,\xi)u(y)\,dy\,d\xi. \tag{13}$$

It is now natural to consider also the quantization

$$\tilde{P}u(x) = (2\pi)^{-n} \int e^{i(x-y)\xi} p(y,\xi)u(y)\,dy\,d\xi. \tag{14}$$

If p is the polynomial in (4), then

$$\tilde{P}u(x) = \sum_{|\alpha|\le m} D^\alpha(c_\alpha(x)u),$$

which is sometimes convenient to use as representation of a differential operator. A compromise between (13) and (14) is given by the Weyl quantization

$$\mathrm{Op}^w(p)u(x) = (2\pi)^{-n} \int e^{i(x-y)\xi} p\left(\frac{x+y}{2},\xi\right) u(y)\,dy\,d\xi. \tag{15}$$

By this correspondence, the L^2-adjoint of the operator $\mathrm{Op}^w(p)$ is given by $\mathrm{Op}^w(\overline{p})$, from which it is clear that when p is a real function one obtains a formally self-adjoint operator.

An operator P of the form (13) or (14) can be written as a Weyl operator by using a new symbol, and there are precise formulas connecting such a Weyl symbol to the former $p(x,\xi)$. In particular, for a classical pseudo-differential operator, the Weyl symbol is again of the form (5) with the same principal part. Weyl quantization will be used in Section 3. We refer to Section 4 for details on the so-called Wick quantization. Note that, when dealing with Wick operators, it is also standard to use a different normalization of the Fourier transform, namely $\int e^{-2\pi i\omega t} f(t)\,dt$, where we now write t and ω for variables and dual variables.

3. Operators with Multiple Characteristics and Lower Bounds

In this section, before treating symbols with limited smoothness, we recall from [43], [40], some recent results about lower bounds for pseudo-differential operators with multiple characteristics.

Precisely, let $X \subset \mathbb{R}^n$ be an open subset of $\mathbb{R}^n$, and $P = P^* \in \mathrm{OPS}^m(X)$ be a pseudo-differential operator with classical Weyl symbol $p \sim \sum_{j\ge 0} p_{m-j}$, where p_{m-j} are homogeneous functions of degree $m-j$ with respect to the variables ξ.

We are interested in estimates of the type

$$(H_\sigma) \qquad (Pu,u)_{L^2(\mathbb{R}^n)} \ge -C_K\|u\|^2_{\frac{m-\sigma}{2}}, \quad \forall u \in C_0^\infty(K),$$

for arbitrary $K \subset\subset X$, where $\|u\|_s$ is the Sobolev norm (8).

We suppose that Σ is a smooth manifold and that P belongs to Sjöstrand-Boutet de Monvel's classes $\mathrm{OPN}^{m,k}(X,\Sigma)$, i.e., for $j<k/2$, $p_{m-j}(x,\xi)$ vanishes to the order $k-2j$ on Σ, k being an even integer. Moreover, the principal symbol p_m of P is assumed non negative and transversally elliptic with respect to Σ, namely it vanishes exactly to order k. Consider now $N\Sigma$, the normal bundle to Σ, and $\rho\in\Sigma$; given any $v\in N_\rho\Sigma$, we take any vector field Ξ on T^*X with $\Xi(\rho)=v$. Then we define the smooth map $p^{(k)}$ by

$$p^{(k)}(\rho,v)=\sum_{0\leq j\leq k/2}\frac{1}{(k-2j)!}\underbrace{\Xi\cdots\Xi}_{k-2j\ \text{times}}\ p_{m-j}(\rho):N\Sigma\to\mathbb{R}.$$

It is well-known (see for example Boutet de Monvel-Grigis-Helffer [10], Mascarello-Rodino [36]) that many properties of the operator P, as hypoellipticity, spectral lower bound, etc., are strictly related to this map.

Results on lower bounds in general are obtained under the requirement that the restriction of the symplectic form to Σ have constant rank. Also the results proved in [43], as well as our subsequent results, depend on the symplectic nature of Σ. More precisely, in [43] the manifold Σ is assumed (regular) involutive as a submanifold of $T^*X\backslash 0$ (i.e., for any $\rho\in\Sigma$, $T_\rho\Sigma^\omega\subset T_\rho\Sigma$, $T_\rho\Sigma^\omega$ being the symplectic orthogonal of $T_\rho\Sigma$, and the canonical 1-form $\sum_{j=1}^n\xi_j dx_j$ does not vanish on $T_\rho\Sigma$) or symplectic (i.e., $T_\rho\Sigma\cap T_\rho\Sigma^\omega=\{0\}$ for every $\rho\in\Sigma$). The case when Σ is non-involutive and non-symplectic will be considered in [39].

Theorem 3.1. [43] *Let $P=P^*\in\mathrm{OPN}^{m,k}(X,\Sigma)$ be transversally elliptic and suppose Σ is involutive. Moreover, suppose that*

$$p^{(k)}(\rho,v)\geq 0,\ \ \forall\rho\in\Sigma,\ \ \forall v\in N_\rho\Sigma, \tag{16}$$

$$p^{(k)}(\rho,v)=0\Longrightarrow v=0, \tag{17}$$

$$p^{(k)}(\rho,0)=0\Longrightarrow\left(d_v^2p^{(k)}\right)(\rho,0)\ \textit{is invertible}. \tag{18}$$

Then (H_σ) holds with $\sigma=k/2+1$.

For operators in the class $\mathrm{OPN}^{m,k}(X,\Sigma)$ with Σ involutive, condition (16) is also necessary to have a gain greater than $k/2$ derivatives (see [42]), whereas once $k>2$, (17) and (18) are not necessary in general. On the other hand one sees easily that, under (16) alone, when $k>2$ estimate (H_σ) with $\sigma=k/2+1$ is, in general, false.

Suppose now that Σ is a symplectic sub-manifold of $T^*X\setminus 0$. Let $\rho\in\Sigma$ and let $\zeta:T^*\mathbb{R}^\nu\to T_\rho\Sigma^\omega$ ($2\nu=\operatorname{codim}\Sigma$) be any linear symplectomorphism. Setting $p_\zeta(y,\eta)=p^{(k)}(\rho,\zeta(y,\eta))$ for $(y,\eta)\in T^*\mathbb{R}^\nu$, we then consider the Weyl quantization $P_{\rho,\zeta}=\mathrm{Op}^w(p_\zeta)(y,D_y):\mathcal{S}(\mathbb{R}^\nu)\to\mathcal{S}(\mathbb{R}^\nu)$. As shown in [42], the spectrum of $P_{\rho,\zeta}$, as an unbounded operator on $L^2(\mathbb{R}^\nu)$, is independent of the parameterization ζ, and it turns out to be discrete and bounded from below. Thus the lowest eigenvalue $\lambda(\rho):=\min\mathrm{Spec}(\mathrm{Op}^w(p_\zeta))$ is a continuous function on Σ, independent of ζ. Similarly, the dimension of the corresponding (finite dimensional) eigenspace $V_{\rho,\zeta}\subset\mathcal{S}(\mathbb{R}^\nu)$ is invariantly defined, as well.

Theorem 3.2. [43] *Let $P = P^* \in \mathrm{OPN}^{m,k}(X,\Sigma)$, P transversally elliptic, and suppose Σ symplectic. Moreover, assume that:*

(i) *for any $\zeta : T^*\mathbb{R}^\nu \to T_\rho\Sigma^\omega$ as above, one has*

$$(\mathrm{Op}^w(p_\zeta)(y,D_y)f,f) \geq 0, \quad \forall f \in \mathcal{S}(\mathbb{R}^\nu),\ \forall \rho \in \Sigma;$$

(ii) *if $\lambda(\rho_0) = 0$, there exists a conic neighborhood $\Gamma \subset \Sigma$ of ρ_0 such that*

$$\dim V_{\rho,\zeta} = \text{const},\ \forall \rho \in \Gamma.$$

Then estimate (H_σ) holds with $\sigma = k/2 + 1/2$.
If we suppose, in addition, that

(iii) *the eigenspace $V_{\rho,\zeta}$ consists of functions which are either all even for every $\rho \in \Gamma$ or all odd for every $\rho \in \Gamma$,*

then estimate (H_σ) holds with $\sigma = k/2 + 1$.

By standard localization techniques it is easy to see that hypothesis (i) is necessary to reach any gain $\sigma > k/2$. In the subsequent paper [44], the authors relaxed hypothesis (ii), whereas in [38] it is shown that hypothesis (iii) cannot be omitted in order to obtain (1.2). In fact, when $k > 2$ there are operators which satisfy (i) and (ii) but do not verify any of the estimates (H_σ) with $\sigma > k/2+1/2$.

We observe that in the double characteristic case treated by Hörmander [28], the eigenspace $V_{\rho,\zeta}$ has dimension 1, hence conditions (ii) and (iii) are automatically satisfied, (see [42]).

Let us now come to the work [40]. Suppose Σ is (regular) involutive, and consider the smooth maps

$$I_j(\rho,v) = \frac{1}{j!}\underbrace{\Xi\cdots\Xi}_{j \text{ times}} p_{m-1}(\rho) : N\Sigma \to \mathbb{R}, \tag{19}$$

with $\Xi(\rho) = v$. Under such assumption, these maps have an invariant meaning for $0 \leq j \leq k-2$, see for example [35], [36].

The main result of [40], which can be regarded as a possible continuation and variation of [43], deals with operators that are not necessarily in the classes $\mathrm{OPN}^{m,k}(X,\Sigma)$ above. More precisely, we have the following lower bound.

Theorem 3.3. *Let $P = P^* \in \mathrm{OPS}^m(X)$ be a classical pseudo-differential operator with p_m vanishing exactly to even order $k \geq 2$ on Σ smooth regular involutive manifold. Let $J \in \{0,1,\ldots,k/2-1\}$. Let us suppose*

$$p_m \geq 0 \text{ on } T^*X \setminus 0, \tag{20}$$

and that for every $\overline{\rho} \in \Sigma$ there exists a neighborhood $\Gamma \subset N\Sigma$ of $(\overline{\rho},0)$ such that

$$\sum_{j=0}^{J} I_j(\rho,v) \geq 0 \quad \forall(\rho,v) \in \Gamma. \tag{21}$$

Then (H_σ) holds with $\sigma = k/(k-J-1)$.

Notice that for $J = 0$ conditions (20) and (21) reduce to

$$\begin{cases} p_m(x,\xi) \geq 0 \quad \forall (x,\xi) \in T^*X \setminus 0, \\ p_{m-1}(x,\xi) \geq 0 \quad \forall (x,\xi) \in \Sigma, \end{cases} \tag{22}$$

which, when Σ is involutive, are also necessary for any lower bound with a gain σ greater than one derivative. This follows at once from Melin's results (see [37, 42]). In general we have $1 < \sigma \leq 2$ and the maximum gain of derivatives $\sigma = 2$ is obtained for $J = k/2 - 1$. Moreover, σ decreases and tends to 1 as $k \to +\infty$, for fixed J.

Finally we observe that (21) is certainly satisfied if p_{m-1} vanishes to the order $J+1$ on Σ, in fact in this case $I_j = 0$, for $j = 0, \dots, J$.

The Case of Symbols with Limited Smoothness

Let us now pass to studying lower bounds for pseudo-differential operators with non-regular symbols. Several papers are devoted to this argument. We recall, in particular, the paper of Bony [8] for a proof of the Sharp-Gårding inequality for paradifferential operators (see also Hérau [25] for an improved version), and the works by Hérau [23, 24, 25], where Melin-type inequalities are proved, with applications to the interior regularity problem. Here we are mostly interested in the recent contribution of Tataru [49], where a version of the Fefferman-Phong inequality for operators with low-regular symbols is presented.

Precisely, as in [49] denote by $C^{k,1}S^m$ the space of symbols $p(x,\xi)$ satisfying

$$|\partial_x^\alpha \partial_\xi^\beta p(x,\xi)| \leq C_{\alpha\beta}(1+|\xi|)^{m-|\beta|}, \quad 0 \leq |\alpha| \leq k+1, \ |\beta| \geq 0. \tag{23}$$

If X is an open subset of $\mathbb{R}^n$, we indicate by $C^{k,1}S^m(X)$ the space of symbols satisfying estimates (23) locally for x in compact subsets of X. Moreover we denote by $\mathrm{Op}^w(p)$ the operator with Weyl symbol $p(x,\xi)$.

Theorem 3.4. [49] *Let $p(x,\xi) \in C^{3,1}S^2$ be a real scalar nonnegative symbol, and $P = \mathrm{Op}^w(p)$. Then there exits a constant $C > 0$ such that*

$$(Pu,u) \geq -C\|u\|_0^2 \quad \forall u \in \mathcal{S}(\mathbb{R}^n).$$

The following result is an extension of a particular case of Theorem 3.3 to low-regular symbols. In the proof we do not need to suppose that Σ is regular involutive.

Theorem 3.5. *Let $P = P^*$ be a pseudo-differential operator with a classical Weyl symbol $p \in C^{k-1,1}S^2(X)$, for an even integer $k > 2$. Let p_2 vanish exactly to order k on the C^k-manifold Σ. Then conditions* (22) *(with $m = 2$) are equivalent to estimate (H_σ) with $\sigma = k/(k-1)$ (and $m = 2$).*

We recall from [40] the following elementary result that will be used in the proof of Theorem 3.5. For convenience of the reader we recall its easy proof.

Lemma 3.6. *Let $p(\zeta) = \sum_{|\alpha|=k} c_\alpha \zeta^\alpha$ be a homogeneous real polynomial in $\mathbb{R}^l$, with $p(\zeta) > c|\zeta|^k$ for $\zeta \neq 0$, $c > 0$, and $q(\zeta)$ a homogeneous real polynomial of degree $s \in \{1, \dots, k-1\}$. Then there exists a constant $C > 0$ such that for every $t > 0$ we have*

$$\min_{\zeta \in \mathbb{R}^l} \{p(\zeta) + tq(\zeta)\} \geq -Ct^{\frac{k}{k-s}}.$$

If the coefficients of p and q vary in a bounded subset of $\mathbb{R}$ but the constant c is uniform, so is C.

Proof of Lemma 3.6. We have

$$\begin{aligned} p(\zeta) + tq(\zeta) &\geq c|\zeta|^k - C_1 t|\zeta|^s \\ &= |\zeta|^s(c|\zeta|^{k-s} - C_1 t); \end{aligned}$$

so $p(\zeta) + tq(\zeta) \geq 0$ when $|\zeta| \geq (C_1 t/c)^{\frac{1}{k-s}}$. Hence, since $p(0) = q(0) = 0$,

$$\begin{aligned} \min_{\zeta \in \mathbb{R}^l} \{p(\zeta) + tq(\zeta)\} &= \min_{|\zeta| \leq (C_1 t/c)^{\frac{1}{k-s}}} \{p(\zeta) + tq(\zeta)\} \\ &\geq -C_1 t (C_1 t/c)^{\frac{s}{k-s}} \\ &= -Ct^{\frac{k}{k-s}}. \end{aligned}$$

This concludes the proof. □

Proof of Theorem 3.5. The proof of the necessity of conditions (22) is exactly the classical one, see [37, 42]. Let us prove that they are sufficient.

Since we have to prove a local estimate, we can suppose that p is compactly supported in the x-variables, hence in the global class $C^{k-1,1}S^2$. For such a symbol it suffices to prove that for every $\overline{\rho} = (\overline{x}, \overline{\xi}) \in T^*X \setminus 0$ (the cotangent bundle with the zero-section removed), with $|\overline{\xi}| = 1$, there exist a conic neighborhood V of $\overline{\rho}$ and a constant $C > 0$ such that

$$p(x,\xi) \geq -C(1+|\xi|^2)^{(2-\sigma)/2}, \quad \forall (x,\xi) \in V. \tag{24}$$

Indeed, granted (24), we obtain at once (by homogeneity and compactness)

$$p(x,\xi) \geq -C(1+|\xi|^2)^{(2-\sigma)/2}, \quad (x,\xi) \in T^*\mathbb{R}^n,$$

whence, since $\sigma < 2$, an application of Theorem 3.4 to the operator with Weyl symbol $p(x,\xi) + C(1+|\xi|^2)^{(2-\sigma)/2} \geq 0$ gives the wanted estimate.

Let us prove (24) first for $\overline{\rho} \notin \Sigma$. Then there exist a constant c and a conic neighborhood V of $\overline{\rho}$ such that $p_2(x,\xi) \geq c|\xi|^2$ for $(x,\xi) \in V$. Since $|p(x,\xi) - p_2(x,\xi)| \leq C|\xi|$ for $|\xi| \geq 1$, it is clear that (24) is satisfied in V.

Consider then $\overline{\rho} \in \Sigma$. There exists a conic neighborhood U of $\overline{\rho}$, with coordinates (u,v) such that u (respectively v) are homogeneous C^k-functions of degree 0 (respectively 1) and $\Sigma \cap U = \{u = 0,\ v \neq 0\}$. For simplicity, we keep the notation $p(u,v)$, $p_2(u,v)$, etc., for the symbol, principal symbol, etc., expressed in the new coordinates (u,v). We emphasize that they have C^k regularity in the u, v variables. Set $\Gamma_\epsilon := \{|u| < \epsilon,\ |v/|v| - v_{\overline{\rho}}/|v_{\overline{\rho}}|| < \epsilon,\ |v| > 1\}$.

By Taylor's formula we can write

$$p_2(u,v) = \sum_{|\alpha|=k} a_\alpha(v)u^\alpha + r(u,v),$$

$$p_1(u,v) = p_1(0,v) + \sum_{|\alpha|=1} c_\alpha(u,v)u^\alpha,$$

where $r(u,v)$ and $c_\alpha(u,v)$ are homogeneous functions with respect to v of degree 2 and 1 respectively, and $|r(u,v)| \le c_\epsilon |u|^k|v|^2$ in Γ_ϵ, with $c_\epsilon \to 0$ as $\epsilon \to 0^+$. Since by the transversal ellipticity assumption we have

$$\sum_{|\alpha|=k} a_\alpha(v)u^\alpha \ge C_1|u|^k|v|^2 \quad \text{in } \Gamma_\epsilon,$$

for a suitable constant $C_1 > 0$, if ϵ is small enough we obtain

$$\begin{aligned} p_2(u,v) + p_1(u,v) &\ge \frac{C_1}{2}|v|^2|u|^k + \sum_{|\alpha|=1} c_\alpha(u,v)u^\alpha = \\ &= |v|^2\left(\frac{C_1}{2}|u|^k + \frac{1}{|v|}\sum_{|\alpha|=1}\frac{c_\alpha(u,v)}{|v|}u^\alpha\right) \\ &\ge -C_2|v|^{2-k/(k-1)}, \end{aligned}$$

where we applied Lemma 3.6 with $t = |v|^{-1}$ and $s = 1$. Since obviously $|p(u,v) - p_2(u,v) - p_1(u,v)| \le C_3$, we obtain (24) and this concludes the proof. □

4. Positivity and Product of Wick Operators

Let us begin with a short historical survey. The classical Wick correspondence has been introduced as a quantization rule in Physics [4] and since then variations have occurred in other fields of Mathematics under names such as Toeplitz operators, wave packets [15, 20]. Recently they have come again into focus in applied mathematics and signal analysis as localization operators and as short-time Fourier transform multipliers [16, 19, 45, 52, 13]. The Wick correspondence is a mapping

$$p \to A_p \tag{25}$$

from a symbol p defined on the phase space $\mathbb{R}^{2n}$ to an operator A_p acting on a subspace of $L^2(\mathbb{R}^n)$. The coherent states, i.e., translated and modulated Gaussian functions, occur implicitly in the classical definition of A_p [47, 5]. The very nice property of the quantization in (25) is the positivity property: *If the symbol p is non-negative on $\mathbb{R}^{2n}$ then the operator A_p is also non-negative on $L^2(\mathbb{R}^n)$.* This condition is not fulfilled by the Weyl quantization, for which the Fefferman-Phong inequality, cf. Theorem 3.4 above, represents the best lower bound for generic operators with non-negative symbols.

The continuity of the mapping (25) has been studied by many mathematicians, both in the PDE and in signal analysis contexts, see, e.g., [5, 7, 34, 51, 52, 19, 13]. In signal processing the analysis and the synthesis of a signal is done by considering suitable windows functions. These functions can be different from the Gaussian and, moreover, the window to analyze the signal can be different from the one to synthesize it, see [12]. In particular, all the pairs (φ_1, φ_2), with φ_i, $i = 1, 2$, being non-zero window functions in the Schwartz space $\mathcal{S}(\mathbb{R}^n)$, are admissible, although rougher windows can be used as well. This leads to the need of a more general formula for the operator in (25), that takes care of different windows functions: the generalized Wick operator $A_p^{\varphi_1,\varphi_2}$, with analysis window φ_1 and synthesis window φ_2 introduced in Definition 4.1 below. Therefore, the study of the linear mapping (25) can be replaced by the investigation of the multilinear map

$$(p, \varphi_1, \varphi_2) \rightarrow A_p^{\varphi_1,\varphi_2}, \tag{26}$$

which involves mutual dependence between symbol and windows: the choice of a rough symbol p must be *balanced* by taking two window functions φ_1, φ_2 smooth enough and viceversa. For more details on the topic we refer to [6, 13, 50].

The next step concerns the study of the product of two Wick operators. It is well-known that, in general, the product of two Wick operators with symbols in certain classes does not yield a Wick operator with symbol in the same class [17]. To circumvent this fact, an approximate formula has been preferred to the exact one [34, 1]. Namely, the product of Wick operators has been written in terms of a sum of Wick operators, modulo a remainder term, expressed in the Weyl form. If we use suitable classes, in particular the Shubin classes G_ρ^m, see (30) below, the symbol of the remainder Weyl term has lower order than the Wick symbols, so one can get rid of it when using the formula.

What happens if we consider products of generalized Wick operators $A_p^{\varphi_1,\varphi_2}$, $A_q^{\varphi_3,\varphi_4}$? Here the main difficulty lies in handling four different un-known window functions (we assume they belong to the Schwartz class), instead of just using only the Gaussian. The answer to the previous question is contained in [14], where a general asymptotic expansion is given by using time-frequency analysis methods. We shall recall here this general formula and show how to derive a different expression (cf. (32), (35)) in a very simple way. Precisely, instead of repeating a long proof, similar to the one of [14, Thm. 1.1], a simple trick turns out to be useful: we compute the adjoint of the generalized Wick operator and then, working with adjoint Weyl and Wick operators, the symmetric product formula can be easily derived from the previous one of [14, Thm. 1.1]. We next introduce the elements and the tools we need to define our operators and refer to [20, 22] for details. The protagonists of signal analysis are the operators of translation and modulation, defined by

$$T_x f(t) = f(t - x) \quad \text{and} \quad M_\omega f(t) = e^{2\pi i \omega \cdot t} f(t) .$$

They are the building blocks of the time-frequency representation we mostly use in the sequel. Namely, given a "window function" g and a "signal" f, the *short-time Fourier transform (STFT)* $V_g f$ is given by

$$V_g f(x,\omega) = \langle f, M_\omega T_x g\rangle = \int_{\mathbb{R}^n} f(t)\,\overline{g(t-x)}\,e^{-2\pi i\omega\cdot t}\,dt\,, \qquad x,\omega,t\in\mathbb{R}^n\,,$$

whenever the integral or the inner product exist, e.g., for $(f,g)\in L^2(\mathbb{R}^n)\times L^2(\mathbb{R}^n)$ or $(f,g)\in\mathcal{S}'(\mathbb{R}^n)\times\mathcal{S}(\mathbb{R}^n)$.

Definition 4.1. *The generalized Wick operator $A_p^{\varphi_1,\varphi_2}$ with symbol p and windows φ_1,φ_2 is defined to be*

$$A_p^{\varphi_1,\varphi_2} f(t) = \int_{\mathbb{R}^{2n}} p(x,\omega)V_{\varphi_1}f(x,\omega)M_\omega T_x\varphi_2(t)\,dxd\omega$$

whenever this vector-valued integral makes sense.

We address the reader to [32, Appendix] for the precise definition of the vector-valued integral. If $\varphi_1(t)=\varphi_2(t)=e^{-\pi t^2}$, then $A_p = A_p^{\varphi_1,\varphi_2}$ is the classical Wick operator [4, 34, 1].

To prove our results we heavily exploit the weak definition of the previous integral, that is

$$\begin{aligned}\langle A_p^{\varphi_1,\varphi_2} f, g\rangle &= \int p(x,\omega)\,V_{\varphi_1}f(x,\omega)\,\langle M_\omega T_x\varphi_2, g\rangle\,dxd\omega\\ &= \langle p, \overline{V_{\varphi_1}f}\,V_{\varphi_2}g\rangle \qquad \text{for } f,g\in\mathcal{S}(\mathbb{R}^n)\,.\end{aligned}$$

Here $\langle\cdot,\cdot\rangle$ denotes the duality B', B (with B a suitable space) that extends the inner product on $L^2(\mathbb{R}^{2n})$. Also the Weyl quantization can be defined by means of a time-frequency representation. In fact, one introduces the cross-Wigner distribution, given by

$$W(f,g)(x,\omega) = \int f(x+\frac{t}{2})\overline{g(x-\frac{t}{2})}e^{-2\pi i\omega t}\,dt \tag{27}$$

(the quadratic expression $W(f,f)$ is usually called the Wigner distribution of f), and the Weyl operator $\mathrm{Op}^w(\sigma)$ of Weyl symbol σ is defined by

$$\langle \mathrm{Op}^w(\sigma)f, g\rangle = \langle\sigma, W(g,f)\rangle, \qquad f,g\in\mathcal{S}(\mathbb{R}^n). \tag{28}$$

Given a localization operator $A_p^{\varphi_1,\varphi_2}$, with symbol $p\in\mathcal{S}'(\mathbb{R}^{2n})$ and non-zero window functions $\varphi_1,\varphi_2\in\mathcal{S}(\mathbb{R}^n)$, a calculation in [6, 20, 47] reveals that $A_p^{\varphi_1,\varphi_2} = \mathrm{Op}^w\left(p * W(\varphi_2,\varphi_1)\right)$, so the (Weyl) symbol of $A_p^{\varphi_1,\varphi_2}$ is given by

$$\sigma = p * W(\varphi_2,\varphi_1)\,. \tag{29}$$

In particular, (28) shows that the Weyl operator $\mathrm{Op}^w(\sigma)$ has a meaning as a mapping from $\mathcal{S}(\mathbb{R}^n)$ to $\mathcal{S}'(\mathbb{R}^n)$, for every $\sigma\in\mathcal{S}'(\mathbb{R}^{2n})$. In this general setting, for example, the symbol $\sigma(x,\xi)=\pi^n\delta(x,\xi)$ gives the operator $\mathrm{Op}^w(\sigma)f(x)=f(-x)$.

In order to simplify this theory, we restrict our attention to symbols in the classical Shubin classes, related to the weight function

$$\langle z\rangle = (1+|z|^2)^{1/2}, \qquad \text{with } z=(x,\omega)\in\mathbb{R}^{2n}.$$

More general weights are considered in [14]. Moreover, the product formula works with other classes of symbols as well [34, 1].

Let $m \in \mathbb{R}$, $0 < \rho \leq 1$. Then the Shubin classes G^m_ρ, in short G^m, are defined by (9), and in the present notation consist of the functions $a(z) \in C^\infty(\mathbb{R}^{2n})$ which satisfy the estimates [1]

$$|\partial^\gamma a(z)| \leq C_\gamma \langle z\rangle^{m-\rho|\gamma|}, \quad \forall \gamma \in \mathbb{Z}^{2n}_+. \tag{30}$$

Under this assumption on the symbols, the product formula in [14, Thm. 1.1] is the following:

Theorem 4.2. *Consider four windows $\varphi_j \in \mathcal{S}(\mathbb{R}^n)$, $j = 1, \dots, 4$, and Shubin symbols $p \in G^{m_1}$, $q \in G^{m_2}$, with $m_j \in \mathbb{R}$, $j = 1, 2$. Let N be a positive integer. If we define the window functions*

$$\Phi_\alpha = \frac{1}{(2\pi i)^{|\alpha_2|}} \sum_{\beta \leq \alpha} \binom{\alpha}{\beta} (-1)^{|\beta_1|} \langle \varphi_3, t^{\alpha_1-\beta_1}\partial^{\alpha_2-\beta_2}\varphi_4\rangle t^{\beta_1}\partial^{\beta_2}\varphi_1, \tag{31}$$

where $\alpha = (\alpha_1, \alpha_2)$, $\beta = (\beta_1, \beta_2) \in \mathbb{Z}^n_+ \times \mathbb{Z}^n_+$, then we have the following composition formula

$$A^{\varphi_1,\varphi_2}_p A^{\varphi_3,\varphi_4}_q = \sum_{|\alpha|=0}^{N-1} \frac{(-1)^{|\alpha|}}{\alpha!} A^{\Phi_\alpha,\varphi_2}_{p\partial^\alpha q} + \mathrm{Op}^w(r), \tag{32}$$

where $\mathrm{Op}^w(r)$ is a Weyl operator with Weyl symbol $r \in G^{m_1+m_2-N}$.

In formula (32), all the derivatives are on the symbol q. If we want to consider derivatives of the symbol p, then we need a symmetric formula. For that, we shall use the formal adjoint formula for a localization operator $A^{\varphi_1,\varphi_2}_p$.

Lemma 4.1. *Let $\varphi_1, \varphi_2 \in \mathcal{S}(\mathbb{R}^n)$, $p \in \mathcal{S}(\mathbb{R}^{2n})$, then the adjoint operator of the generalized Wick operator $A^{\varphi_1,\varphi_2}_p$ is given by*

$$(A^{\varphi_1,\varphi_2}_p)^* = A^{\varphi_2,\varphi_1}_{\bar{p}}, \tag{33}$$

where $\bar{p}$ is the complex conjugate of the symbol p.

Proof. Let $f, g \in \mathcal{S}(\mathbb{R}^n)$, since by assumption symbol and window functions live in the Schwartz class, the integral written below are all absolute convergent integrals and this allows to interchange the order of integration. If we introduce the *time-frequency shift operators*:

$$\pi(z)f = M_\omega T_x f,$$

[1] $\gamma = (\gamma_1, \dots, \gamma_{2n}) \in \mathbb{Z}^{2n}_+$, $t^\gamma = t_1^{\gamma_1} \cdots t_{2n}^{\gamma_{2n}}$, $\partial^\gamma = \partial_{t_1}^{\gamma_1} \cdots \partial_{t_{2n}}^{\gamma_{2n}}$.

by the weak definition of $A_p^{\varphi_1,\varphi_2}$ we have

$$\begin{aligned}\langle A_p^{\varphi_1,\varphi_2} f, g\rangle &= \langle pV_{\varphi_1} f, V_{\varphi_2} g\rangle = \int_{\mathbb{R}^{2n}} p(z)\left(\int_{\mathbb{R}^n} f(t)\overline{\pi(z)\varphi_1}\,dt\right)\overline{V_{\varphi_2} g(z)}\,dz\\ &= \int_{\mathbb{R}^n} f(t)\left(\int_{\mathbb{R}^{2n}} p(z)\overline{\pi(z)\varphi_1 V_{\varphi_2} g(z)}\,dz\right)dt\\ &= \langle f, \left(\int_{\mathbb{R}^{2n}} \overline{p(z)} V_{\varphi_2} g(z)\pi(z)\varphi_1\,dz\right)\rangle = \langle f, A_{\bar p}^{\varphi_2,\varphi_1} g\rangle.\end{aligned}$$

By a density argument one can consider every functions $f, g \in L^2(\mathbb{R}^n)$ and (33) follows. □

Proposition 4.3. *Under the assumptions of Theorem 4.2, we define the window functions*

$$\Psi_\alpha = \frac{1}{(2\pi i)^{|\alpha_2|}} \sum_{\beta\le\alpha} \binom{\alpha}{\beta} (-1)^{|\beta_1|}\langle\varphi_2, t^{\alpha_1-\beta_1}\partial^{\alpha_2-\beta_2}\varphi_1\rangle t^{\beta_1}\partial^{\beta_2}\varphi_4, \tag{34}$$

then the symmetric *composition formula is given by*

$$A_p^{\varphi_1,\varphi_2} A_q^{\varphi_3,\varphi_4} = \sum_{|\alpha|=0}^{N-1} \frac{(-1)^{|\alpha|}}{\alpha!} A_{(\partial^\alpha p)q}^{\varphi_3,\Psi_\alpha} + \mathrm{Op}^w(\bar r). \tag{35}$$

Formula (35) is referred to be the *symmetric* formula of (32), because it is quite similar to (32), but the localization symbols $p\partial^\alpha q$ in (32) are replaced in (35) by $(\partial^\alpha p)q$.

Proof. We use the formal adjoint formula for a localization operator $A_p^{\varphi_1,\varphi_2}$ given by (33). We recall from Section 2 that the formal adjoint of a Weyl operator $\mathrm{Op}^w(r)$ is given by $\mathrm{Op}^w(\bar r)$. Then, using formula (32) for the product $A_{\bar q}^{\varphi_4,\varphi_3} A_{\bar p}^{\varphi_2,\varphi_1}$ we get the result. In fact, we have

$$\begin{aligned} A_p^{\varphi_1,\varphi_2} A_q^{\varphi_3,\varphi_4} &= (A_p^{\varphi_1,\varphi_2} A_q^{\varphi_3,\varphi_4})^{**} = [(A_q^{\varphi_3,\varphi_4})^*(A_p^{\varphi_1,\varphi_2})^*]^*\\ &= [A_{\bar q}^{\varphi_4,\varphi_3} A_{\bar p}^{\varphi_2,\varphi_1}]^* = \left[\sum_{|\alpha|=0}^{N-1} \frac{(-1)^{|\alpha|}}{\alpha!} A_{\bar q\partial^\alpha \bar p}^{\Psi_\alpha,\varphi_3} + \mathrm{Op}^w(r)\right]^*\\ &= \sum_{|\alpha|=0}^{N-1} \frac{(-1)^{|\alpha|}}{\alpha!} A_{(\partial^\alpha p)q}^{\varphi_3,\Psi_\alpha} + \mathrm{Op}^w(\bar r), \end{aligned}$$

that is (35). □

Let us finally point out that the positivity property of the classical Wick operators holds true in our context if the two window functions coincide. In fact, if $\varphi_1 = \varphi_2 = \varphi$ and $p \ge 0$, we have

$$\langle A_p^{\varphi,\varphi} u, u\rangle = \langle p, \overline{V_\varphi u} V_\varphi u\rangle = \langle p, |V_\varphi u|^2\rangle \ge 0. \tag{36}$$

The estimate (36) can be used to obtain lower bounds for Weyl operators. In short: starting from P with non-negative principal Weyl symbol p_m, we may represent

$$P = A^{\varphi,\varphi}_{p_m} + \mathrm{Op}^w(r)$$

for some lower order symbol r. Applying (36) to $A^{\varphi,\varphi}_{p_m}$ we deduce the Sharp-Gårding inequality, cf. [5],[34],[47]. We also observe that the Fefferman-Phong inequality can be re-obtained by using the FBI transform, which is related to the Wick quantization, see [49] for details.

Finally, concerning (36), let us emphasize that $A^{\varphi,\varphi}_p \geq 0$ does not imply in general $p \geq 0$. In fact, for Wick operators, as well as for Weyl operators, the problem of characterizing operators which are non-negative, or satisfy (H_σ) of Section 3, is largely open, and challenging.

Acknowledgments. We are very indebted to the referee for many detailed remarks. His invaluable comments led to a substantial improvement of the work, as regards the English, as regards presenting things in the proper order, and as regards precision.

References

[1] H. Ando and Y. Morimoto, Wick calculus and the Cauchy problem for some dispersive equations, *Osaka J. Math.* **39** (2002), 123–147.

[2] R. Beals, A general calculus of pseudo-differential operators, *Duke Math. J.* **42** (1975), 1–42.

[3] R. Beals and C. Fefferman, Spatially inhomogeneous pseudo-differential operators I, *Comm. Pure Appl. Math.* **27** (1974), 1–24.

[4] F.A. Berezin, Wick and anti-Wick symbols of operators, *Math. Sb. (N.S.)* **86**(128) (1971), 578–610.

[5] P. Boggiatto, E. Buzano and L. Rodino, *Global Hypoellipticity and Spectral Theory*, Akademie Verlag, Berlin, 1996.

[6] P. Boggiatto, E. Cordero and K. Gröchenig, Generalized anti-Wick operators with symbols in distributional Sobolev spaces, *Integral Equations Operator Theory*, to appear.

[7] P. Boggiatto and L. Rodino, Quantization and pseudo-differential operators, *Cubo Mat Ed.* **5** (2003), 237–272.

[8] J.-M. Bony, Calcul symbolique et propagations des singularités pour les équations aux dérivées partielles non linéaires, *Ann. Sci. École Norm. Sup.* **14** (1981), 209–246.

[9] J.-M. Bony, Sur l'inégalité de Fefferman-Phong, in *Séminaire sur les Équations aux Dérivé Partielles*, 1998-1999, Exp. No. III, École Polytech., Palaiseau, 1998.

[10] L. Boutet de Monvel, A. Grigis and B. Helffer, Paramétrixes d'opérateurs pseudo-différentiels à caractéristiques multiples, *Astérique* **34–35** (1976), 93–121.

[11] A.P. Calderón and R. Vaillancourt, A class of bounded pseudo-differential operators, *Proc. Nat. Acad. Sci. USA* **69** (1972), 1185–1187.

[12] L. Cohen, *Time-Frequency Analysis*, Prentice Hall, Englewood Cliffs, NJ, 1995.

[13] E. Cordero and K. Gröchenig, Time-frequency analysis of localization operators, *J. Funct. Anal.* **205** (2003), 107–131

[14] E. Cordero and L. Rodino, Wick calculus: a time-frequency approach, *Osaka J. Math.*, to appear.

[15] A. Córdoba and C. Fefferman, Wave packets and Fourier integral operators, *Comm. Partial Differential Equations* **3** (1978), 979–1005, 1978.

[16] I. Daubechies, Time-frequency localization operators: a geometric phase space approach, *IEEE Trans. Inform. Theory* **34** (1988), 605–612.

[17] J. Du and M.W. Wong, A product formula for localization operators, *Bull. Korean Math. Soc.* **37** (2000), 77–84.

[18] C. Fefferman and D.H. Phong, On positivity of pseudo-differential operators, *Proc. Natl. Acad. Sci. USA* **75** (1978), 4673–4674.

[19] H.G. Feichtinger and K. Nowak, A first survey of Gabor multipliers, in *Advances in Gabor Analysis*, Editors: H. G. Feichtinger and T. Strohmer, Birkhäuser, Boston, 2002, 99–128.

[20] G.B. Folland, *Harmonic Analysis in Phase Space*, Princeton Univ. Press, Princeton, NJ, 1989.

[21] L. Gårding, Dirichlet's problem for linear elliptic partial differential equations, *Math. Scand.* **1** (1953), 55–72.

[22] K. Gröchenig, *Foundations of Time-Frequency Analysis*, Birkhäuser, Boston, 2001.

[23] F. Hérau, *Opérateurs Pseudo-Differentieles Semi-Bornés*, Ph.D. Dissertation, University of Rennes, 1999.

[24] F. Hérau, Melin-Hörmander inequality in a Wiener type pseudo-differential algebra, *Ark. Mat.* **39** (2001), 311–338.

[25] F. Hérau, Melin inequality for paradifferential operators and applications, *Comm. Partial Differential Equations* **27** (2002), 1659–1680.

[26] L. Hörmander, Pseudo-differential operators, *Comm. Pure Appl. Math.* **18** (1965), 501–517.

[27] L. Hörmander, Pseudo-differential operators and hypoelliptic equations, in Singular Integrals, *Proc. Sympos. Pure Math.*, Amer. Math. Soc. **10** (1966), 138–183.

[28] L. Hörmander, Fourier integral operators I, *Acta Math.* **127** (1971), 79–183.

[29] L. Hörmander, The Cauchy problem for differential equations with double characteristics, *J. Analyse Math.* **32** (1977), 118–196.

[30] L. Hörmander, The Weyl calculus of pseudo-differential operators, *Comm. Pure Appl. Math.* **32** (1979), 359–443.

[31] L. Hörmander, *The Analysis of Linear Partial Differential Operators* I–IV, Springer-Verlag, Berlin, 1983–1985.

[32] Y. Katznelson, *An Introduction to Harmonic Analysis*, John Wiley & Sons, New York, 1968.

[33] J.J. Kohn and L. Nirenberg, On the algebra of pseudo-differential operators, *Comm. Pure Appl. Math.* **18** (1965), 269–305.

[34] N. Lerner, The Wick calculus of pseudo-differential operators and energy estimates, in *New Trends in Microlocal Analysis*, Springer, 1997, 23–37.

[35] O. Liess and L. Rodino, Linear partial differential equations with multiple involutive characteristics, in *Microlocal Analysis and Spectral Theory*, Editor: L. Rodino, Kluwer Academic Publishers, 1997, 227–250.

[36] M. Mascarello and L. Rodino, *Linear Partial Differential Operators with Multiple Characteristics*, Wiley-Akademie Verlag, Berlin, 1997.

[37] A. Melin, Lower bounds for pseudo-differential operators, *Ark. Mat.* **9** (1971), 117–140.

[38] M. Mughetti and F. Nicola, A counterexample to a lower bound for a class of pseudo-differential operators, *Proc. Amer. Math. Soc.*, to appear.

[39] M. Mughetti and F. Nicola, *A generalization of Hörmander's inequality II*, in preparation.

[40] F. Nicola and L. Rodino, *Remarks on lower bounds for pseudo-differential operators*, preprint.

[41] C. Parenti and A. Parmeggiani, Lower bounds for pseudo-differential operators, in *Microlocal Analysis and Spectral Theory*, Editor: L. Rodino, Kluwer Academic Publishers, 1997, 227–250.

[42] C. Parenti and A. Parmeggiani, Some remarks on almost-positivity of pseudo-differential operators, *Boll. Un. Mat. It.*1-B **8** (1998), 187–215.

[43] C. Parenti and A. Parmeggiani, A generalization of Hörmander's inequality I, *Comm. Partial Differential Equations* **25** (2000), 457–506.

[44] C. Parenti and A. Parmeggiani, Lower bounds for systems with double characteristics, *J. Analyse. Math.* **86** (2002), 49–91.

[45] J. Ramanathan and P. Topiwala, Time-frequency localization via the Weyl correspondence, *SIAM J. Math. Anal.* **24** (1993), 1378–1393.

[46] L. Schwartz, *Théorie des Distributions*, Hermann, Paris, 1966.

[47] M.A. Shubin, *Pseudo-Differential Operators and Spectral Theory*, Second Edition, Springer-Verlag, Berlin, 2001.

[48] J. Sjöstrand, Parametrices for pseudo-differential operators with multiple characteristics, *Ark. Mat.* **12** (1974), 85–130.

[49] D. Tataru, On the Fefferman-Phong inequality and related problems, *Comm. Partial Differential Equations* **27** (2002), 2101–2138.

[50] J. Toft, *Modulation spaces and pseudo-differential operators*, preprint.

[51] M.W. Wong, *Localization Operators*, Seoul National University, Research Institute of Mathematics, Global Analysis Research Center, Seoul, 1999.

[52] M.W. Wong, *Wavelet Transforms and Localization Operators*, Birkhäuser, Basel, 2002.

E. Cordero, F. Nicola and L. Rodino
Department of Mathematics
University of Torino
Via Carlo Alberto 10, 10123 Torino
Italy
e-mail: cordero@dm.unito.it
nicola@dm.unito.it
rodino@dm.unito.it

Operator Theory:
Advances and Applications, Vol. 155, 19–64

The Conormal Symbolic Structure of Corner Boundary Value Problems

T. Krainer and B.-W. Schulze

Abstract. Ellipticity of operators on manifolds with conical singularities or parabolicity on space-time cylinders are known to be linked to parameter-dependent operators (conormal symbols) on a corresponding base manifold. We introduce the conormal symbolic structure for the case of corner manifolds, where the base itself is a manifold with edges and boundary. The specific nature of parameter-dependence requires a systematic approach in terms of meromorphic functions with values in edge-boundary value problems. We develop here a corresponding calculus, and we construct inverses of elliptic elements.

Mathematics Subject Classification (2000). Primary 35J30, 35J55; Secondary 47G30, 58J32.

Keywords. Pseudo-differential operators, boundary value problems, corner manifolds, parameter-dependent edge calculus, symbolic hierarchies.

Introduction

Boundary value problems for elliptic differential operators on a manifold with corners (or parabolic operators on a spatial configuration of that kind) can be studied in terms of symbolic structures that contain parameter-dependent operators on base manifolds of corresponding local cones (or cylinders). The analysis of such so called conormal symbols is the first essential step for establishing parametrices or regularity and asymptotics of solutions in weighted Sobolev spaces, as is done in Kontratyev's work [10] for the case of conical singularities with a smooth base.

The classical theory of parabolic boundary value problems in connection with parameter-dependent ellipticity is developed in Agranovich and Vishik [1]. New applications concern the construction of inverses of parabolic operators in infinite cylinders, cf. [11], the characterisation of resolvents of elliptic operators as meromorphic inverses with a specific dependence on parameters, cf. [27], or Maniccia and Schulze [15], and the evaluation of Shapiro-Lopatinskij elliptic edge conditions for elliptic operators on manifolds with edges, cf. [25], [29], or Nazaikinskij, Savin, Schulze, and Sternin [16].

Conormal symbols also play an important role in the recent development of the index theory on manifolds with geometric singularities, e.g., in analytic index formulas, cf. Schulze, Sternin, and Shatalov [32], or Fedosov, Schulze, and Tarkhanov [5], [6]. Another interesting aspect is the spectral flow of families of conormal symbols associated with edge singularities, cf. [17].

The main purpose of the present paper is to establish a new calculus of boundary value problems on a manifold W with edges with meromorphy in a complex parameter.

Such parameter-dependent operators should be contained in a future algebra of boundary value problems on manifolds with higher corner singularities as a component of a corresponding symbolic hierarchy.

Moreover, if the manifold W plays the role of a cross section of an infinite space-time cylinder our calculus (in a corresponding anisotropic form) can be related to iterated long-time asymptotics of solutions to parabolic equations, similarly as the author's joint paper [12] for the simpler case of conical singularities and without boundary.

Here we study isotropic parameter-dependent operators associated with the elliptic theory of boundary value problems on manifolds with corners and base W. The calculus as a whole contains interesting substructures, e.g., parameter-dependent operators that are flat in the cone and corner axial variables $r, t \in \mathbb{R}$, cf. [18], or smoothing operators with iterated asymptotic information for r and t tending to zero, cf. [3].

This paper is organised as follows.

Chapter 1 starts from manifolds with corner points where the base spaces are manifolds with edges and boundary. Configurations of that kind can be described by corresponding 'corner metrics'; the associated Laplacians are then corner degenerate. For such differential operators, we observe how corner conormal symbols appear as parameter-dependent edge degenerate differential operators. Moreover, we consider operators with the transmission property at the boundary for the case of manifolds with conical exits to infinity. Then we formulate the cone algebra on an infinite cone with discrete asymptotics at the tip of the cone.

Chapter 2 develops the calculus of boundary value problems on a manifold with edges in parameter-dependent form. We introduce edge amplitude functions taking values in boundary value problems on the infinite model cone. Discrete asymptotic data are formulated in terms of the meromorphic structure of subordinate cone conormal symbols and of the mapping properties of Green symbols. The edge algebra itself will be formulated with continuous asymptotic types, based on vector- and operator-valued analytic functionals in the complex plane of the Mellin covariable belonging to the axial variable of the model cone. We investigate parameter-dependent ellipticity of edge boundary value problems and obtain invertibility of operators in weighted edge Sobolev spaces for large absolute values of the parameter (Theorem 2.16).

An essential technical tool is the kernel cut-off procedure which generates elements that are holomorphic in the parameter (Theorem 2.6), cf. also [24], [28], [26], and [13].

In Chapter 3 we investigate 'corner conormal symbols', i.e., families of edge boundary value problems meromorphically depending on a parameter. Kernel cut-off produces a rich space of such families (Theorem 3.3), and parameter-dependent ellipticity is preserved in this process.

We finally show that the space of corner conormal symbols is closed under inversion of elliptic elements (Theorem 3.10).

1. Parameter-Dependent Boundary Value Problems

1.1. Differential Operators on Manifolds with Edges

By a manifold W with edge Y and boundary we understand a topological space such that $W \setminus Y$ is a C^∞ manifold with boundary, Y a C^∞ manifold, and every $y \in Y$ has a neighbourhood U in W that is modelled on a wedge $X^\Delta \times \Omega$ with $X^\Delta := (\overline{\mathbb{R}}_+ \times X)/(\{0\} \times X)$ for a compact C^∞ manifold X with boundary, and an open set $\Omega \subseteq \mathbb{R}^q$, $q = \dim Y$. In addition we require a specific behaviour of transition maps belonging to different 'singular charts' $U \to X^\Delta \times \Omega$.

It will be convenient to first pass to the double $\tilde{W} = 2W$ which is a manifold with edge Y (without boundary). The space $\tilde{W}$ may be described by a stretched manifold $\tilde{\mathbb{W}}$ associated with $\tilde{W}$ which is a C^∞ manifold with boundary $\partial\tilde{\mathbb{W}}$, and $\partial\tilde{\mathbb{W}}$ is an $\tilde{X}$-bundle over Y for $\tilde{X} = 2X$, the double of X. Then $\tilde{W}$ itself follows by squeezing down the fibres $(\partial\tilde{\mathbb{W}})_y$ for every $y \in Y$ to the single point y. This gives us the local structure of $\tilde{W}$ near a point of Y as $\tilde{X}^\Delta \times \Omega$, and transition maps are induced by the ones for a collar neighbourhood of $\partial\tilde{\mathbb{W}}$ in $\tilde{\mathbb{W}}$, represented by diffeomorphisms $\overline{\mathbb{R}}_+ \times \tilde{X} \times \Omega \to \overline{\mathbb{R}}_+ \times \tilde{X} \times \Omega$ that restrict to corresponding transition maps of the $\tilde{X}$-bundle $\partial\tilde{\mathbb{W}}$ on the boundary.

By construction, there is then a projection

$$\tilde{\pi} : \tilde{\mathbb{W}} \to \tilde{W},$$

defined as the map that restricts to the bundle projection $\partial\tilde{\mathbb{W}} \to Y$ on the boundary and to the identity map on $\operatorname{int}\tilde{\mathbb{W}}$. Let us write

$$\tilde{\mathbb{W}}_{\text{sing}} := \partial\tilde{\mathbb{W}}, \ \tilde{\mathbb{W}}_{\text{reg}} := \operatorname{int}\tilde{W}.$$

In the following, for convenience, we assume $\partial\tilde{\mathbb{W}}$ to be a trivial $\tilde{X}$-bundle on Y, i.e., $\partial\tilde{\mathbb{W}} = \tilde{X} \times Y$. The general case requires more comment on invariance properties of our operators below which is not the main intention of the present paper.

For references below, we identify neighbourhoods $\tilde{\mathbb{U}}$ near $\tilde{\mathbb{W}}_{\text{sing}}$ with $[0,1) \times \tilde{X} \times G$, where G is a coordinate neighbourhood on Y, and we fix diffeomorphisms

$$\chi : \tilde{\mathbb{U}} \to \overline{\mathbb{R}}_+ \times \tilde{X} \times \Omega, \quad \kappa : G \to \Omega \tag{1.1}$$

for an open set $\Omega \subseteq \mathbb{R}^q$, such that $\chi(r,x,y) = (r,x,\kappa(g))$ for $0 \le r \le \frac{1}{2}$. Now $\tilde{\mathbb{W}}$ can be regarded as the double of another stretched space $\mathbb{W}$ associated with our W, with a projection

$$\pi := \tilde{\pi}|_{\mathbb{W}} : \mathbb{W} \to W.$$

To make the idea more transparent we consider an example, namely

$$\tilde{\mathbb{W}} := \overline{\mathbb{R}}_+ \times \tilde{X} \times \Omega, \ \mathbb{W} := \overline{\mathbb{R}}_+ \times X \times \Omega.$$

This shows in what sense $\tilde{\mathbb{W}}$ is equal to $2\mathbb{W}$ for $\tilde{X} = 2X$. Here $\pi : \overline{\mathbb{R}}_+ \times X \times \Omega \to X^\Delta \times \Omega = W$ is induced by $\tilde{\pi} : \overline{\mathbb{R}}_+ \times \tilde{X} \times \Omega \to \tilde{X}^\Delta \times \Omega = \tilde{W}$. The invariance of the latter construction under transition maps for a general stretched manifold $\tilde{\mathbb{W}}$ gives us correct global definitions also for $\mathbb{W}$. Let us set

$$\mathbb{W}_{\mathrm{sing}} := \pi^{-1}Y \quad \text{and} \quad \mathbb{W}_{\mathrm{reg}} := \mathbb{W} \setminus \mathbb{W}_{\mathrm{sing}}.$$

Then (under our assumption on $\partial\tilde{\mathbb{W}}$) we have $\mathbb{W}_{\mathrm{sing}} = X \times Y$, and $\mathbb{W}_{\mathrm{reg}}$ is a C^∞ manifold with boundary.

Let us now pass to a corner $W^\Delta = (\overline{\mathbb{R}}_+ \times W)/(\{0\} \times W)$ with base W, and consider the associated stretched corner $\mathbb{R}_+ \times \mathbb{W}$. We also look at the doubles $\tilde{W}^\Delta$ and $\mathbb{R}_+ \times \tilde{\mathbb{W}}$, respectively. Differential operators on $\mathbb{R}_+ \times \mathbb{W}$ may be defined in terms of restrictions of corresponding operators on $\mathbb{R}_+ \times \tilde{\mathbb{W}}$ to $\mathbb{R}_+ \times \mathbb{W}$. Typical examples are Laplace-Beltrami operators to corner metrics of the form

$$dt^2 + t^2 g_{\tilde{\mathbb{W}}}(t), \tag{1.2}$$

where $t \in \mathbb{R}_+$ is the corner axis variable and $g_{\tilde{\mathbb{W}}}(t)$ a family of edge-metrics on $\tilde{\mathbb{W}}$, smoothly dependent on t up to 0. By an edge metric we understand a Riemannian metric on $\tilde{\mathbb{W}}_{\mathrm{reg}}$ that is close to $\tilde{\mathbb{W}}_{\mathrm{sing}}$ in the splitting of the variables $(r,x,y) \in \mathbb{R}_+ \times \tilde{X} \times \Omega$ of the form

$$dr^2 + r^2 g_{\tilde{X}}(r) + g_\Omega(r), \tag{1.3}$$

where $g_{\tilde{X}}(r)$ and $g_\Omega(r)$ are families of Riemannian metrics on $\tilde{X}$ and Ω, respectively, smoothly dependent on r up to 0. In other words, (1.2) takes the form

$$dt^2 + t^2(dr^2 + r^2 g_{\tilde{X}}(t,r) + g_\Omega(t,r)) \tag{1.4}$$

with $g_{\tilde{X}}$ and g_Ω being smooth in $t, r \in \mathbb{R}_+ \times \mathbb{R}_+$ up to $(0,0)$.

Let $\mathrm{Diff}^\mu(M)$ for a C^∞ manifold M denote the space of all differential operators on M (with smooth coefficients) of order μ; this is a Fréchet space in a natural way (all manifolds in this paper are assumed to be paracompact and locally compact). Moreover, let $\mathrm{Diff}^\mu_{\mathrm{edge}}(\tilde{\mathbb{W}})$ be the space of all elements of $\mathrm{Diff}^\mu(\tilde{\mathbb{W}}_{\mathrm{reg}})$ that have close to $\tilde{\mathbb{W}}_{\mathrm{sing}}$ in the splitting of variables $(r,x,y) \in \mathbb{R}_+ \times \tilde{X} \times \Omega$ the form

$$r^{-\mu} \sum_{j+|\alpha| \le \mu} a_{j\alpha}(r,y)\Big(-r\frac{\partial}{\partial r}\Big)^j (rD_y)^\alpha \tag{1.5}$$

with coefficients $a_{j\alpha} \in C^\infty(\overline{\mathbb{R}}_+ \times \Omega, \mathrm{Diff}^{\mu-(j+|\alpha|)}(\tilde{X}))$. Also $\mathrm{Diff}^\mu_{\mathrm{edge}}(\tilde{\mathbb{W}})$ is a Fréchet space in a canonical way. Then the Laplace-Beltrami operator associated with an edge metric on $\tilde{\mathbb{W}}_{\mathrm{reg}}$ belongs to $\mathrm{Diff}^2_{\mathrm{edge}}(\tilde{\mathbb{W}})$.

Moreover, the Laplace-Beltrami operator associated with a corner metric (1.2) has the form

$$\tilde{A} = t^{-\mu} \sum_{k=0}^{\mu} b_k(t)(-t\frac{\partial}{\partial t})^k \tag{1.6}$$

(for $\mu = 2$) with coefficients $b_k \in C^\infty(\overline{\mathbb{R}}_+, \mathrm{Diff}^{\mu-k}_{\mathrm{edge}}(\tilde{\mathbb{W}}))$. For r near 0 it follows that

$$\tilde{A} = t^{-\mu} r^{-\mu} \sum_{k+j+|\alpha| \le \mu} c_{kj\alpha}(t,r,y)(-rt\frac{\partial}{\partial t})^k(-r\frac{\partial}{\partial r})^j (rD_y)^\alpha \tag{1.7}$$

with coefficients $c_{kj\alpha} \in C^\infty(\overline{\mathbb{R}}_+ \times \overline{\mathbb{R}}_+ \times \Omega, \mathrm{Diff}^{\mu-(k+j+|\alpha|)}(\tilde{X}))$.

The corner conormal symbol for the operator $\tilde{A}$ in our terminology is the operator function

$$\sigma_{\mathrm{c}}(\tilde{A})(w) = \sum_{k=0}^{\mu} b_k(0) w^k \tag{1.8}$$

with a complex variable w varying in $\mathbb{C}$ or on a weight line

$$\Gamma_\beta := \{w \in \mathbb{C} : \mathrm{Re}\, w = \beta\}$$

for some real β.

In other words, the Fuchs type derivative $-t\frac{\partial}{\partial t}$ is replaced by the covariable w in the Mellin transform $Mu(w) = \int_0^\infty t^{w-1} u(t)dt$. We also employ the weighted Mellin transform $(M_\gamma u)(w) = M(t^{-\gamma}u)(w+\gamma)$, $\gamma \in \mathbb{R}$ (such that $M = M_0$), with the inverse

$$M_\gamma^{-1} g(t) = (2\pi i)^{-1} \int_{\Gamma_{\frac{1}{2}-\gamma}} t^{-w} g(w) dw.$$

Setting

$$\mathrm{op}_M^\gamma(f)u(t) = (2\pi)^{-1} \iint \left(\frac{t}{t'}\right)^{-(\frac{1}{2}-\gamma+i\tau)} f(t,t',w)u(t')dt'd\tau,$$

$w = \frac{1}{2} - \gamma + i\tau$, which has the meaning of a Mellin pseudo-differential operator $M_\gamma^{-1} f M_\gamma$ with amplitude function $f(t,t',w)$ (to be specified below), we have $M_\gamma^{-1} w M_\gamma = -t\frac{\partial}{\partial t}$. More generally, for $f(t,w) = \sum_{k=0}^{\mu} b_k(t) w^k$ it follows that

$$\tilde{A} = t^{-\mu}\, \mathrm{op}_M^\gamma(f).$$

Here $\gamma \in \mathbb{R}$ is arbitrary when $\tilde{A}$ is applied to argument functions with compact support in $t \in \mathbb{R}$, otherwise, for extensions of the operator to weighted Sobolev spaces we have to specify the number γ, cf. [30].

Let $\tilde{\mathbb{W}}$ be compact. Writing $f(\tau) := \sigma_{\mathrm{c}}(\tilde{A})(\beta + i\tau)$ for any fixed $\beta \in \mathbb{R}$ we obtain a τ-dependent family of operators in the edge algebra on the space $\tilde{\mathbb{W}}$, acting in weighted edge Sobolev spaces $\mathcal{W}^{s,\gamma}(\tilde{\mathbb{W}})$, cf. [29], and Definition 1.1 below.

Operators in the edge algebra are characterised by a principal symbolic hierarchy which is particularly simple for the subalgebra generated by $\mathrm{Diff}^{\mu}_{\mathrm{edge}}(\tilde{\mathbb{W}})$, the edge-degenerate differential operators. We shall formulate symbols at once in parameter-dependent form with parameter $\tau \in \mathbb{R}$.

First, we have the standard homogeneous principal symbol of order μ

$$\sigma_\psi(f)(\tilde{x}, \tilde{\xi}, \tau) \in C^\infty((T^*\tilde{\mathbb{W}}_{\text{reg}} \times \mathbb{R}) \setminus 0)$$

which is independent of β. For points $\tilde{x} \in \tilde{\mathbb{W}}_{\text{reg}}$ close to $\tilde{\mathbb{W}}_{\text{sing}}$ in the splitting of variables $\tilde{x} = (r, x, y) \in \mathbb{R}_+ \times \Sigma \times \Omega$, $\Sigma \subseteq \mathbb{R}^n$, $\Omega \subseteq \mathbb{R}^q$ open, $n = \dim \tilde{X}$, $q = \dim Y$, with covariables $(\varrho, \xi, \eta) \in \mathbb{R}^{1+n+q}$, we can write

$$\sigma_\psi(f)(r, x, y, \varrho, \xi, \eta, \tau) = r^{-\mu} \tilde{\sigma}_\psi(f)(r, x, y, r\varrho, \xi, r\eta, r\tau)$$

for a function

$$\tilde{\sigma}_\psi(f)(r, x, y, \tilde{\varrho}, \xi, \tilde{\eta}, \tilde{\tau}),$$

homogeneous in $(\tilde{\varrho}, \xi, \tilde{\eta}, \tilde{\tau}) \neq 0$ of order μ and smooth in r up to $r = 0$. This is immediate from the definition.

Furthermore, we have a so called homogeneous principal edge symbol

$$\sigma_\wedge(f)(y, \eta, \tau)$$

which is an operator valued function on $(T^*Y \times \mathbb{R}) \setminus 0$, acting between weighted Sobolev spaces on the infinite (stretched) model cone $\tilde{X}^\wedge$ of the local wedges that characterise $\tilde{\mathbb{W}}$ near $\tilde{\mathbb{W}}_{\text{sing}}$. Representing $f(\tau)$ locally in the form

$$f(\tau) = r^{-\mu} \sum_{k+j+|\alpha| \leq \mu} c_{kj\alpha}(0, r, y)(-r(\beta + i\tau))^k (-r\frac{\partial}{\partial r})^j (rD_y)^\alpha$$

we have

$$\begin{aligned} \sigma_\wedge(f)(y, \eta, \tau) &= r^{-\mu} \sum_{k+j+|\alpha| \leq \mu} c_{kj\alpha}(0, 0, y)(-ir\tau)^k (-r\frac{\partial}{\partial r})^j (r\eta)^\alpha \\ &: \mathcal{K}^{s,\gamma}(\widetilde{X}^\wedge) \to \mathcal{K}^{s-\mu,\gamma-\mu}(\widetilde{X}^\wedge). \end{aligned}$$

Let us briefly recall the definition of the involved spaces. Given a C^∞ manifold N, by $L^\mu_{\text{cl}}(N; \mathbb{R}^l)$ we denote the space of all classical parameter-dependent pseudo-differential operators of order μ on N, with parameter $\lambda \in \mathbb{R}^l$. Recall that local (left-) amplitude functions $a(x, \xi, \lambda)$ are classical symbols in the covariables $(\xi, \lambda) \in \mathbb{R}^{n+l}$, $n = \dim N$, while $L^{-\infty}(N; \mathbb{R}^l)$ is identified with $A(\lambda) \in \mathcal{S}(\mathbb{R}^l, L^{-\infty}(N))$. Let N be compact. Then it is known that for every $s \in \mathbb{R}$ there exists an element $R^s(\lambda) \in L^s_{\text{cl}}(N; \mathbb{R}^l)$ that induces isomorphisms $H^r(N) \to H^{r-s}(N)$ for all $r \in \mathbb{R}$. Let $\mathcal{H}^{s,\gamma}(N^\wedge)$ for $s, \gamma \in \mathbb{R}$, $N^\wedge = \mathbb{R}_+ \times N$, denote the completion of the space $C_0^\infty(N^\wedge)$ with respect to the norm

$$\big\{(2\pi i)^{-1} \int_{\Gamma_{\frac{n+1}{2}-\gamma}} \|R^s(\operatorname{Im} w)(Mu)(w)\|^2_{L^2(N)} dw\big\}^{1/2},$$

where $R^s(\tau) \in L^s(N; \mathbb{R})$ is any choice of an order reducing family.

In this paper, by a cut-off function on the half-axis we understand any real-valued $\omega \in C_0^\infty(\overline{\mathbb{R}}_+)$ that is equal to 1 in a neighbourhood of 0. We then define the space

$$\mathcal{K}^{s,\gamma}(N^\wedge) = \{\omega u + (1-\omega)v : u \in \mathcal{H}^{s,\gamma}(N^\wedge),\ v \in H^s_{\mathrm{cone}}(N^\wedge)\}$$

for any cut-off function ω. Here $H^s_{\mathrm{cone}}(N^\wedge)$ for the unit sphere $N := S^n$ in $\mathbb{R}^{n+1}$ is defined as the subspace of all $v \in H^s_{\mathrm{loc}}(\mathbb{R}^{n+1} \setminus \{0\})$ such that $(1-\omega)v \in H^s(\mathbb{R}^{n+1})$, otherwise, for general N we define $H^s_{\mathrm{cone}}(N^\wedge)$ by a simple reduction to the previous case by a localisation on sets $\mathbb{R}_+ \times U$ for coordinate neighbourhoods U on N, cf. [29].

Both $\mathcal{H}^{s,\gamma}(N^\wedge)$ and $\mathcal{K}^{s,\gamma}(N^\wedge)$ are Hilbert spaces with suitable scalar products, where $\mathcal{K}^{0,0}(N^\wedge) = \mathcal{H}^{0,0}(N^\wedge) = t^{-\frac{n}{2}}L^2(\mathbb{R}_+ \times N)$ with L^2 referring to the measure $dtdx$.

If N is equal to the double $\tilde{X} = 2X$ of a manifold X with boundary, consisting of two copies $X_\pm$ of X (and X identified with X_+) we set $\mathcal{K}_0^{s,\gamma}(X_\pm^\wedge) := \{\tilde{u} \in \mathcal{K}^{s,\gamma}(\tilde{X}^\wedge) : \operatorname{supp}\tilde{u} \subset X_\pm^\wedge\}$ which is a closed subspace of $\mathcal{K}^{s,\gamma}(\tilde{X}^\wedge)$, and

$$\mathcal{K}^{s,\gamma}(X^\wedge) := \left\{\tilde{u}|_{\mathrm{int}\,X^\wedge} : \tilde{u} \in \mathcal{K}^{s,\gamma}(\tilde{X}^\wedge)\right\}$$

isomorphic to $\mathcal{K}^{s,\gamma}(X^\wedge) \cong \mathcal{K}^{s,\gamma}(\tilde{X}^\wedge)/\mathcal{K}_0^{s,\gamma}(X_-^\wedge)$ and endowed with the corresponding Hilbert space structure.

For references below we set

$$\mathcal{S}^\gamma(X^\wedge) = \{\omega u + (1-\omega)v : u \in \mathcal{K}^{\infty,\gamma}(X^\wedge),\ v \in \mathcal{S}(\overline{\mathbb{R}}_+, C^\infty(X))\}. \tag{1.9}$$

Given a Hilbert space E together with a strongly continuous group of isomorphisms $\kappa_\delta : E \to E, \delta \in \mathbb{R}_+$, such that $\kappa_\delta\kappa_{\delta'} = \kappa_{\delta\delta'}$ for all $\delta, \delta' \in \mathbb{R}_+$, we say that E is equipped with the group action $\kappa = \{\kappa_\delta\}_{\delta\in\mathbb{R}_+}$. More generally, if a Fréchet space E is written as a projective limit of Hilbert spaces $(E^j)_{j\in\mathbb{N}}$ with continuous embeddings $E^{j+1} \hookrightarrow E^j \hookrightarrow \ldots \hookrightarrow E^0$ for all j, and if κ is a group action on E^0 that restricts to group actions on E^j for all j, we say that E is endowed with the group action κ.

We apply this terminology to $E = \mathcal{K}^{s,\gamma}(N^\wedge)$ for $(\kappa_\delta u)(t,x) = \delta^{\frac{n+1}{2}}u(\delta t, x)$, $\delta \in \mathbb{R}_+, n = \dim N$, and later on to a variety of Fréchet subspaces.

Definition 1.1. 1. Let E be a Hilbert space with group action $\kappa = \{\kappa_\delta\}_{\delta\in\mathbb{R}_+}$. Then $\mathcal{W}^s(\mathbb{R}^q, E)$, $s = \mathbb{R}$, denotes the completion of $\mathcal{S}(\mathbb{R}^q, E)$ with respect to the norm

$$\|u\|_{\mathcal{W}^s(\mathbb{R}^q,E)} = \left\{\int \langle\eta\rangle^{2s}\|\kappa^{-1}_{\langle\eta\rangle}\hat{u}(\eta)\|^2_E d\eta\right\}^{1/2},$$

where $\hat{u}(\eta) = Fu(\eta)$ is the Fourier transform in $\mathbb{R}^q$ and $\langle\eta\rangle = (1+|\eta|^2)^{1/2}$.

2. If $E = \varprojlim_{j\in\mathbb{N}} E^j$ is a Fréchet space with group action, we denote by $\mathcal{W}^s(\mathbb{R}^q, E)$, $s \in \mathbb{R}$, the projective limit of the spaces $\mathcal{W}^s(\mathbb{R}^q, E^j)$, $j \in \mathbb{N}$.

The space $\mathcal{W}^s(\mathbb{R}^q, E)$ in the case (ii) is Fréchet with $\|\cdot\|_{\mathcal{W}^s(\mathbb{R}^q,E^j)}$, $j \in \mathbb{N}$, as a system of norms.

There is also an analogue of standard 'comp' and 'loc' spaces in the present vector-valued case

$$W^s_{\mathrm{comp}}(\Omega, E) \quad \text{and} \quad W^s_{\mathrm{loc}}(\Omega, E)$$

for any open set $\Omega \subseteq \mathbb{R}^q$.

Applying Definition 1.1 to $E = \mathcal{K}^{s,\gamma}(N^\wedge)$ we obtain the spaces

$$\mathcal{W}^{s,\gamma}(N^\wedge \times \mathbb{R}^q) := \mathcal{W}^s(\mathbb{R}^q, \mathcal{K}^{s,\gamma}(N^\wedge))$$

for every $s, \gamma \in \mathbb{R}$. For $N = \tilde{X}$ the space $\mathcal{W}^{s,\gamma}(\tilde{\mathbb{W}})$ is defined as the subspace of all $u \in H^s_{\mathrm{loc}}(\tilde{\mathbb{W}}_{\mathrm{reg}})$ such that $\omega\varphi u = \chi^* v$ for some $v \in \mathcal{W}^{s,\gamma}(\tilde{X}^\wedge \times \mathbb{R}^q)$ for every χ of the kind (1.1) and any cut-off function $\omega(r)$ vanishing for $r > \frac{1}{2}$, and $\varphi \in C_0^\infty(G)$, cf. also Section 2.2 below.

We set

$$\mathcal{W}^{s,\gamma}(\tilde{\mathbb{W}}) := \big\{\omega u + (1-\omega)v : u \in \mathcal{W}^{s,\gamma}(\tilde{X} \times \mathbb{R}^q),\ v \in H^s_{\mathrm{loc}}(\tilde{\mathbb{W}}_{\mathrm{reg}})\big\}$$

for any cut-off function $\omega(r)$, with r referring to the local splitting of variables (r, x, y) near $\tilde{\mathbb{W}}_{\mathrm{sing}}$. For simplicity, pull backs under charts are suppressed in the latter notation; also below we shall identify a neighbourhood of $\tilde{\mathbb{W}}_{\mathrm{sing}}$ with $\overline{\mathbb{R}}_+ \times \tilde{X} \times \mathbb{R}^q$. Now let us set

$$\mathcal{W}^{s,\gamma}_0(\mathbb{W}_\pm) = \{\tilde{u} \in \mathcal{W}^{s,\gamma}(\tilde{\mathbb{W}}) : \operatorname{supp} \tilde{u} \subseteq \mathbb{W}_\pm\}$$

with $\mathbb{W}_\pm$, where the two copies of $\mathbb{W}$ constitute the double $\tilde{\mathbb{W}}$ (with $\mathbb{W}$ being identified with $\mathbb{W}_+$), and

$$\mathcal{W}^{s,\gamma}(\mathbb{W}) = \big\{\tilde{u}|_{\mathrm{int}\,\mathbb{W}_{\mathrm{reg}}} : \tilde{u} \in \mathcal{W}^{s,\gamma}(\tilde{\mathbb{W}})\big\}.$$

The conormal symbol (1.8) can be written in the form

$$\sigma_{\mathrm{c}}(\tilde{A})(w) = r^{-\mu}\omega(r) \sum_{k+j+|\alpha|\le\mu} c_{kj\alpha}(0, r, y)(rw)^k\big(-r\frac{\partial}{\partial r}\big)^j (rD_y)^\alpha$$
$$+ (1-\omega)r^{-\mu}\sum_{k=0}^{\mu} b_k(0)w^j.$$

Proposition 1.2. (1.8) *represents a holomorphic family of continuous operators*

$$\sigma_{\mathrm{c}}(\tilde{A})(w) : \mathcal{W}^{s,\gamma}(\tilde{\mathbb{W}}) \to \mathcal{W}^{s-\mu,\gamma-\mu}(\tilde{\mathbb{W}}). \tag{1.10}$$

Corollary 1.3. *The restriction of* (1.10) *to* $\mathrm{int}\,\mathbb{W}_{\mathrm{reg}}$ *gives us a holomorphic family of continuous operators*

$$\sigma_{\mathrm{c}}(A)(w) : \mathcal{W}^{s,\gamma}(\mathbb{W}) \to \mathcal{W}^{s-\mu,\gamma-\mu}(\mathbb{W}). \tag{1.11}$$

for all $s, \gamma \in \mathbb{R}$.

The continuity of (1.10) for every fixed $w \in \mathbb{C}$ is a consequence of the fact that (1.10) near $\tilde{\mathbb{W}}_{\text{sing}}$ can be regarded as a pseudo-differential operator with an operator-valued symbol of a specific kind. The general definition is as follows.

Let E and $\tilde{E}$ be Hilbert spaces with group actions $\kappa = \{\kappa_\lambda\}_{\lambda \in \mathbb{R}_+}$ and $\tilde{\kappa} = \{\tilde{\kappa}_\lambda\}_{\lambda \in \mathbb{R}_+}$, respectively. Then

$$S^\mu(U \times \mathbb{R}^q; E, \tilde{E}) \tag{1.12}$$

for $\mu \in \mathbb{R}, U \subseteq \mathbb{R}^p$ open, is defined as the set of all $a(y,\eta) \in C^\infty(U \times \mathbb{R}^q, \mathcal{L}(E, \tilde{E}))$ such that

$$\sup_{y \in K, \eta \in \mathbb{R}^q} \langle \eta \rangle^{-\mu+|\beta|} \| \tilde{\kappa}^{-1}_{\langle \eta \rangle} \{ D^\alpha_y D^\beta_\eta a(y,\eta) \} \kappa_{\langle \eta \rangle} \|_{\mathcal{L}(E,\tilde{E})} < \infty \tag{1.13}$$

for every $K \subset\subset U, \alpha \in \mathbb{N}^p, \beta \in \mathbb{N}^q$. The expressions (1.13) form a semi-norm system which makes (1.12) a Fréchet space.

Elements of (1.12) are called operator-valued symbols of order μ.

We also have operator-valued symbols in the case when E or $\tilde{E}$ are Fréchet spaces, written as projective limits of Hilbert spaces, where the respective group actions extend to group actions in the former sense in all Hilbert spaces of the projective limits, cf. [29].

Let $S^\mu_{\text{cl}}(U \times \mathbb{R}^q; E, \tilde{E})$ denote the space of classical symbols, that is, for $a(y,\eta)$ there exist homogeneous components $a_{(\mu-j)}(y,\eta) \in C^\infty(U \times (\mathbb{R}^q \setminus \{0\}), \mathcal{L}(E, \tilde{E}))$, $j \in \mathbb{N}$, i.e.,

$$a_{(\mu-j)}(y, \delta\eta)) = \delta^{\mu-j} \tilde{\kappa}_\delta a_{(\mu-j)}(y,\eta) \kappa^{-1}_\delta$$

for all $\delta \in \mathbb{R}_+$, such that

$$a(y,\eta) - \chi(\eta) \sum_{j=0}^{N} a_{(\mu-j)}(y,\eta) \in S^{\mu-(N+1)}(U \times \mathbb{R}^q; E, \tilde{E})$$

for all $N \in \mathbb{N}$ and any excision function χ. Also $S^\mu_{\text{cl}}(U \times \mathbb{R}^q; E, \tilde{E})$ is a Fréchet space in a natural way.

If a relation holds both for classical or general symbols, we write '(cl)' as subscript.

The spaces $S^\mu_{(\text{cl})}(\mathbb{R}^q; E, \tilde{E})$ of y-independent symbols are closed in $S^\mu_{(\text{cl})}(U \times \mathbb{R}^q; E, \tilde{E})$.

Theorem 1.4. *Let $U = \Omega \times \Omega$ for $\Omega \subseteq \mathbb{R}^q$ open, and let $a(y, y', \eta) \in S^\mu(\Omega \times \Omega \times \mathbb{R}^q; E, \widetilde{E})$.*

Then, setting $\text{Op}(a)u(y) = \iint e^{i(y-y')\eta} a(y,y',\eta) u(y') dy' đ\eta$, $đ\eta = (2\pi)^{-q} d\eta$, the operator

$$\text{Op}(a) : C^\infty_0(\Omega, E) \to C^\infty(\Omega, \widetilde{E})$$

extends to a continuous operator

$$\text{Op}(a) : W^s_{\text{comp}}(\Omega, E) \to W^{s-\mu}_{\text{loc}}(\Omega, \widetilde{E})$$

for every $s \in \mathbb{R}$.

1.2. Boundary Value Problems with the Transmission Property

Let X be a smooth manifold with boundary ∂X. In this section we discuss a few basic constructions on parameter-dependent pseudo-differential operators on X with the transmission property at the boundary. For simplicity, we assume X to be compact. The non-compact case will also be of interest in a variety of cases. We will tacitly use the corresponding generalisations, unless special precautions are necessary; those will be separately described.

Let $\mathrm{Vect}(\cdot)$ denote the set of all smooth complex vector bundles on the manifold in brackets. The manifolds in question are assumed to be equipped with Riemannian metrics and the vector bundles with Hermitian metrics. Operators will refer to Sobolev spaces $H^s(X,E)$ of distributional sections in vector bundles $E \in \mathrm{Vect}(X)$.

Parameter-dependent boundary value problems in our set-up will be families of continuous operators

$$\mathcal{A}(\lambda) : \begin{matrix} H^s(X,E) \\ \oplus \\ H^{s-\frac{1}{2}}(\partial X, J_-) \end{matrix} \to \begin{matrix} H^{s-\mu}(X,F) \\ \oplus \\ H^{s-\mu-\frac{1}{2}}(\partial X, J_+) \end{matrix} \tag{1.14}$$

for $E, F \in \mathrm{Vect}(X)$, $J_-, J_+ \in \mathrm{Vect}(\partial X)$, $\mu \in \mathbb{Z}$, given in the form

$$\mathcal{A}(\lambda) = \begin{pmatrix} \mathrm{r}^+ A(\lambda) \mathrm{e}^+ & 0 \\ 0 & 0 \end{pmatrix} + \mathcal{G}(\lambda), \tag{1.15}$$

with the following ingredients.

We assume that

$$A(\lambda) \in L^\mu_{\mathrm{cl}}(\tilde{X}; \tilde{E}, \tilde{F}; \mathbb{R}^l) \tag{1.16}$$

for vector bundles $\tilde{E}, \tilde{F} \in \mathrm{Vect}(\tilde{X})$ on the double $\tilde{X}$ such that $\tilde{E}|_X = E, \tilde{F}|_X = F$. In other words, (1.16) is a family of classical parameter-dependent pseudo-differential operators on $\tilde{X}$ ($\tilde{E}, \tilde{F}$ in (1.16) means that the operators act between distributional sections of corresponding vector bundles). In addition we require the operators $A(\lambda)$ to have the transmission property at the interface ∂X, cf. [19], or [29]. Let $L^\mu_{\mathrm{cl}}(\tilde{X}; \tilde{E}, \tilde{F}; \mathbb{R}^l)_{\mathrm{tr}}$ denote the subspace of all elements of (1.16) with the transmission property. Moreover,

$$\mathrm{e}^+ : H^s(X,E) \to H^s(\tilde{X}, \tilde{E})$$

is the operator of extension by zero from int X to the double $\tilde{X}$ of X, $s > -\frac{1}{2}$, and

$$\mathrm{r}^+ : H^s(\tilde{X}, \tilde{E}) \to H^s(X,E)$$

the operator of restriction, i.e., $\mathrm{r}^+ u := u|_{\mathrm{int}\, X}$.

To explain $\mathcal{G}(\lambda)$ on the right hand side of (1.15) we first introduce some other notation. Choose any differential operator $T \in \mathrm{Diff}^1(X; E, E)$ on X with smooth coefficients up to the boundary (acting between sections in E) which is of the form $T = \frac{\partial}{\partial x_n} \otimes \mathrm{id}_E$ in the splitting of variables (x', x_n) in a collar neighbourhood $\cong \partial X \times [0,1)$ of ∂X.

Moreover, let $\mathcal{B}^{-\infty,0}(X;\boldsymbol{v})$ for $\boldsymbol{v} = (E,F;J_-,J_+)$ denote the space of all operators

$$\mathcal{C} \in \bigcap_{\substack{s,s'\in\mathbb{R}\\ s>-\frac{1}{2}}} \mathcal{L}(H^s(X,E)\oplus H^{s'}(\partial X,J_-),\ C^\infty(X,F)\oplus C^\infty(\partial X,J_+))$$

such that the formal adjoint $\mathcal{C}^*$ with respect to the respective L^2-scalar products represents an operator

$$\mathcal{C}^* \in \bigcap_{\substack{s,s'\in\mathbb{R}\\ s>-\frac{1}{2}}} \mathcal{L}(H^s(X,F)\oplus H^{s'}(\partial X,J_+),\ C^\infty(X,E)\oplus C^\infty(\partial X,J_-)).$$

The space $\mathcal{B}^{-\infty,0}(X;\boldsymbol{v})$ is Fréchet in a canonical way, and we set $\mathcal{B}^{-\infty,0}(X;\boldsymbol{v};\mathbb{R}^l) = \mathcal{S}(\mathbb{R}^l,\mathcal{B}^{-\infty,0}(X;\boldsymbol{v};\mathbb{R}^l))$. We then form the space $\mathcal{B}^{-\infty,d}(X;\boldsymbol{v};\mathbb{R}^l)$ of all operator families

$$\mathcal{C}(\lambda) = \mathcal{C}_0(\lambda) + \sum_{j=1}^{d} \mathcal{C}_j(\lambda)\operatorname{diag}(T^j,0) \tag{1.17}$$

for arbitrary $\mathcal{C}_j(\lambda) \in \mathcal{B}^{-\infty,0}(X;\boldsymbol{v};\mathbb{R}^l)$.

Let us endow the spaces $L^2(\mathbb{R}_+,\mathbb{C}^N)\oplus\mathbb{C}^M$ and $\mathcal{S}(\overline{\mathbb{R}}_+,\mathbb{C}^N)\oplus\mathbb{C}^M$ for $\mathcal{S}(\overline{\mathbb{R}}_+,\mathbb{C}^N) := \varprojlim_{j\in\mathbb{N}}\langle x_n\rangle^{-j}H^j(\mathbb{R}_+,\mathbb{C}^N)$ with the group action $\varphi(x_n)\oplus c \to (\delta^{\frac{1}{2}}\varphi(\delta x_n))\oplus c$, $\delta\in\mathbb{R}_+$. Then we can form operator-valued symbols

$$g(x',\xi',\lambda) \in S^\mu_{\mathrm{cl}}(\Omega\times\mathbb{R}^{n-1}; L^2(\mathbb{R}_+,\mathbb{C}^e)\oplus\mathbb{C}^{j_-}, \mathcal{S}(\overline{\mathbb{R}}_+,\mathbb{C}^f)\oplus\mathbb{C}^{j_+})$$

with e, f, j_- and j_+ in the meaning of fibre dimensions of the bundles E, F, J_- and J_+, respectively, $\Omega\subseteq\mathbb{R}^{n-1}$ open, such that the pointwise adjoints $g^*(x',\xi',\lambda)$ in the sense of $(g(x',\xi',\lambda)u,v)_{L^2(\mathbb{R}_+,\mathbb{C}^f)\oplus\mathbb{C}^{j_+}} = (u,g^*(x',\xi',\lambda)v)_{L^2(\mathbb{R}_+,\mathbb{C}^e)\oplus\mathbb{C}^{j_-}}$ have the property

$$g^*(x',\xi',\lambda) \in S^\mu_{\mathrm{cl}}(\Omega\times\mathbb{R}^{n-1+l};\ L^2(\mathbb{R}_+,\mathbb{C}^f)\oplus\mathbb{C}^{j_+},\ \mathcal{S}(\overline{\mathbb{R}}_+,\mathbb{C}^e)\oplus\mathbb{C}^{j_-}).$$

Note that then

$$b(x',\xi',\lambda) := g_0(x',\xi',\lambda) + \sum_{j=1}^{d} g_j(x',\xi',\lambda)\operatorname{diag}(\frac{\partial^j}{\partial x_n^j},0)$$

for arbitrary $g_j(x',\xi',\lambda)$ of the abovementioned structure, of order $\mu-j$, represents a symbol

$$b(x',\xi',\lambda) \in S^\mu_{\mathrm{cl}}(\Omega\times\mathbb{R}^{n-1+l}; H^s(\mathbb{R}_+,\mathbb{C}^e)\oplus\mathbb{C}^{j_-}, \mathcal{S}(\overline{\mathbb{R}}_+,\mathbb{C}^f)\oplus\mathbb{C}^{j_+}) \tag{1.18}$$

for all $s > d-\frac{1}{2}$. Also

$$\operatorname{diag}(1,\langle\xi',\lambda\rangle^{\frac{1}{2}})b(x',\xi',\lambda)\operatorname{diag}(1,\langle\xi',\lambda\rangle^{-\frac{1}{2}}) =: h(x',\xi',\lambda) \tag{1.19}$$

is a classical operator-valued symbol, although with slightly modified group actions in the involved spaces, namely $\varphi(x_n)\oplus c \to \delta^{\frac{1}{2}}\varphi(\delta x_n)\oplus\delta^{\frac{1}{2}}c$, $\delta\in\mathbb{R}_+$, instead of the previous ones.

Choose an open covering of the collar neighbourhood $\partial X \times [0,1)$ of the boundary by charts $\chi_k : U'_k \times [0,1) \to \Omega \times \overline{\mathbb{R}}_+$, $\Omega \subseteq \mathbb{R}^{n-1}$ open, $k = 1,\dots,N$, such that $\{U'_1,\dots,U'_N\}$ is an open covering of ∂X by induced charts $\chi'_k : U'_k \to \Omega$. Let $\{\varphi_1,\dots,\varphi_N\}$, $\{\psi_1,\dots,\psi_N\}$ be systems of functions $\varphi_k, \psi_k \in C_0^\infty(U'_l \times [0,1))$ such that $\sum \varphi_k = 1$ in a neighbourhood of ∂X and $\psi_k \equiv 1$ on $\operatorname{supp}\varphi_k$. Set $\varphi'_k = \varphi_k|_{\partial X}$, $\psi'_k = \psi_k|_{\partial X}$. Then, using symbols $h_k(x', \xi', \lambda)$ of the kind (1.19), we can pass to operator pull backs $\mathcal{H}_k(\lambda)$ of pseudo-differential operators $\operatorname{Op}_{x'}(h_k)(\lambda)$ to $U'_k \times (0,1)$ (that also take into account the transition functions of involved bundles) and form

$$\mathcal{H}(\lambda) := \sum_{k=1}^{N} \operatorname{diag}(\varphi_k, \varphi'_k)\mathcal{H}_k(\lambda)\operatorname{diag}(\psi_k, \psi'_k). \tag{1.20}$$

The operator $\mathcal{G}(\lambda)$ in (1.15) is assumed to be of the form

$$\mathcal{G}(\lambda) = \mathcal{H}(\lambda) + \mathcal{C}(\lambda) \tag{1.21}$$

for arbitrary $\mathcal{H}(\lambda)$ as in (1.20) and $\mathcal{C}(\lambda) \in \mathcal{B}^{-\infty,d}(X; \boldsymbol{v}; \mathbb{R}^l)$.

Definition 1.5. Let $\mathcal{B}^{\mu,d}(X; \boldsymbol{v}; \mathbb{R}^l)$ for $\mu \in \mathbb{Z}$, $d \in \mathbb{N}$, $\boldsymbol{v} = (E, F; J_-, J_+)$ denote the space of all operator families of the form (1.15) for arbitrary $A(\lambda) \in L^\mu_{\mathrm{cl}}(\widetilde{X}; \widetilde{E}, \widetilde{F}; \mathbb{R}^l)_{\mathrm{tr}}$ and $\mathcal{G}(\lambda) \in \mathcal{B}^{\mu,d}_G(X; \boldsymbol{v}; \mathbb{R}^l)$, the subspace of all elements of the kind (1.21). Given any $\mathcal{A} \in \mathcal{B}^{\mu,d}(X; \boldsymbol{v}; \mathbb{R}^l)$ we also write $d = d_{\mathcal{A}}$, called the type of $\mathcal{A}$. Moreover, let

$$\mathcal{B}^\mu(X; \boldsymbol{v}; \mathbb{R}^l) = \bigcup_{d \in \mathbb{N}} \mathcal{B}^{\mu,d}(X; \boldsymbol{v}; \mathbb{R}^l),$$

and, similarly, $\mathcal{B}^\mu_G(X; \boldsymbol{v}; \mathbb{R}^l) = \bigcup_{d\in\mathbb{N}} \mathcal{B}^{\mu,d}_G(X; \boldsymbol{v}; \mathbb{R}^l)$.

The operator families $\mathcal{A}(\lambda) \in \mathcal{B}^\mu(X; \boldsymbol{v}; \mathbb{R}^l)$ have a parameter-dependent principal symbolic structure

$$\sigma(\mathcal{A}) = (\sigma_\psi(\mathcal{A}), \sigma_\partial(\mathcal{A}))$$

with the homogeneous principal interior and boundary symbols $\sigma_\psi(\mathcal{A})$ and $\sigma_\partial(\mathcal{A})$, respectively. The interior symbol of $\mathcal{A}(\lambda)$ is the restriction of the parameter-dependent interior symbol of (1.16) to $(T^*X \times \mathbb{R}^l) \setminus 0$, where 0 stands for $(\xi, \lambda) = 0$. The boundary symbol, expressed in a collar neighbourhood of ∂X in the variables (x', x_n), contains an ingredient from $\mathrm{r}^+ A(\lambda)\mathrm{e}^+$, namely

$$\sigma_\partial(\mathrm{r}^+ A\mathrm{e}^+))(x', \xi', \lambda) := \mathrm{r}^+ \operatorname{op}_{x_n}(\sigma_\psi(\mathcal{A})\big|_{x_n=0})(x', \xi', \lambda)\mathrm{e}^+,$$

regarded as a family of maps

$$\sigma_\partial(\mathrm{r}^+ A\mathrm{e}^+)(x', \xi', \lambda) : H^s(\mathbb{R}_+) \to H^{s-\mu}(\mathbb{R}_+)$$

for $s > -\frac{1}{2}$, invariantly defined as an operator family parametrised by $(T^*(\partial X) \times \mathbb{R}^l) \setminus 0$, with 0 being interpreted as $(\xi', \lambda) = 0$.

Another ingredient is defined by the summand $\mathcal{G}(\lambda)$ in (1.21), more precisely by $\mathcal{H}(\lambda)$ in (1.21), locally given by

$$\sigma_\partial(\mathcal{G})(x', \xi', \lambda) = \operatorname{diag}(1, |\xi', \lambda|^{\frac{1}{2}}) b_{(\mu)}(x', \xi', \lambda) \operatorname{diag}(1, |\xi', \lambda|^{-\frac{1}{2}})$$

for $(\xi', \lambda) \neq 0$, cf. (1.18), (1.19), and

$$b_{(\mu)}(x', \xi', \lambda) = g_{0,(\mu)}(x', \xi', \lambda) + \sum_{j=1}^{d} g_{j,(\mu-j)}(x', \xi', \lambda) \operatorname{diag}\Big(\frac{\partial^j}{\partial x_n^j}, 0\Big),$$

with $g_{j,(\mu-j)}(x', \xi', \lambda)$ as the homogeneous principal components of the classical operator-valued symbols $g_j(x', \xi', \lambda)$ of order $\mu - j$, $j = 0, \ldots, d$. Then

$$\sigma_\partial(\mathcal{A})(x', \xi', \lambda) := \operatorname{diag}\big(\sigma_\partial(\mathrm{r}^+ A \mathrm{e}^+)(x', \xi', \lambda), 0\big) + \sigma_\partial(\mathcal{G})(x', \xi', \lambda)$$

represents an invariantly defined family of operators

$$\sigma_\partial(\mathcal{A})(x', \xi', \lambda) : \begin{array}{c} H^s(\mathbb{R}_+) \\ \oplus \\ \mathbb{C}^{j_-} \end{array} \rightarrow \begin{array}{c} H^{s-\mu}(\mathbb{R}_+) \\ \oplus \\ \mathbb{C}^{j_+} \end{array} \tag{1.22}$$

parametrised by $(T^*(\partial X) \times \mathbb{R}^l) \setminus 0$, with homogeneity

$$\sigma_\partial(\mathcal{A})(x', \delta\xi', \delta\lambda) = \delta^\mu \begin{pmatrix} \kappa_\delta & 0 \\ 0 & \delta^{\frac{1}{2}} \end{pmatrix} \sigma_\partial(\mathcal{A}(x', \xi', \lambda) \begin{pmatrix} \kappa_\delta & 0 \\ 0 & \delta^{\frac{1}{2}} \end{pmatrix}^{-1}$$

for all $\delta \in \mathbb{R}_+$.

Given a Fréchet space F and an open set $U \subseteq \mathbb{C}$ we denote by $\mathcal{A}(U, F)$ the space of all holomorphic functions in U with values in F. By

$$\mathcal{B}^{\mu,d}(X; \boldsymbol{v}; \mathbb{C} \times \mathbb{R}^q) \tag{1.23}$$

we will denote the space of all operator families

$$f(z, \eta) \in \mathcal{A}(\mathbb{C}, \mathcal{B}^{\mu,d}(X; \boldsymbol{v}; \mathbb{R}^q))$$

such that

$$f(\beta + i\varrho, \eta) \in \mathcal{B}^{\mu,d}(X; \boldsymbol{v}; \mathbb{R}^{1+q}_{\varrho,\eta})$$

for every $\beta \in \mathbb{R}$, uniformly in $c \leq \beta \leq c'$ for every $c \leq c'$. For $q = 0$ we simply write $\mathcal{B}^{\mu,d}(X; \boldsymbol{v}; \mathbb{C})$. Moreover, we set

$$\mathcal{B}^{\mu}(X; \boldsymbol{v}; \mathbb{C} \times \mathbb{R}^q) = \bigcup_{d \in \mathbb{N}} \mathcal{B}^{\mu,d}(X; \boldsymbol{v}; \mathbb{C} \times \mathbb{R}^q). \tag{1.24}$$

The space (1.23) Fréchet in a natural way.

Remark 1.6. As is well known, cf. [22], [23], or [9], the space $\mathcal{B}^{\mu,d}(X; \boldsymbol{v}; \mathbb{C} \times \mathbb{R}^q)$ is rich in the sense that for every operator family $p \in \mathcal{B}^{\mu,d}(X; \boldsymbol{v}; \mathbb{R}^{1+q}_{\varrho,\eta})$ there exists an $f(z, \eta) \in \mathcal{B}^{\mu,d}(X; \boldsymbol{v}; \mathbb{C} \times \mathbb{R}^q)$ such that

$$p(\varrho, \eta) = f(i\varrho, \eta) \bmod \mathcal{B}^{-\infty,d}(X; \boldsymbol{v}; \mathbb{R}^{1+q}).$$

The construction of f is based on a kernel cut-off construction, cf. Section 2.1 below.

The spaces $\mathcal{B}^{\mu,d}(X;\boldsymbol{v};\mathbb{R}^l)$ as well as (1.23) have a straightforward generalisation to the case of a non-compact C^∞ manifold X with boundary. This will tacitly be used below. Instead of (1.14) we then talk about continuous maps between Sobolev spaces with subscript 'comp' or 'loc'.

An interesting situation with a specific control of the non-compactness at infinity is the case of an infinite stretched cone $X^\wedge = \mathbb{R}_+ \times X \ni (r,x)$ for a compact C^∞ manifold X with boundary. Let us formulate a few basic notions for this case. We first discuss operators far from $r=0$; operators close to $r=0$ will be studied in the following section.

The typical situation is the half-space $\overline{\mathbb{R}^{n+1}_+} = \{\tilde{x} = (\tilde{x}', \tilde{x}_{n+1}) \in \mathbb{R}^{n+1} : \tilde{x}_{n+1} \geq 0\}$, $\tilde{x}' = (\tilde{x}_1, \dots, \tilde{x}_n)$; in this case X is equal to $S^n_+ = S^n \cap \overline{\mathbb{R}^{n+1}_+}$ the upper half of the unit sphere. A simple localisation on conical subsets Γ of $\overline{\mathbb{R}^{n+1}_+}$, using invariance of our constructions under transition maps $\kappa : \Gamma \to \tilde{\Gamma}$ that satisfy $\kappa(\lambda\tilde{x}) = \lambda\kappa(\tilde{x})$ for $|\tilde{x}| \geq C$ for some $C > 0$ and $\lambda \geq 1$, allows us to pass to operators on $X^\wedge$. In other words, we mainly look at $\overline{\mathbb{R}^{n+1}_+}$.

In this case we define a global calculus of operators of the class

$$\mathcal{B}^{\mu,d;\varrho}(\overline{\mathbb{R}^{n+1}_+};\boldsymbol{v}) \tag{1.25}$$

with a weight $\varrho \in \mathbb{R}$ at infinity. We do not need any parameter-dependent variant here.

Similarly to $\mathcal{B}^{\mu,d}(X;\boldsymbol{v})$, cf. Definition 1.5, the elements of (1.25) consist of 2×2 block matrices of continuous operators

$$\mathcal{A} : \begin{array}{c} H^{s;\delta}(\mathbb{R}^{n+1}_+, \mathbb{C}^e) \\ \oplus \\ H^{s-\frac{1}{2};\delta}(\mathbb{R}^n, \mathbb{C}^{j_-}) \end{array} \longrightarrow \begin{array}{c} H^{s-\mu;\delta-\varrho}(\mathbb{R}^{n+1}_+, \mathbb{C}^f) \\ \oplus \\ H^{s-\mu-\frac{1}{2};\delta-\varrho}(\mathbb{R}^n, \mathbb{C}^{j_+}) \end{array}$$

for $s > d - \frac{1}{2}$. Here $H^{s;\delta}(\mathbb{R}^{n+1}_+) := \langle\tilde{x}\rangle^{-\delta} H^s(\mathbb{R}^{n+1}_+)$, $H^{s';\delta}(\mathbb{R}^n) := \langle\tilde{x}'\rangle^{-\delta} H^{s'}(\mathbb{R}^n)$. For simplicity, we consider the case $e = f = 1$, $j_- = j_+ = 0$. The constructions for the general case are straightforward and left to the reader.

First, on $\mathbb{R}^{n+1}$ we have the standard calculus of pseudo-differential operators with exit conditions. Let $S^{\mu;\varrho}(\mathbb{R}^{2(n+1)}_{\tilde{x},\tilde{\xi}})$ for $\mu, \varrho \in \mathbb{R}$ denote the set of all $a(\tilde{x},\tilde{\xi}) \in C^\infty(\mathbb{R}^{2(n+1)})$ such that

$$\sup_{(\tilde{x},\tilde{\xi})\in\mathbb{R}^{2(n+1)}} \langle\tilde{x}\rangle^{-\varrho+|\alpha|} \langle\tilde{\xi}\rangle^{-\mu+|\beta|} |D^\alpha_{\tilde{x}} D^\beta_{\tilde{\xi}} a(\tilde{x},\tilde{\xi})| < \infty \tag{1.26}$$

for all $\alpha, \beta \in \mathbb{N}^{n+1}$. Observe that $S^{\mu;\varrho}(\mathbb{R}^{2(n+1)})$ for $\varrho \geq 0$ and $\mu \geq 0$, respectively contains the subspaces $S^\mu_{(\mathrm{cl})}(\mathbb{R}^{n+1}_{\tilde{\xi}})$ and

$S^\varrho_{(\mathrm{cl})}(\mathbb{R}^{n+1}_{\tilde{x}})$ 'with constant coefficients' ('(cl)' means classical or non-classical in the respective variables, treated as covariables). Note that $S^\mu_{\mathrm{cl}}(\mathbb{R}^{n+1})$ is nuclear in the natural Fréchet topology. Let us define

$$S^{\mu;\varrho}_{\mathrm{cl}}(\mathbb{R}^{2(n+1)}) := S^\mu_{\mathrm{cl}}(\mathbb{R}^{n+1}_{\tilde{\xi}}) \hat{\otimes}_\pi S^\varrho_{\mathrm{cl}}(\mathbb{R}^{n+1}_{\tilde{x}})$$

which is the set of all $a(\tilde{x},\tilde{\xi}) \in S^{\mu;\varrho}(\mathbb{R}^{2(n+1)})$ that are classical both in $\tilde{\xi}$ and $\tilde{x}$. Moreover, let $S^{\mu;\varrho}_{\mathrm{cl}}(\mathbb{R}^{2(n+1)})_{\mathrm{tr}}$ be the subspace of all elements with the transmission property at $\tilde{x}_{n+1} = 0$.

Let us set

$$L^{\mu;\varrho}_{\mathrm{cl}}(\mathbb{R}^{n+1})_{\mathrm{tr}} = \{\mathrm{Op}_{\tilde{x}}(a) : a(\tilde{x},\tilde{\xi}) \in S^{\mu;\varrho}_{\mathrm{cl}}(\mathbb{R}^{2(n+1)})_{\mathrm{tr}}\}. \tag{1.27}$$

Then $\mathrm{r}^+ A\mathrm{e}^+$ for $A \in L^{\mu;\varrho}_{\mathrm{cl}}(\mathbb{R}^{n+1})_{\mathrm{tr}}$ (with e^+ being the operator of extension by zero from $\mathbb{R}^{n+1}_+$ to $\mathbb{R}^{n+1}$ and r^+ the restriction from $\mathbb{R}^{n+1}$ to $\mathbb{R}^{n+1}_+$) induces continuous operators

$$\mathrm{r}^+ A\mathrm{e}^+ : H^{s;\delta}(\mathbb{R}^{n+1}_+) \to H^{s-\mu;\delta-\varrho}(\mathbb{R}^{n+1}_+)$$

for all $s > -\frac{1}{2}$; $\delta \in \mathbb{R}$.

Symbols $a(\tilde{x},\tilde{\xi})$ that are classical in $\tilde{\xi}$ of order μ and $\tilde{x}$ of order ϱ have a triple of principal components, consisting of

$$\sigma_\psi(a)(\tilde{x},\tilde{\xi}) \in C^\infty(\mathbb{R}^{n+1}_{\tilde{\xi}} \setminus \{0\}, S^{\varrho}_{\mathrm{cl}}(\mathbb{R}^{n+1}_{\tilde{x}})),$$

homogeneous of order μ in $\tilde{\xi} \neq 0$,

$$\sigma_{\mathrm{e}}(a)(\tilde{x},\tilde{\xi}) \in C^\infty(\mathbb{R}^{n+1}_{\tilde{x}} \setminus \{0\}, S^{\mu}_{\mathrm{cl}}(\mathbb{R}^{n+1}_{\tilde{\xi}})),$$

homogeneous of order ϱ in $\tilde{x} \neq 0$, and

$$\sigma_{\psi,\mathrm{e}}(a)(\tilde{x},\tilde{\xi}) \in C^\infty((\mathbb{R}^{n+1}_{\tilde{x}} \setminus \{0\}) \times (\mathbb{R}^{n+1}_{\tilde{\xi}} \setminus \{0\})),$$

homogeneous in $\tilde{\xi} \neq 0$ of order μ and $\tilde{x} \neq 0$ of order ϱ. For $A = \mathrm{Op}(a)$ we then set

$$\sigma_\psi(\mathrm{r}^+ A\mathrm{e}^+) = \sigma_\psi(a),\ \sigma_{\mathrm{e}}(\mathrm{r}^+ A\mathrm{e}^+) = \sigma_{\mathrm{e}}(a),\ \sigma_{\psi,\mathrm{e}}(\mathrm{r}^+ A\mathrm{e}^+) = \sigma_{\psi,\mathrm{e}}(a).$$

There is also a variant of operator-valued symbols with exit conditions in $\mathbb{R}^n_{\tilde{x}'}$, acting between Hilbert spaces E and $\tilde{E}$ with group action, cf. (1.12) and (1.13). Let $S^{\mu;\varrho}(\mathbb{R}^{2n}; E, \tilde{E})$ be the set of all $a(\tilde{x}',\tilde{\xi}') \in C^\infty(\mathbb{R}^{2n}, \mathcal{L}(E,\tilde{E}))$ such that

$$\sup_{(\tilde{x}',\tilde{\xi}')\in\mathbb{R}^{2n}} \langle\tilde{x}'\rangle^{-\varrho+|\alpha|}\langle\tilde{\xi}'\rangle^{-\mu+|\beta|}\|\tilde{\kappa}^{-1}_{\langle\xi'\rangle}\{D^\alpha_{\tilde{x}'}D^\beta_{\tilde{\xi}'}a(\tilde{x}',\tilde{\xi}')\}\kappa_{\langle\xi'\rangle}\|_{\mathcal{L}(E,\tilde{E})} < \infty$$

for all $\alpha, \beta \in \mathbb{N}^n$.

Also in the operator-valued case there is a natural notion of classical symbols in both variables $\tilde{x}'$ and $\tilde{\xi}'$, for more details, cf. [9, Chapter 3]. Let $S^{\mu;\varrho}_{\mathrm{cl}}(\mathbb{R}^{2n}; E, \tilde{E})$ denote the corresponding subspace of $S^{\mu;\varrho}(\mathbb{R}^{2n}; E, \tilde{E})$. Finally, we can generalise such symbol spaces to the case of Fréchet spaces $E, \tilde{E}$ with group actions.

Classical operator-valued symbols $a(\tilde{x}',\tilde{\xi}')$ in $(\tilde{x}',\tilde{\xi}')$ of order μ in $\tilde{\xi}'$ and ϱ in $\tilde{x}'$ have also a triple of principal symbols, namely

$$\sigma_\partial(a)(\tilde{x}') \in C^\infty(\mathbb{R}^n_{\tilde{\xi}'} \setminus \{0\}, S^{\varrho}_{\mathrm{cl}}(\mathbb{R}^n_{\tilde{x}'}; E, \tilde{E})),$$

homogeneous of order μ in $\tilde{\xi}' \neq 0$ in the sense

$$\sigma_\partial(a)(\tilde{x}',\lambda\tilde{\xi}') = \lambda^\mu\tilde{\kappa}_\lambda\sigma_\partial(a)(\tilde{x}',\tilde{\xi}')\kappa^{-1}_\lambda \tag{1.28}$$

for all $\lambda \in \mathbb{R}_+$,

$$\sigma_{e'}(a)(\tilde{x}', \tilde{\xi}') \in C^\infty(\mathbb{R}^n_{\tilde{x}'} \setminus \{0\}, S^\mu_{\mathrm{cl}}(\mathbb{R}^n_{\tilde{\xi}'}; E, \tilde{E})),$$

homogeneous of order ϱ in $\tilde{x}' \neq 0$ in the sense

$$\sigma_{e'}(a)(\delta\tilde{x}', \tilde{\xi}') = \delta^\varrho \sigma_{e'}(a)(\tilde{x}', \tilde{\xi}') \tag{1.29}$$

for all $\delta \in \mathbb{R}_+$, and a corresponding mixed term

$$\sigma_{\partial,e'}(a)(\tilde{x}', \tilde{\xi}') \in C^\infty((\mathbb{R}^n_{\tilde{x}'} \setminus \{0\}) \times (\mathbb{R}^n_{\tilde{\xi}'} \setminus \{0\}), \mathcal{L}(E, \tilde{E})),$$

homogeneous in $\tilde{\xi}' \neq 0$ of order μ as (1.28) and in $\tilde{x}' \neq 0$ of order ϱ as (1.29).

In particular, we can talk about so called Green symbols $g(\tilde{x}', \tilde{\xi}')$ of type 0 in the half-space, defined by the conditions

$$g(\tilde{x}', \tilde{\xi}'), g^*(\tilde{x}', \tilde{\xi}') \in S^{\mu;\varrho}_{\mathrm{cl}}(\mathbb{R}^n \times \mathbb{R}^n; L^2(\mathbb{R}_+), \mathcal{S}(\overline{\mathbb{R}}_+))$$

with g^* being the pointwise adjoint with respect to the $L^2(\mathbb{R}_+)$-scalar product. Moreover, Green symbols $g(\tilde{x}', \tilde{\xi}')$ of type $d \in \mathbb{N}$ are defined by

$$g(\tilde{x}', \tilde{\xi}') = \sum_{j=0}^{d} g_j(\tilde{x}', \tilde{\xi}') \frac{\partial^j}{\partial \tilde{x}^j_{n+1}}$$

with Green symbols $g_j(\tilde{x}', \tilde{\xi}')$ of type 0 and order $(\mu - j, \varrho)$, $j = 0, \ldots, d$.

Let $\mathcal{B}^{-\infty,d;-\infty}(\overline{\mathbb{R}^{n+1}_+})$ denote the set of all operators

$$C = \sum_{j=0}^{d} C_j \frac{\partial^j}{\partial \tilde{x}^j_{n+1}},$$

where C_j are integral operators with kernels in

$$\mathcal{S}(\overline{\mathbb{R}^{n+1}_+ \times \mathbb{R}^{n+1}_+}) = \mathcal{S}(\mathbb{R}^{2(n+1)})|_{\overline{\mathbb{R}^{n+1}_+ \times \mathbb{R}^{n+1}_+}}.$$

An element $\psi \in C^\infty(\overline{\mathbb{R}^{n+1}_+})$ is called an admissible cut-off function, if it has the following properties:

(i) There are constants $R < R'$ such that

$$\psi = 1 \quad \text{for} \quad \tilde{x} \in L_R\ , \ \psi = 0 \quad \text{for} \quad \tilde{x} \notin L_{R'}$$

for $L_c := \{\tilde{x} \in \mathbb{R}^{n+1} : |\tilde{x}_{n+1}| \leq c|\tilde{x}'|\} \cup \{\tilde{x} \in \mathbb{R}^{n+1} : |\tilde{x}_{n+1}| \leq c\}$, $c > 0$.

(ii) $\psi(\delta\tilde{x}) = \psi(\tilde{x})$ for all $\delta \geq 1$, $|\tilde{x}| \geq \tilde{c}$ for a sufficiently large $\tilde{c} > 0$.

Now $\mathcal{B}^{\mu,d;\varrho}(\overline{\mathbb{R}^{n+1}_+})$ is defined as the set of all operators

$$A := \mathrm{r}^+ \operatorname{Op}_{\tilde{x}}(a) \mathrm{e}^+ + \psi_0 \operatorname{Op}_{\tilde{x}'}(g) \psi_1 + C \tag{1.30}$$

for arbitrary $a(\tilde{x}, \tilde{\xi}) \in S^{\mu;\varrho}_{\mathrm{cl}}(\mathbb{R}^{2(n+1)})_{\mathrm{tr}}$, a Green symbol $g(\tilde{x}', \tilde{\xi}')$ of type d, admissible cut-off functions ψ_0, ψ_1, and a smoothing operator $C \in \mathcal{B}^{-\infty,d;-\infty}(\overline{\mathbb{R}^{n+1}_+})$.

Let N be a closed compact C^∞ manifold, $n = \dim N$, and form a C^∞ manifold $N_\asymp \cong \mathbb{R} \times N \ni (r, x)$ with conical exits for $r \to \pm\infty$ (i.e., on $N_\asymp$ we fix a Riemannian metric that has the form $dr^2 + r^2 g_N$ for $|r| > R$ for some $R > 0$ where

g_N is a Riemannian metric on N). Moreover, let $H^{s;\delta}(N_{\asymp})$ for $s, \delta \in \mathbb{R}$ denote the subspace of all $u \in H^s_{\mathrm{loc}}(N_{\asymp})$ such that $(1-\omega)u, (1-\omega)u^{\vee} \in \langle r \rangle^{-\delta} H^s_{\mathrm{cone}}(N^{\wedge})$ for $u^{\vee}(r,x) = u(-r,x)$ and any cut-off function $\omega(r)$ on the positive half-axis. On $N_{\asymp}$ we then have a calculus of pseudo-differential operators $L^{\mu;\varrho}(N_{\asymp}), \mu, \varrho \in \mathbb{R}$, the local symbols of which on subsets of $U_{\asymp} \cong \mathbb{R} \times U$ (for any coordinate neighbourhood U on N) satisfy the symbolic estimates of the form (1.26) in coordinates $\tilde{x} \in \mathbb{R}^{n+1}$ (the corresponding charts $\chi : U_{\asymp} \to \Gamma$ are assumed to be homogeneous in the sense $\chi(\lambda r, x) = \lambda \tilde{x}$ for $\lambda \geq 1, |\tilde{x}| > R$ for some $R > 0$). In this calculus the smoothing operators have kernels in $\mathcal{S}(N_{\asymp} \times N_{\asymp}) \cong \mathcal{S}(\mathbb{R} \times \mathbb{R}, C^{\infty}(N \times N))$.

Taking classical local symbols in $(\tilde{x}, \tilde{\xi})$ we get the subspace $L^{\mu;\varrho}_{\mathrm{cl}}(N_{\asymp})$ of classical operators with exit behaviour for $|r| \to \infty$.

Given vector bundles $\tilde{E}, \tilde{F} \in \mathrm{Vect}(N_{\asymp})$ we have similarly the spaces

$$L^{\mu;\varrho}_{(\mathrm{cl})}(N_{\asymp}; \tilde{E}, \tilde{F})$$

of operators $\tilde{A}$ acting between corresponding spaces of distributional sections

$$\tilde{A} : H^{s;\delta}(N_{\asymp}, \tilde{E}) \to H^{s-\mu,\delta-\varrho}(N_{\asymp}, \tilde{F}).$$

Applying this picture to $N := 2X$, the double of a compact C^{∞} manifold X with boundary, there is a subspace $L^{\mu;\varrho}_{\mathrm{cl}}((2X)_{\asymp}; \tilde{E}, \tilde{F})_{\mathrm{tr}}$ of operators which have the transmission property at $(\partial X)_{\asymp}$. Then, if $\mathrm{r}^+ : H^{s;\delta}((2X)_{\asymp}, \tilde{E}) \to H^{s;\delta}(X_{\asymp}, E)$ (for $E = \tilde{E}|_{X_{\asymp}}$) is the operator of restriction to $\mathrm{int}\, X_{\asymp}$, $\mathrm{e}^+ : H^{s;\delta}(\mathrm{int}\, X_{\asymp}, E) \to H^{s;\delta}((2X)_{\asymp}, \tilde{E})$ the extension by zero to the opposite side of $\mathrm{int}\, X_{\asymp}$ in $(2X)_{\asymp}$, every $\tilde{A} \in L^{\mu;\varrho}_{\mathrm{cl}}((2X)_{\asymp}; \tilde{E}, \tilde{F})_{\mathrm{tr}}$ gives rise to continuous operators

$$\mathrm{r}^+ \tilde{A} \mathrm{e}^+ : H^{s;\delta}(X_{\asymp}, E) \to H^{s-\mu;\delta-\varrho}(X_{\asymp}, F)$$

for every $s, \delta \in \mathbb{R}, s > -\frac{1}{2}$.

There is also a generalisation of Green and smoothing operators of the class $\mathcal{B}^{\mu,d;\varrho}(\overline{\mathbb{R}}^{n+1}_+)$ to the case of a smooth manifold $X_{\asymp}$ with boundary and conical exits for $|r| \to \infty$, including the aspect of operators between sections of vector bundles and additional trace and potential operators, cf. also [9, Chapter 3]. This gives us spaces of boundary value problems

$$\mathcal{B}^{\mu,d;\varrho}(X_{\asymp}; \boldsymbol{v}) \quad \text{for} \quad \boldsymbol{v} = (E, F; J_-, J_+), \tag{1.31}$$

$E, F \in \mathrm{Vect}(X_{\asymp}),\ J_{\pm} \in \mathrm{Vect}((\partial X)_{\asymp})$. The operators in (1.31) have the form

$$\mathcal{A} = \begin{pmatrix} \mathrm{r}^+ \tilde{A} \mathrm{e}^+ & 0 \\ 0 & 0 \end{pmatrix} + \mathcal{G},$$

where $\mathcal{G}$ is a 2×2 block matrix of Green, trace and potential operators with exit behaviour for $|r| \to \infty$, and $\tilde{A} \in L^{\mu;\varrho}_{\mathrm{cl}}((2X)_{\asymp}; E, F)_{\mathrm{tr}}$. The operators $\mathcal{A}$ in (1.31) are continuous in the sense

$$\begin{matrix} H^{s;\delta}(X_{\asymp}, E) & & H^{s-\mu;\delta-\varrho}(X_{\asymp}, F) \\ \oplus & \to & \oplus \\ H^{s-\frac{1}{2};\delta}((\partial X)_{\asymp}, J_-) & & H^{s-\mu-\frac{1}{2};\delta-\varrho}((\partial X)_{\asymp}, J_+) \end{matrix} \tag{1.32}$$

for all $s > d - \frac{1}{2}$. Let us set

$$\mathcal{B}^{\mu,d;\varrho}(X^\wedge; \boldsymbol{v}) = \{\mathcal{A}|_{X^\wedge} : \mathcal{A} \in \mathcal{B}^{\mu,d;\varrho}(X_\asymp; \boldsymbol{v})\}; \tag{1.33}$$

in this notation $X^\wedge$ is regarded as a subset of $X_\asymp$ in a canonical way, and bundles on $X_\asymp$ $((\partial X)_\asymp)$ and their restrictions to $X^\wedge$ $((\partial X)^\wedge)$ are denoted by the same letters.

Operators $\mathcal{A} \in \mathcal{B}^{\mu,d;\varrho}(X_\asymp; \boldsymbol{v})$ with $A := \mathrm{r}^+ \widetilde{A} \mathrm{e}^+$ for $\widetilde{A} \in L_{\mathrm{cl}}^{\mu;\varrho}((2X)_\asymp; E, F)_{\mathrm{tr}}$ in the upper left corner have a principal symbolic structure

$$\sigma(\mathcal{A}) = (\sigma_\psi(\mathcal{A}), \sigma_{\mathrm{e}}(\mathcal{A}), \sigma_{\psi,\mathrm{e}}(\mathcal{A}); \sigma_\partial(\mathcal{A}), \sigma_{\mathrm{e}'}(\mathcal{A}), \sigma_{\partial,\mathrm{e}'}(\mathcal{A})).$$

Here $\sigma_\psi(\mathcal{A})$ is the standard homogeneous principal symbol of A of order μ, further $\sigma_{\mathrm{e}}(\mathcal{A})$ is the homogeneous principal exit symbol of A of order ϱ (by definition, this concerns homogeneity in the variable r for $r \to \pm\infty$) and $\sigma_{\psi,\mathrm{e}}(\mathcal{A})$ is the homogeneous principal part of $\sigma_{\mathrm{e}}(\mathcal{A})$ of order μ in the $X_\asymp$-covariables. Moreover, $\sigma_\partial(\mathcal{A})$ is the standard homogeneous principal symbol of $\mathcal{A}$ of order μ, further $\sigma_{\mathrm{e}'}(\mathcal{A})$ is the homogeneous principal exit symbol of $\mathcal{A}$ of order ϱ (which refers again to homogeneity of order ϱ in the variable r for $r \to \pm\infty$), and $\sigma_{\partial,\mathrm{e}'}(\mathcal{A})$ is the homogeneous principal part of $\sigma_{\mathrm{e}'}(\mathcal{A})$ of order μ in the $\partial X_\asymp$-covariables.

An operator $\mathcal{A} \in \mathcal{B}^{\mu,d;\varrho}(X_\asymp; \boldsymbol{v})$ is said to be elliptic, if all components of $\sigma(\mathcal{A})$ are bijective on the respective sets of variables and covariables.

Theorem 1.7. *An operator $\mathcal{A} \in \mathcal{B}^{\mu,d;\varrho}(X_\asymp; \boldsymbol{v})$ is elliptic if and only if $\mathcal{A}$ defines a Fredholm operator* (1.32) *for any $s \in \mathbb{R}, s > \max(\mu, d) - \frac{1}{2}$.*

If $\mathcal{A} \in \mathcal{B}^{\mu,d;\varrho}(X_\asymp; \boldsymbol{v})$ is elliptic, there is a parametrix $\mathcal{P} \in \mathcal{B}^{-\mu,h;-\varrho}(X_\asymp; \boldsymbol{v}^{-1})$ for $\boldsymbol{v}^{-1} := (F, E; J_+, J_-)$ and $h = \max(d - \mu, 0)$ such that

$$\mathcal{I} - \mathcal{P}\mathcal{A} \in \mathcal{B}^{-\infty,d_{\mathrm{l}};0}(X_\asymp; \boldsymbol{v}_{\mathrm{l}}), \quad \mathcal{I} - \mathcal{A}\mathcal{P} \in \mathcal{B}^{-\infty,d_{\mathrm{r}};0}(X_\asymp; \boldsymbol{v}_{\mathrm{r}})$$

for $d_{\mathrm{l}} = \max(\mu, d), d_{\mathrm{r}} = \max(d - \mu, 0)$, and $\boldsymbol{v}_{\mathrm{l}} = (E, E; J_-, J_-), \boldsymbol{v}_{\mathrm{r}} = (F, F; J_+, J_+)$.

Boundary value problems with the transmission property on a manifold with exits to infinity have been studied systematically by Schrohe [21] where one can find, in particular, the necessity of the ellipticity for the Fredholm property of (1.32). In the present paper we refer to the corresponding calculus of [9, Chapter 3].

1.3. The Cone Algebra

We now turn to boundary value problems of the classes $\mathcal{B}^{\mu,d}$ globally on $X^\wedge = \mathbb{R}_+ \times X \ni (r, x)$. Close to $r = 0$ we impose the structure of the cone algebra. For $r \to \infty$ the calculus corresponds to operators on a manifold with conical exit to infinity as formulated in the preceding section.

Let $\omega_0(r), \omega_1(r), \omega_2(r)$ be cut-off functions on the half-axis, $\omega_1 \equiv 1$ on $\operatorname{supp} \omega_0$, and $\omega_0 \equiv 1$ on $\operatorname{supp} \omega_2$, and set

$$\mathcal{A} = \omega_0 \mathcal{A}_{\mathrm{cone}} \omega_1 + (1 - \omega_0) \mathcal{A}_{\mathrm{exit}} (1 - \omega_2) + \mathcal{C}. \tag{1.34}$$

Here $\mathcal{A}_{\text{exit}} \in \mathcal{B}^{\mu,d;0}(X^\wedge;\boldsymbol{v})$, and $\mathcal{A}_{\text{cone}}$ is of the form

$$\mathcal{A}_{\text{cone}} = r^{-\mu}\{\text{op}_M^{\gamma-\frac{n}{2}}(h) + \sum_{j=0}^{k} r^j\big[\text{op}_M^{\delta_j-\frac{n}{2}}(f_j) + \text{op}_M^{\tilde{\delta}_j-\frac{n}{2}}(\tilde{f}_j)\big]\} \tag{1.35}$$

for some $k \in \mathbb{N}$ and weights

$$\delta_0 = \tilde{\delta}_0 = \gamma \text{ and } \delta_j = \gamma - \frac{1}{3},\ \tilde{\delta}_j = \gamma - \frac{2}{3} \text{ for } 1 \le j \le k. \tag{1.36}$$

Note that in (1.36) we could fix any other weights $\delta_j, \tilde{\delta}_j$ such that $\gamma - j \le \delta_j,\ \tilde{\delta}_j \le \gamma$ and $\delta_j \ne \tilde{\delta}_j$ for $j \ge 1$.

We assume

$$h(r,z) \in C^\infty(\overline{\mathbb{R}}_+, \mathcal{B}^{\mu,d}(X;\boldsymbol{v};\mathbb{C})). \tag{1.37}$$

In order to define f_j and $\tilde{f}_j$ we first introduce so called discrete and continuous asymptotic types of Mellin symbols.

Given a set $A \subseteq \mathbb{C}$ we define $A^I := \{\lambda z_1 + (1-\lambda)z_2 : z_1, z_2 \in A, \operatorname{Re} z_1 = \operatorname{Re} z_2, 0 \le \lambda \le 1\}$. Let $\mathcal{V}$ denote the system of all closed subsets V of $\mathbb{C}$ such that $V \cap \{c \le \operatorname{Re} z \le c'\}$ is compact for every $c \le c'$, and $V^I = V$. A V-excision function is any $\chi \in C^\infty(\mathbb{C})$ such that $\chi(z) = 0$ for $\operatorname{dist}(z,V) < \varepsilon_0$, $(z) = 1$ for $\operatorname{dist}(z,V) > \varepsilon_1$ for certain $0 < \varepsilon_0 < \varepsilon_1$.

For every $V \in \mathcal{V}$ we form the space

$$\mathcal{B}_V^{-\infty,d}(X;\boldsymbol{v};\mathbb{C}) \tag{1.38}$$

of all $f(z) \in \mathcal{A}(\mathbb{C} \setminus V, \mathcal{B}^{-\infty,d}(X;\boldsymbol{v}))$ such that for every V-excision function $\chi(z)$, we have

$$\chi(z)f(z)|_{z=\beta+i\varrho} \in \mathcal{B}^{-\infty,d}(X;\boldsymbol{v};\mathbb{R}_\varrho)$$

for every $\beta \in \mathbb{R}$, uniformly in $c \le \beta \le c'$ for every $c \le c'$. The space (1.38) is Fréchet in a canonical way.

Consider, in particular, a compact set, and let $\mathcal{A}'(K,E)$ be the space of all analytic functionals, carried by K, with values in a Fréchet space E. Then, for $K \in \mathcal{V}$ we have

$$f(z) := M_{\delta,r\to z}\omega(r)\langle\zeta, r^{-z}\rangle \in \mathcal{B}_K^{-\infty,d}(X;\boldsymbol{v};\mathbb{C})$$

for every $\zeta \in \mathcal{A}'(K, \mathcal{B}^{-\infty,d}(X;\boldsymbol{v}))$ and a cut-off function ω. Here $\delta \in \mathbb{R}$ is any weight such that $K \subset \{z : \operatorname{Re} z < \frac{1}{2} - \delta\}$. Moreover,

$$\mathcal{B}_K^{-\infty,d}(X;\boldsymbol{v};\mathbb{C}) = \{M_\delta\omega\langle\zeta, r^{-z}\rangle + f_0 : \zeta \in \mathcal{A}'(K, \mathcal{B}^{-\infty,d}(X;\boldsymbol{v})),$$
$$f_0 \in \mathcal{B}^{-\infty,d}(X;\boldsymbol{v};\mathbb{C})\}.$$

For $V, \tilde{V} \in \mathcal{V}$, $V + \tilde{V} := (V \cup \tilde{V})^I$, we have the relation

$$\mathcal{B}_{V+\tilde{V}}^{-\infty,d}(X;\boldsymbol{v};\mathbb{C}) = \mathcal{B}_V^{-\infty,d}(X;\boldsymbol{v};\mathbb{C}) + \mathcal{B}_{\tilde{V}}^{-\infty,d}(X;\boldsymbol{v};\mathbb{C}) \tag{1.39}$$

as a non-direct sum of Fréchet spaces. The sets $V \in \mathcal{V}$ will be called continuous asymptotic types for Mellin symbols of the cone algebra.

A useful property of decompositions like (1.39) is that for every $V_1 \in \mathcal{V}$ and $\delta, \tilde{\delta} \in \mathbb{R}$, $\delta \neq \tilde{\delta}$, there exist sets $V, \tilde{V} \in \mathcal{V}$ such that

$$V_1 = V + \tilde{V} \text{ and } \Gamma_\delta \cap V = \Gamma_{\tilde{\delta}} \cap \tilde{V} = \emptyset. \tag{1.40}$$

In other words, Mellin symbols l with continuous asymptotic types can be decomposed as sums $l = f + \tilde{f}$, where $f \in \mathcal{B}_V^{-\infty,d}(X;\boldsymbol{v};\mathbb{C})$ and $\tilde{f} \in \mathcal{B}_{\tilde{V}}^{-\infty}(X;\boldsymbol{v};\mathbb{C})$ are holomorphic in a neighbourhood of Γ_δ and $\Gamma_{\delta'}$, respectively.

Now the Mellin symbols f_j and $\tilde{f}_j$ in (1.35) are assumed as

$$f_j(z) \in \mathcal{B}_{V_j}^{-\infty,d}(X;\boldsymbol{v};\mathbb{C}) \quad \text{and} \quad \tilde{f}_j(z) \in \mathcal{B}_{\tilde{V}_j}^{-\infty,d}(X;\boldsymbol{v};\mathbb{C}), \tag{1.41}$$

respectively, with sets $V_j, \tilde{V}_j \in \mathcal{V}$ such that $\Gamma_{\frac{n+1}{2}-\delta_j} \cap V_j = \Gamma_{\frac{n+1}{2}-\tilde{\delta}_j} \cap \tilde{V}_j = \emptyset$ for all j.

It is interesting also to consider Mellin symbols, where the asymptotic information is supported by a discrete set $\{p_j\}_{j\in\mathbb{Z}} \subset \mathbb{C}$.

A sequence of triples

$$R = \{(p_j, m_j, L_j)\}_{j\in\mathbb{Z}}$$

for $p_j \in \mathbb{C}, m_j \in \mathbb{N}$, is called a discrete asymptotic type for Mellin symbols if

(i) $\pi_{\mathbb{C}} R := \{p_j\}_{j\in\mathbb{N}}$ has a finite intersection with $\{z : c \leq \operatorname{Re} z \leq c'\}$ for every $c \leq c'$,

(ii) $L_j \subset \mathcal{B}^{-\infty,d}(X;\boldsymbol{v})$ is a finite-dimensional subspace of operators of finite rank, for every $j \in \mathbb{N}$.

We denote by $\mathcal{B}_R^{-\infty,d}(X;\boldsymbol{v};\mathbb{C})$ the subspace of all $f(z) \in \mathcal{B}_{\pi_{\mathbb{C}}R}^{-\infty,d}(X;\boldsymbol{v};\mathbb{C})$ that are meromorphic with poles at the points p_j of multiplicity $m_j + 1$ and Laurent coefficients at $(z-p_j)^{-(k+1)}$ belonging to L_j for all $0 \leq k \leq m_j$, $j \in \mathbb{N}$.

It remains to explain the nature of operators $\mathcal{C}$ in the expression (1.34). To this end we first recall the notion of continuous and discrete asymptotics in the spaces $\mathcal{K}^{s,\gamma}(X^\wedge)$.

First, if $\Theta = (\vartheta, 0]$, $-\infty \leq \vartheta < 0$, is a weight interval, we set

$$\mathcal{K}_\Theta^{s,\gamma}(X^\wedge) := \varprojlim_{\varepsilon>0} \mathcal{K}^{s,\gamma-\vartheta-\varepsilon}(X^\wedge)$$

which is a Fréchet space in a canonical way. Let Θ be finite, consider any $V \in \mathcal{V}$, $V \subset \{z : \operatorname{Re} z < \frac{n+1}{2} - \gamma\}$, and set $K := V \cap \{z : \frac{n+1}{2} - \gamma - \vartheta \leq \operatorname{Re} z \leq \frac{n+1}{2} - \gamma\}$ which is a compact set. Then

$$\mathcal{E}_K(X^\wedge) := \{\omega(r)\langle\zeta, r^{-z}\rangle : \zeta \in \mathcal{A}'(K, C^\infty(X))\}$$

is a Fréchet subspace of $\mathcal{K}^{\infty,\gamma}(X^\wedge)$, and we define

$$\mathcal{K}_P^{s,\gamma}(X^\wedge) := \mathcal{K}_\Theta^{s,\gamma}(X^\wedge) + \mathcal{E}_K(X^\wedge) \tag{1.42}$$

in the Fréchet topology of the non-direct sum. Subscript 'P' has the meaning of a so called continuous asymptotic type associated with V and weight data $\boldsymbol{g} = (\gamma, \Theta)$. We interpret P as the quotient space $\mathcal{A}'(K, C^\infty(X))/\sim$, where equivalence $\zeta \sim \zeta'$ means that $\omega(r)\langle\zeta - \zeta', r^{-z}\rangle$ belongs to $\mathcal{K}_\Theta^{\infty,\gamma}(X^\wedge)$. Let us set

$$\pi_{\mathbb{C}} P := K \cap \{z : \frac{n+1}{2} - \gamma + \vartheta < \operatorname{Re} z\}. \tag{1.43}$$

For $\Theta = (-\infty, 0]$ we choose a sequence $\{\vartheta_j\}_{j\in\mathbb{N}}$, $\vartheta_{j+1} < \vartheta_j < 0$ for all j and $\vartheta_j \to -\infty$ for $j \to \infty$. Then, if P_j is the continuous asymptotic type, associated with V and weight data $\boldsymbol{g}_j = (\gamma, (\vartheta_j, 0])$, there are continuous embeddings

$$\mathcal{K}^{s,\gamma}_{P_{j+1}}(X^\wedge) \hookrightarrow \mathcal{K}^{s,\gamma}_{P_j}(X^\wedge) \tag{1.44}$$

for all j. We then set $\mathcal{K}^{s,\gamma}_P(X^\wedge) = \varprojlim_{j\in\mathbb{N}} \mathcal{K}^{s,\gamma}_{P_j}(X^\wedge)$, where P is represented by the sequence of P_j, $j \in \mathbb{N}$, and is regarded as a continuous asymptotic type associated with V and $\boldsymbol{g} = (\gamma, (-\infty, 0])$. This construction is independent of the specific choice of the sequence $\{\vartheta_j\}_{j\in\mathbb{N}}$. Let $\operatorname{As}(X, \boldsymbol{g})$ denote the set of all continuous asymptotic types P, belonging to any $V \in \mathcal{V}$ and $\boldsymbol{g} = (\gamma, \Theta)$. Similarly to (1.43) we write

$$\pi_{\mathbb{C}} P = V,$$

if P belongs to $V \in \mathcal{V}$.

Let us write

$$P_1 \subseteq P_2 \quad \text{for} \quad \pi_{\mathbb{C}} P_1 \subseteq \pi_{\mathbb{C}} P_2. \tag{1.45}$$

A sequence

$$P = \{(p_j, m_j, L_j)\}_{0 \le j \le N},$$

with $p_j \in \mathbb{C}, m_j \in \mathbb{N}$, $N = N(P) \le \infty$, is called a discrete asymptotic type for weighted distributions on $X^\wedge$ near $r = 0$, associated with weight data $\boldsymbol{g} = (\gamma, \Theta)$, $\Theta = (\vartheta, 0]$, if

(i) $\pi_{\mathbb{C}} P = \{p_j\}_{0\le j\le N} \subset \{z : \frac{n+1}{2} - \gamma + \vartheta < \operatorname{Re} z < \frac{n+1}{2} - \gamma\}$, $N(P) < \infty$ for finite Θ, $\pi_{\mathbb{C}} P \cap \{z : c \le \operatorname{Re} z \le c'\}$ finite for every $c \le c'$.

(ii) $m_j \in \mathbb{N}$ and $L_j \subset C^\infty(X, E)$ (for a given $E \in \operatorname{Vect}(X)$) is a finite-dimensional subspace, for every $j \in \mathbb{N}$.

Let $\operatorname{As}(X, \boldsymbol{g}^\bullet)$ denote the set of all discrete asymptotic types in that sense. For $P \in \operatorname{As}(X, \boldsymbol{g}^\bullet)$, $\boldsymbol{g} = (\gamma, \Theta)$, Θ finite, the space

$$\mathcal{E}_P(X^\wedge) := \{\omega(r) \sum_{j=0}^{N} \sum_{k=0}^{m_j} c_{jk}(x) r^{-p_j} \log^k r : c_{jk} \in L_j \\ \text{for } 0 \le k \le m_j,\ 0 \le j \le N\}$$

(for any fixed cut-off function ω) is finite-dimensional, and we have $\mathcal{E}_P(X^\wedge) \subset \mathcal{K}^{\infty,\gamma}(X^\wedge)$ and $\mathcal{E}_P(X^\wedge) \cap \mathcal{K}^{\infty,\gamma}_\Theta(X^\wedge) = \{0\}$. We then set

$$\mathcal{K}^{s,\gamma}_P(X^\wedge) := \mathcal{K}^{s,\gamma}_\Theta(X^\wedge) + \mathcal{E}_P(X^\wedge) \tag{1.46}$$

in the Fréchet topology of the direct sum.

For $\Theta = (-\infty, 0]$ we choose a sequence $\{\vartheta_j\}_{j\in\mathbb{N}}$ as above, and form $P_j := \{(p, m, L) \in P : \operatorname{Re} p > \frac{n+1}{2} - \gamma + \vartheta_j\} \in \operatorname{As}(X, \boldsymbol{g}_j^\bullet)$ for $\boldsymbol{g}_j = (\gamma, \Theta_j)$, $\Theta_j = (\vartheta_j, 0]$. Then we have continuous embeddings (1.44) for all j, and we set again

$$\mathcal{K}^{s,\gamma}_P(X^\wedge) = \varprojlim_{j\in\mathbb{N}} \mathcal{K}^{s,\gamma}_{P_j}(X^\wedge).$$

Moreover, let us set

$$\mathcal{S}_P^{\gamma}(X^\wedge) = \{\omega u + (1-\omega)v : u \in \mathcal{K}_P^{\infty,\gamma}(X^\wedge),\ v \in \mathcal{S}(\overline{\mathbb{R}}_+, C^\infty(X))\}. \tag{1.47}$$

Remark 1.8. The spaces $\mathcal{K}_P^{s,\gamma}(X^\wedge)$ and $\mathcal{S}_P^{\gamma}(X^\wedge)$ for $P \in \mathrm{As}(X,\boldsymbol{g})$ or $P \in \mathrm{As}(X,\boldsymbol{g}^\bullet)$ are Fréchet spaces with group action $\kappa = \{\kappa_\delta\}_{\delta\in\mathbb{R}_+}$, $(\kappa_\delta u)(r,x) = \delta^{\frac{n+1}{2}} u(\delta r, x)$ (cf. the notation of Section 1.1).

An operator

$$\mathcal{C} : \begin{array}{c}\mathcal{K}^{s,\gamma}(X^\wedge,E)\\ \oplus \\ \mathcal{K}^{s',\gamma-\frac{1}{2}}((\partial X)^\wedge, J_-)\end{array} \to \begin{array}{c}\mathcal{K}^{\infty,\gamma-\mu}(X^\wedge,F)\\ \oplus \\ \mathcal{K}^{\infty,\gamma-\mu-\frac{1}{2}}((\partial X)^\wedge, J_+)\end{array}$$

which is continuous for $s > -\frac{1}{2}, s' \in \mathbb{R}$, is called a Green operator of type 0 in the cone algebra on $X^\wedge$, if $\mathcal{C}$ and its formal adjoint $\mathcal{C}^*$ induce continuous operators

$$\mathcal{C} : \begin{array}{c}\mathcal{K}^{s,\gamma}(X^\wedge,E)\\ \oplus \\ \mathcal{K}^{s',\gamma-\frac{1}{2}}((\partial X)^\wedge, J_-)\end{array} \to \begin{array}{c}\mathcal{S}_P^{\gamma-\mu}(X^\wedge,F)\\ \oplus \\ \mathcal{S}_{P'}^{\gamma-\mu-\frac{1}{2}}((\partial X)^\wedge, J_+)\end{array} \tag{1.48}$$

and

$$\mathcal{C}^* : \begin{array}{c}\mathcal{K}^{s,-\gamma+\mu}(X^\wedge,F)\\ \oplus \\ \mathcal{K}^{s',-\gamma+\mu-\frac{1}{2}}((\partial X)^\wedge, J_+)\end{array} \to \begin{array}{c}\mathcal{S}_{Q'}^{-\gamma-\frac{1}{2}}((\partial X^\wedge,E)\\ \oplus \\ \mathcal{S}_{Q'}^{-\gamma-\frac{1}{2}}((\partial X)^\wedge, J_-)\end{array}, \tag{1.49}$$

respectively, for all $s > -\frac{1}{2}, s' \in \mathbb{R}$, with asymptotic types

$$(P,P') \in \mathrm{As}(X,(\gamma-\mu,\Theta)) \times \mathrm{As}(\partial X,(\gamma-\mu-\frac{1}{2},\Theta)) \tag{1.50}$$

and

$$(Q,Q') \in \mathrm{As}(X,(-\gamma,\Theta)) \times \mathrm{As}(\partial X,(-\gamma-\frac{1}{2},\Theta)), \tag{1.51}$$

depending on $\mathcal{C}$ (not on s,s'). The formal adjoint is defined in terms of the respective $\mathcal{K}^{0,0}(X^\wedge,\cdot) \oplus \mathcal{K}^{0,-\frac{1}{2}}((\partial X)^\wedge,\cdot)$-scalar products (where dots stand for the corresponding bundles).

An operator of the form

$$\mathcal{C} = \mathcal{C}_0 + \sum_{j=1}^{d} \mathcal{C}_j \operatorname{diag}(D^j, 0)$$

for Green operators $\mathcal{C}_j$ of type 0, $0 \le j \le d$, and D^j as in (1.17), is called a Green operator of type d.

Definition 1.9. Let $\mathcal{C}^{\mu,d}(X^\wedge,\boldsymbol{g};\boldsymbol{v})$ for $\boldsymbol{g} = (\gamma,\gamma-\mu,\Theta)$, $\Theta = (-(k+1),0]$, $\boldsymbol{v} = (E,F;J_-,J_+)$, denote the space of all operators of the form (1.34), such that $\mathcal{A}_{\mathrm{exit}} \in \mathcal{B}^{\mu,d;0}(X^\wedge;\boldsymbol{v})$, moreover, $\mathcal{A}_{\mathrm{cone}}$ given by (1.34), (1.37), (1.41), and a Green operator $\mathcal{C}$ of type d. For $\Theta = (-\infty, 0]$ we set $\mathcal{C}^{\mu,d}(X, \boldsymbol{g}; \boldsymbol{v}) = \bigcap_{k\in\mathbb{N}} \mathcal{C}^{\mu,d}(X; \boldsymbol{g}_k; \boldsymbol{v})$ for $\boldsymbol{g}_k := (\gamma, \gamma - \mu, (-(k+1), 0])$.

Theorem 1.10. *Every $\mathcal{A} \in \mathcal{C}^{\mu,d}(X^\wedge, \boldsymbol{g}; \boldsymbol{v})$ induces continuous operators*

$$\begin{matrix} \mathcal{K}^{s,\gamma}(X^\wedge, E) & & \mathcal{K}^{s-\mu,\gamma-\mu}(X^\wedge, F) \\ \oplus & \to & \oplus \\ \mathcal{K}^{s-\frac{1}{2},\gamma-\frac{1}{2}}((\partial X)^\wedge, J_-) & & \mathcal{K}^{s-\mu-\frac{1}{2},\gamma-\mu-\frac{1}{2}}((\partial X)^\wedge, J_+) \end{matrix} \tag{1.52}$$

as well as

$$\begin{matrix} \mathcal{K}_P^{s,\gamma}(X^\wedge, E) & & \mathcal{K}_Q^{s-\mu,\gamma-\mu}(X^\wedge, F) \\ \oplus & \to & \oplus \\ \mathcal{K}_{P'}^{s-\frac{1}{2},\gamma-\frac{1}{2}}((\partial X)^\wedge, J_-) & & \mathcal{K}_{Q'}^{s-\mu-\frac{1}{2},\gamma-\mu-\frac{1}{2}}((\partial X)^\wedge, J_+) \end{matrix} \tag{1.53}$$

for every $s \in \mathbb{R}, s > d - \frac{1}{2}$, and every pair of asymptotic types (1.50) *with some resulting* (1.51).

This theorem may be found in [9, Section 2.1.7], cf. also [29, Theorem 2.3.55] for the case without boundary.

Operators $\mathcal{A} \in \mathcal{C}^{\mu,d}(X^\wedge, \boldsymbol{g}; \boldsymbol{v})$ have a principal symbolic hierarchy

$$\sigma(\mathcal{A}) = (\sigma_\psi(\mathcal{A}), \sigma_\partial(\mathcal{A}), \sigma_c(\mathcal{A}), \sigma_E(\mathcal{A})),$$

where $\sigma_\psi(\mathcal{A})$ and $\sigma_\partial(\mathcal{A})$ are the homogeneous interior and boundary symbols of $\mathcal{A}$, regarded as an element of $\mathcal{B}^{\mu,d}(X^\wedge; \boldsymbol{v})$. Moreover, $\sigma_c(\mathcal{A})$ is the principal conormal symbol, defined as

$$\sigma_c(\mathcal{A})(z) = h(0, z) + f_0(z), \tag{1.54}$$

cf. (1.35) (without loss of generality, we assume $\tilde{f}_0 = 0$, otherwise we have to replace the summand $f_0(z)$ in (1.54) by $f_0(z) + \tilde{f}_0(z)$). By definition,

$$\sigma_c(\mathcal{A}) : \begin{matrix} H^s(X, E) & & H^{s-\mu}(X, F) \\ \oplus & \to & \oplus \\ H^{s-\frac{1}{2}}(\partial X, J_-) & & H^{s-\mu-\frac{1}{2}}(\partial X, J_+) \end{matrix} \tag{1.55}$$

is an element of $\mathcal{B}_V^{\mu,d}(X; \boldsymbol{v}; \mathbb{C})$ for some $V \in \mathcal{V}, V \cap \Gamma_{\frac{n+1}{2}-\gamma} = \emptyset$.

Finally, $\sigma_E(\mathcal{A})$ is the tuple of exit symbolic components

$$\sigma_E(\mathcal{A}) = (\sigma_e(\mathcal{A}), \sigma_{\psi,e}(\mathcal{A}), \sigma_{e'}(\mathcal{A}), \sigma_{\partial,e'}(\mathcal{A}))$$

for $r \to \infty$, cf. the notation of the preceding section.

An operator $\mathcal{A} \in \mathcal{C}^{\mu,d}(X^\wedge, \boldsymbol{g}; \boldsymbol{v})$ is called elliptic (with respect to the weight $\gamma \in \mathbb{R}$), if all components of $\sigma(\mathcal{A})$ are bijective; for (1.54) this means that (1.55) is a family of isomorphisms for all $z \in \Gamma_{\frac{n+1}{2}-\gamma}$ and any $s \in \mathbb{R}, s > \max(\mu, d) - \frac{1}{2}$ (this condition is independent of the choice of s).

Theorem 1.11. *An operator $\mathcal{A} \in \mathcal{C}^{\mu,d}(X^\wedge, \boldsymbol{g}; \boldsymbol{v})$ is elliptic if and only if* (1.52) *is a Fredholm operator for some $s > \max(\mu, d) - \frac{1}{2}$.*

Ellipticity entails the Fredholm property of (1.52) *for every* $s > \max(\mu, d) - \frac{1}{2}$, *and the operator* $\mathcal{A}$ *has a paramatrix* $\mathcal{P} \in \mathcal{C}^{-\mu,h}(X^\wedge, \boldsymbol{g}^{-1}; \boldsymbol{v}^{-1})$ *for* $h = \max(d - \mu, 0)$, $\boldsymbol{g}^{-1} = (\gamma - \mu, \gamma, \Theta)$, $\boldsymbol{v}^{-1} = (F, E; J_+, J_-)$, *i.e.,*

$$\mathcal{I} - \mathcal{P}\mathcal{A} \in \mathcal{C}_G^{d_l}(X^\wedge, \boldsymbol{g}_e; \boldsymbol{v}_e), \quad \mathcal{I} - \mathcal{A}\mathcal{P} \in \mathcal{C}_G^{d_r}(X^\wedge, \boldsymbol{g}_r; \boldsymbol{v}_r)$$

for $d_l = \max(\mu, d)$, $\boldsymbol{g}_l = (\gamma, \gamma; \Theta)$, $\boldsymbol{v}_l = (E, E; J_-, J_-)$, *and* $d_r = \max(d - \mu, 0)$, $\boldsymbol{g}_r = (\gamma - \mu, \gamma - \mu, \Theta)$, $\boldsymbol{v}_r = (F, F; J_+, J_+)$.

The proof of the Fredholm property of (1.52) follows from a parametrix construction, combining a local parametrix near the tip of the cone from [20] with a parametrix far from the tip up to infinity, using the second part of Theorem 1.7. The necessity of the ellipticity for the Fredholm property can be obtained by writing $\mathcal{A} = \mathcal{A}_0 + \mathcal{A}_{\mathrm{int}} + \mathcal{A}_\infty$ modulo a Green operator, where $\mathcal{A}_0$ is localised near the tip of the cone, $\mathcal{A}_{\mathrm{int}}$ far from the tip as well as far from ∞, and $\mathcal{A}_\infty$ localised near ∞. Now different variants of 'Gohberg's lemma' allows us to treat the principal symbolic components separately, cf. analogously [26, Section 2.2.1] for $\mathcal{A}_0$, [19, Section 3.1.1.1] for $\mathcal{A}_{\mathrm{int}}$, and [21] for $\mathcal{A}_\infty$.

2. The Edge Algebra with Parameters

2.1. Edge-Amplitude Functions

Edge-amplitude functions $a(y, \eta)$ as they will be defined in this section are particular operator-valued symbols in the sense of (1.12), with E and $\tilde{E}$ being of the form

$$\mathcal{K}^{s,\gamma}(X^\wedge, E) \oplus \mathcal{K}^{s-\frac{1}{2},\gamma-\frac{1}{2}}((\partial X)^\wedge, J_-) \oplus \mathbb{C}^{l_-}$$

and

$$\mathcal{K}^{s-\mu,\gamma-\mu}(X^\wedge, F) \oplus \mathcal{K}^{s-\mu-\frac{1}{2},\gamma-\mu-\frac{1}{2}}((\partial X)^\wedge, J_+) \oplus \mathbb{C}^{l_+},$$

respectively, cf. the notation of Section 1.3. The group actions in both spaces are

$$\kappa_\delta : u(r, x) \oplus u'(r, x') \oplus c \to \delta^{\frac{n+1}{2}}(u(\delta r, x) \oplus u'(\delta r, x') \oplus c), \tag{2.1}$$

$\delta \in \mathbb{R}_+$. The 2×2 upper left corners of the operator functions $a(y, \eta)$ take values in the cone algebra on $X^\wedge$, and there is a specific dependence on the parameter η.

These amplitude functions will constitute a space

$$\mathcal{R}^{\mu,d}(U \times \mathbb{R}^q, \boldsymbol{g}; \boldsymbol{w}), \tag{2.2}$$

$d \in \mathbb{N}$, with weight data $\boldsymbol{g} = (\gamma, \gamma - \mu, \Theta)$, for a finite weight interval $\Theta = (-(k+1), 0]$, and tuples $\boldsymbol{w} = (E, F; J_-, J_+; l_-, l_+)$, with vector bundles $E, F \in \mathrm{Vect}(X)$, $J_\pm \in \mathrm{Vect}(\partial X)$. The numbers $l_\pm$ are the fibre dimensions of bundles $L_\pm \in \mathrm{Vect}(Y)$ in the global calculus below.

Moreover, we will single out a subspace of so called Green edge-symbols

$$\mathcal{R}_G^{\mu,d}(U \times \mathbb{R}^q, \boldsymbol{g}; \boldsymbol{w}). \tag{2.3}$$

Then the elements $a(y, \eta)$ of (2.2) will have the form

$$a(y, \eta) = \mathrm{diag}(b(y, \eta), 0) + g(y, \eta) \tag{2.4}$$

for arbitrary $g(y,\eta) \in \mathcal{R}_G^{\mu,d}(U \times \mathbb{R}^q, \boldsymbol{g}; \boldsymbol{w})$ and $b(y,\eta) \in \mathcal{R}^{\mu,d}(U \times \mathbb{R}^q, \boldsymbol{g}; \boldsymbol{v})$, $\boldsymbol{v} := (E, F; J_-, J_+)$ (the latter space corresponds to (2.2) for the case $l_- = l_+ = 0$).

Let us now turn to more details. The structure of $b(y,\eta)$ is as follows. Choose arbitrary cut-off functions $\sigma(r), \tilde{\sigma}(r)$ and $\omega(r), \tilde{\omega}(r), \tilde{\tilde{\omega}}(r)$, where we assume $\tilde{\omega} = 1$ on $\operatorname{supp}\omega$, and $\omega = 1$ on $\operatorname{supp}\tilde{\tilde{\omega}}$. Let us set $\omega_\eta(r) := \omega(r[\eta])$ for some strictly positive function $[\eta] \in C^\infty(\mathbb{R}^q)$ such that $[\eta] = |\eta|$ for $|\eta| > c$ for some constant $c > 0$.

(i) We choose an arbitrary element

$$\tilde{p}(r, y, \tilde{\varrho}, \tilde{\eta}) \in C^\infty(\overline{\mathbb{R}}_+ \times U, \mathcal{B}^{\mu,d}(X; \boldsymbol{v}; \mathbb{R}^{1+q}_{\tilde{\varrho},\tilde{\eta}})) \tag{2.5}$$

and form the family of operators $\operatorname{op}_r(p)(y,\eta)$ for

$$p(r, \varrho, \eta) := \tilde{p}(r, r\varrho, r\eta). \tag{2.6}$$

(ii) Let $\tilde{h}(r, y, z, \tilde{\eta}) \in C^\infty(\overline{\mathbb{R}}_+ \times U, \mathcal{B}^{\mu,d}(X; \boldsymbol{v}; \mathbb{C} \times \mathbb{R}^q_{\tilde{\eta}}))$ be an element such that

$$\operatorname{op}_r(p)(y,\eta) = \operatorname{op}_M^\beta(h)(y,\eta) \bmod C^\infty(U, \mathcal{B}^{-\infty,d}(X^\wedge; \boldsymbol{v}; \mathbb{R}^q_\eta))$$

for

$$h(r, y, z, \eta) := \tilde{h}(r, y, z, r\eta). \tag{2.7}$$

(iii) We set

$$m(y,\eta) = \omega_\eta r^{-\mu} \sum_{j+|\alpha|\le k} r^j \{\operatorname{op}_M^{\delta_j - \frac{n}{2}}(f_{j\alpha})(y) + \operatorname{op}_M^{\tilde{\delta}_j - \frac{n}{2}}(\tilde{f}_{j\alpha})(y)\} \eta^\alpha \tilde{\omega}_\eta \tag{2.8}$$

with weights $\delta_j, \tilde{\delta}_j$ as in (1.36) and functions

$$f_{j\alpha} \in C^\infty(U, \mathcal{B}^{-\infty,d}_{V_{j\alpha}}(X; \boldsymbol{v}; \mathbb{C})),\ \tilde{f}_{j\alpha} \in C^\infty(U, \mathcal{B}^{-\infty,d}_{\tilde{V}_{j\alpha}}(X; \boldsymbol{v}; \mathbb{C})) \tag{2.9}$$

for certain $V_{j\alpha}, \tilde{V}_{j\alpha} \in \mathcal{V}$, satisfying the conditions

$$\Gamma_{\frac{n+1}{2} - \delta_j} \cap V_{j\alpha} = \emptyset, \quad \Gamma_{\frac{n+1}{2} - \tilde{\delta}_j} \cap \tilde{V}_{j\alpha} = \emptyset \tag{2.10}$$

for all j, α.

(iv) The operator function $b(y,\eta)$ in (2.4) has the form

$$\begin{aligned} b(y,\eta) = \sigma r^{-\mu} \{ &\omega_\eta \operatorname{op}_M^{\gamma - \frac{n}{2}}(h)(y,\eta) \tilde{\omega}_\eta \\ &+ (1 - \omega_\eta) \operatorname{op}_r(p)(y,\eta)(1 - \tilde{\tilde{\omega}}_\eta)\} \tilde{\sigma} + m(y,\eta) \end{aligned} \tag{2.11}$$

for p, h and m as in (i), (ii) and (iii), respectively.

Let us set, for abbreviation,

$$\boldsymbol{E}^{0,\gamma} := \mathcal{K}^{0,\gamma}(X^\wedge, E) \oplus \mathcal{K}^{0,\gamma-\frac{1}{2}}((\partial X)^\wedge, J_-) \oplus \mathbb{C}^{l_-},$$

$$\boldsymbol{\mathcal{S}}^{\gamma-\mu}_{\mathcal{P}} := \mathcal{S}^{\gamma-\mu}_P(X^\wedge, F) \oplus \mathcal{S}^{\gamma-\mu-\frac{1}{2}}_{P'}((\partial X)^\wedge, J_+) \oplus \mathbb{C}^{l_+}$$

for asymptotic types $P \in \mathrm{As}(X,(\gamma,\Theta))$, $P' \in \mathrm{As}(X,(\gamma-\frac{1}{2},\Theta))$, $\mathcal{P}=(P,P')$, and

$$\boldsymbol{F}^{0,-\gamma+\mu} := \mathcal{K}^{0,\gamma-\mu}(X^\wedge,F)\oplus\mathcal{K}^{0,\gamma-\mu-\frac{1}{2}}((\partial X)^\wedge,J_+)\oplus\mathbb{C}^{l_+},$$

$$\boldsymbol{\mathcal{S}}_{\mathcal{Q}}^{-\gamma} := \mathcal{S}_Q^{-\gamma}(X^\wedge,E)\oplus\mathcal{S}_{Q'}^{-\gamma-\frac{1}{2}}((\partial X)^\wedge,J_-)\oplus\mathbb{C}^{l_-}$$

for asymptotic types $Q\in\mathrm{As}(X,(-\gamma+\mu,\Theta))$, $Q'\in\mathrm{As}((\partial X)^\wedge,(-\gamma+\mu-\frac{1}{2},\Theta))$, $\mathcal{Q}=(Q,Q')$.

A Green symbol $g(y,\eta)$ of type $d=0$ in the sense of (2.3) is defined as a function

$$g(y,\eta)=\mathrm{diag}(1,\langle\eta\rangle^{\frac{1}{2}},\langle\eta\rangle^{\frac{n+1}{2}})g_0(y,\eta)\,\mathrm{diag}(1,\langle\eta\rangle^{\frac{1}{2}},\langle\eta\rangle^{\frac{n+1}{2}})^{-1}$$

with an operator-valued symbol

$$g_0(y,\eta)\in S^\mu_{\mathrm{cl}}(U\times\mathbb{R}^q;\boldsymbol{E}^{0,\gamma},\boldsymbol{\mathcal{S}}_{\mathcal{P}}^{\gamma-\mu})_{(\kappa^0,\kappa^0)}, \tag{2.12}$$

cf. notation (2.16) below, such that their pointwise formal adjoint $g_0^*(y,\eta)$ in the sense $(g_0u,v)_{\boldsymbol{E}^{0,0}}=(u,g_0^*v)_{\boldsymbol{F}^{0,0}}$ for all $u\in C_0^\infty(X^\wedge,E)\oplus C_0^\infty((\partial X)^\wedge,J_-)\oplus\mathbb{C}^{l_-}$, $v\in C_0^\infty(X^\wedge,F)\oplus C_0^\infty((\partial X)^\wedge,J_+)\oplus\mathbb{C}^{l_+}$ represents an element

$$g_0^*(y,\eta)\in S^\mu_{\mathrm{cl}}(U\times\mathbb{R}^q;\boldsymbol{F}^{0,-\gamma+\mu},\boldsymbol{\mathcal{S}}_{\mathcal{Q}}^{-\gamma})_{(\kappa^0,\kappa^0)}. \tag{2.13}$$

The asymptotic types

$$\mathcal{P}=(P,P')\in\mathrm{As}(X,(\gamma-\mu,\Theta))\times\mathrm{As}(\partial X,(\gamma-\mu-\frac{1}{2},\Theta)), \tag{2.14}$$

$$\mathcal{Q}=(Q,Q')\in\mathrm{As}(X,(-\gamma,\Theta))\times\mathrm{As}(\partial X,(-\gamma-\frac{1}{2},\Theta)) \tag{2.15}$$

depend on g_0. Subscripts '(κ^0,κ^0)' in the relations (2.12) and (2.13) mean that the spaces of symbols refer to the group actions

$$\kappa^0_\delta: u(r,x)\oplus u'(r,x')\oplus c\to\delta^{\frac{n+1}{2}}u(\lambda r,x)\oplus\delta^{\frac{n}{2}}u'(\lambda r,x)\oplus c, \tag{2.16}$$

$\delta\in\mathbb{R}_+$, in contrast to (2.1). In this way we avoid distinguishing matrices of Douglis-Nirenberg orders for symbols and their adjoints. Clearly, the Green symbols $g(y,\eta)$ themselves are operator-valued symbols with respect to the group actions (2.1), cf. Remark 2.3 below.

Now a Green symbol $g(y,\eta)$ in (2.3) of type $d\in\mathbb{N}$ is defined as a linear combination

$$g(y,\eta)=h_0(y,\eta)+\sum_{j=1}^d h_j(y,\eta)\,\mathrm{diag}(T^j,0,0) \tag{2.17}$$

with T^j being as in (1.17) and $h_j\in\mathcal{R}^{\mu,0}_G(U\times\mathbb{R}^q,\boldsymbol{g};\boldsymbol{w})$ of type zero, $j=0,\dots,d$.

Let

$$\mathcal{R}^{\mu,d}_{M+G}(U\times\mathbb{R}^q,\boldsymbol{g};\boldsymbol{w}) \tag{2.18}$$

be the subspace of all elements of (2.2) of the form

$$a(y,\eta)=m(y,\eta)+g(y,\eta) \tag{2.19}$$

for arbitrary $m(y,\eta)$ as in (iii) above, and a Green symbol $g(y,\eta)$ of type d.

Remark 2.1. The space $\mathcal{R}^{\mu,d}_{M+G}(U\times\mathbb{R}^q,\boldsymbol{g};\boldsymbol{w})$ is equal to the subspace of all $a(y,\eta)\in\mathcal{R}^{\mu,d}(U\times\mathbb{R}^q,\boldsymbol{g};\boldsymbol{w})$ such that $\tilde{p}(r,y,\tilde{\varrho},\tilde{\eta})\in C^\infty(\overline{\mathbb{R}}_+\times U,\mathcal{B}^{-\infty,d}(X;\boldsymbol{v};\mathbb{R}^{1+q}_{\tilde{\varrho},\tilde{\eta}}))$.

For the aspect of holomorphic dependence of symbols in (2.18) on a further parameter it is interesting to single out Fréchet subspaces that are parametrised by the type $d\in\mathbb{N}$ and asymptotic types involved in (2.8) as well as $\mathcal{P}$ and $\mathcal{Q}$ in the Green summands $g(y,\eta)$. For this consideration it is not essential to fix the carrier sets $V_{j\alpha},\tilde{V}_{j\alpha}$ particularly small. It suffices to choose tuples

$$(V_0,V_1,\mathcal{P},\mathcal{Q})=:R \tag{2.20}$$

where $V_0,V_1\in\mathcal{V}$, $V_0\cap\Gamma_{\frac{n+1}{2}-\gamma}=\emptyset$, and to assume that the sets $V_{j\alpha},\tilde{V}_{j\alpha}$ in (2.8) satisfy the condition

$$V_{j\alpha},\tilde{V}_{j\alpha}\subseteq V_1$$

for all $0<j+|\alpha|\le k$ (together with the relations (2.10)).

Lemma 2.2. *For every $V_1\in\mathcal{V}$ there exist asymptotic types $\mathcal{P}_1,\mathcal{Q}_1$ of the kind* (2.14), (2.15) *such that for every sequence*

$$h_{j\alpha}\in C^\infty(U,\mathcal{B}^{-\infty,d}_{V_1}(X;\boldsymbol{v};\mathbb{C})) \tag{2.21}$$

$0<j+|\alpha|\le k$, *and decompositions*

$$h_{j\alpha}=f_{j\alpha}+\tilde{f}_{j\alpha}$$

into Mellin symbols (2.9) *for an arbitrary choice of $V_{j\alpha},\tilde{V}_{j\alpha}\subseteq V_1$, the operator $m(y,\eta)$ given by* (2.8) *is uniquely determined by* (2.21)*, modulo a Green symbol $g_1(y,\eta)$ with asymptotic types $\mathcal{P}_1$ and $\mathcal{Q}_1$.*

Let $\mathbf{As}_{M+G}(X,\boldsymbol{g})$ denote the set of all tuples (2.20) for arbitrary V_0,V_1 as mentioned before, and $\mathcal{P}\supseteq\mathcal{P}_1,\mathcal{Q}\supseteq\mathcal{Q}_1$ (the latter inclusions correspond to the inclusions of carrier sets, cf. the relation (1.45)).

Moreover, let

$$\mathcal{R}^{\mu,d}_{M+G}(U\times\mathbb{R}^q,\boldsymbol{g};\boldsymbol{w})_R \tag{2.22}$$

for $R\in\mathbf{As}_{M+G}(X,\boldsymbol{g})$ be the space of all $m(y,\eta)+g(y,\eta)\in\mathcal{R}^{\mu,d}_{M+G}(U\times\mathbb{R}^q,\boldsymbol{g};\boldsymbol{w})$ of type d, with arbitrary Mellin symbols $f_{j\alpha},\tilde{f}_{j\alpha}$ linked to certain $h_{j\alpha}$ as in (2.21), $f_0\in C^\infty(U,\mathcal{B}^{-\infty,d}_{V_0}(X;\boldsymbol{v};\mathbb{C}))$, and Green symbols $g(y,\eta)$ with asymptotic types $\mathcal{P},\mathcal{Q}$. The space (2.22) is Fréchet in a natural way. It is clear that

$$\mathcal{R}^{\mu,d}_{M+G}(U\times\mathbb{R}^q,\boldsymbol{g};\boldsymbol{w})=\bigcup_{R\in\mathbf{As}_{M+G}(X,\boldsymbol{g})}\mathcal{R}^{\mu,d}_{M+G}(U\times\mathbb{R}^q;\boldsymbol{g};\boldsymbol{w})_R.$$

Remark 2.3. Let us set

$$\boldsymbol{E}^{s,\gamma}:=\mathcal{K}^{s,\gamma}(X^\wedge,E)\oplus\mathcal{K}^{s-\frac12,\gamma-\frac12}((\partial X)^\wedge,J_-)\oplus\mathbb{C}^{l_-}, \tag{2.23}$$

$$\boldsymbol{E}^{s,\gamma}_{\mathcal{P}}:=\mathcal{K}^{s,\gamma}_{P}(X^\wedge,E)\oplus\mathcal{K}^{s-\frac12,\gamma-\frac12}_{P'}((\partial X)^\wedge,J_-)\oplus\mathbb{C}^{l_-}, \tag{2.24}$$

$$\boldsymbol{S}^{\gamma-\mu}:=\mathcal{S}^{\gamma-\mu}(X^\wedge,F)\oplus\mathcal{S}^{\gamma-\mu-\frac12}((\partial X)^\wedge,J_+)\oplus\mathbb{C}^{l_+} \tag{2.25}$$

for

$$\mathcal{S}^{\gamma-\mu}(X^\wedge, F) := \{\omega u + (1-\omega)v : u \in \mathcal{K}^{\infty,\gamma-\mu}(X,F), v \in \mathcal{S}(\overline{\mathbb{R}}_+, C^\infty(X,F))\},$$

cf. also the formula (1.9). Every $a(y,\eta) \in \mathcal{R}^{\mu,d}_{M+G}(U\times\mathbb{R}^q, \boldsymbol{g}; \boldsymbol{w})$ represents operator-valued symbols

$$a(y,\eta) \in S^\mu_{\mathrm{cl}}(U \times \mathbb{R}^q; \boldsymbol{E}^{s,\gamma}, \boldsymbol{\mathcal{S}}^{\gamma-\mu}) \tag{2.26}$$

and

$$a(y,\eta) \in S^\mu_{\mathrm{cl}}(U \times \mathbb{R}^q; \boldsymbol{E}^{s,\gamma}_{\mathcal{P}}, \boldsymbol{\mathcal{S}}^{\gamma-\mu}_{\mathcal{Q}}) \tag{2.27}$$

of order μ with respect to (the first two components of) (2.1), $s > d - \frac{1}{2}$. The relation (2.27) is valid for every pair $\mathcal{P}$ of asymptotic types with some resulting $\mathcal{Q}$ depending on a and $\mathcal{P}$.

Note that in the second components of (2.23), (2.24) instead of $s - \frac{1}{2}$ we can insert any other $s' \in \mathbb{R}$.

Theorem 2.4. *The space $\mathcal{R}^{\mu,d}(U \times \mathbb{R}^q, \boldsymbol{g}; \boldsymbol{w})$ can equivalently be described as the space of all operator functions* (2.4) *where $g(y,\eta)$ is a Green symbol as before, and*

$$\begin{aligned} b(y,\eta) = \sigma r^{-\mu}\{\omega \operatorname{op}_M^{\gamma-\frac{n}{2}}(h)(y,\eta)\tilde{\omega} \\ + (1-\omega)\operatorname{op}_r(p)(y,\eta)(1-\tilde{\tilde{\omega}}\}\tilde{\sigma} + m(y,\eta). \end{aligned} \tag{2.28}$$

This result is a corollary of a corresponding theorem [9, Section 4.6.4] that the η-dependent cut-off functions in (2.4) may be replaced by η-independent ones, modulo a Green symbol with trivial asymptotic types; concerning the boundaryless case, see [7].

Let $\boldsymbol{E}^{s,\gamma}$ and $\boldsymbol{E}^{s,\gamma}_{\mathcal{P}}$ be as in Remark 2.3, and set

$$\boldsymbol{F}^{s-\mu} := \mathcal{K}^{s-\mu,\gamma-\mu}(X^\wedge, F) \oplus \mathcal{K}^{s-\mu-\frac{1}{2},\gamma-\mu-\frac{1}{2}}((\partial X)^\wedge, J_+) \oplus \mathbb{C}^{l_+},$$

$$\boldsymbol{F}^{s-\mu}_{\mathcal{Q}} := \mathcal{K}^{s-\mu,\gamma-\mu}_{Q}(X^\wedge, F) \oplus \mathcal{K}^{s-\mu-\frac{1}{2},\gamma-\mu-\frac{1}{2}}_{Q'}((\partial X)^\wedge, J_+) \oplus \mathbb{C}^{l_+},$$

$\mathcal{P} = (P, P')$, $\mathcal{Q} = (Q, Q')$, cf. the formulas (2.14), (2.15).

Theorem 2.5. *Every $a(y,\eta) \in \mathcal{R}^{\mu,d}(U \times \mathbb{R}^q; \boldsymbol{g}; \boldsymbol{w})$ represents operator-valued symbols*

$$a(y,\eta) \in S^\mu(U \times \mathbb{R}^q; \boldsymbol{E}^{s,\gamma}, \boldsymbol{F}^{s-\mu,\gamma-\mu}) \tag{2.29}$$

and

$$a(y,\eta) \in S^\mu(U \times \mathbb{R}^q; \boldsymbol{E}^{s,\gamma}_{\mathcal{P}}, \boldsymbol{F}^{s-\mu,\gamma-\mu}_{\mathcal{Q}}) \tag{2.30}$$

of order μ with respect to (2.1), *for every $s > d - \frac{1}{2}$.* (2.30) *is valid for every pair $\mathcal{P}$ of asymptotic types with some resulting $\mathcal{Q}$ depending on a and $\mathcal{P}$ (not on s). We also get operator-valued symbols with respect to* (2.16) *with a scheme of Douglis-Nirenberg orders as in Remark* 2.3.

It is now essential for the applications below that the space (2.2) has a variant with a complex parameter $w \in \mathbb{C}$ as an additional covariable, such that the elements $a(y, \eta, w)$ are holomorphic in w.

Let

$$\mathcal{R}^{\mu,d}(U \times \mathbb{R}^q \times \mathbb{C}, \boldsymbol{g}; \boldsymbol{w}) \tag{2.31}$$

denote the corresponding space. Ingredients of (2.31) have been investigated in [18] and [3], namely

$$\sigma r^{-\mu}\{\omega \operatorname{op}_M^{\gamma-\frac{n}{2}}(h)(y,\eta,w)\tilde{\omega} + (1-\omega)\operatorname{op}_r(p)(y,\eta,w)(1-\tilde{\tilde{\omega}})\}\tilde{\sigma}, \tag{2.32}$$

the holomorphic analogue of the first summand on the right hand side of (2.28), and the holomorphic analogues

$$g(y,\eta,w) \quad \text{and} \quad m(y,\eta,w) \tag{2.33}$$

of $g(y, \eta)$ in (2.4) and $m(y, \eta)$ in (2.11), respectively. In order to make the ingredients of the operator-valued symbols of the class (2.31) more transparent we now discuss the kernel cut-off constructions.

Kernel cut-off only concerns covariables. Therefore, to simplify considerations, we assume for a while that symbols have constant coefficients.

Let us first consider Green and smoothing Mellin symbols in the covariables $(\eta, \tau) \in \mathbb{R}^{q+1}$. These are operator-valued symbols

$$a(\eta,\tau) \in S^{\mu}_{\mathrm{cl}}(\mathbb{R}^{q+1}; E, \tilde{E}) \tag{2.34}$$

with E and $\tilde{E}$ running over specific scales of Hilbert spaces. The constructions may be performed for any fixed Hilbert spaces $E, \tilde{E}$; then they are valid also for the projective limits involved in the definition of Green symbols.

In the following we admit symbols to be classical or non-classical. Let $a(\eta, \tau) \in S^{\mu}_{(\mathrm{cl})}(\mathbb{R}^{q+1}_{\eta,\tau}; E, \tilde{E})$, and set

$$k(a)(\eta,\varrho) := \int_{\mathbb{R}} e^{i\varrho\tau} a(\eta,\tau) đ\tau. \tag{2.35}$$

Then, for every $\psi(\varrho) \in C_0^\infty(\mathbb{R})$ the function

$$h(\psi)(a)(\eta,\tau) := \int_{\mathbb{R}} e^{-i\varrho\tau}\psi(\varrho)k(a)(\eta,\varrho)d\varrho \tag{2.36}$$

has a holomorphic extension $h(\psi)(a)(\eta, \zeta)$ into the complex plane of the variable $\zeta = \tau + i\delta$.

Theorem 2.6. *Let $a(\eta, \tau) \in S^{\mu}_{(\mathrm{cl})}(\mathbb{R}^{q+1}; E, \tilde{E})$ and $\psi(\varrho) \in C_0^\infty(\mathbb{R})$.*

1. *We have*

$$h(\psi)(a)(\eta,\zeta) \in \mathcal{A}(\mathbb{C}_\zeta, C^\infty(\mathbb{R}^q, \mathcal{L}(E,\tilde{E}))),$$

and

$$h(\psi)(a)(\eta,\tau+i\delta) \in S^{\mu}_{(\mathrm{cl})}(\mathbb{R}^{q+1}_{\eta,\tau}; E, \tilde{E}) \tag{2.37}$$

for every $\delta \in \mathbb{R}$, uniformly in $c \le \delta \le c'$ for arbitrary $c \le c'$.

2. *The map* $\psi \to h(\psi)(a)(\tau + i\delta)$ *given by* (2.37) *defines a continuous operator*

$$C_0^\infty(\mathbb{R}) \to S^\mu_{(\mathrm{cl})}(\mathbb{R}^{q+1}; E, \tilde{E})$$

for every $\delta \in \mathbb{R}$, *uniformly continuous in* $c \le \delta \le c'$ *for arbitrary* $c \le c'$.

3. *If* $\psi(\varrho)$ *is equal to* 1 *in a neigbourhood of* $\varrho = 0$, *we have*

$$g(\eta, \tau) = h(\psi)(a)(\eta, \tau) \bmod S^{-\infty}(\mathbb{R}^{q+1}; E, \tilde{E}). \qquad (2.38)$$

The map

$$h(\psi) : S^\mu_{(\mathrm{cl})}(\mathbb{R}^{q+1}; E, \tilde{E}) \to S^\mu_{(\mathrm{cl})}(\mathbb{R}^{q+1}; E, \tilde{E})$$

will be called a kernel cut-off operator; the notation is motivated by the relation between (2.35) and the distributional kernel $k(a)(\eta, t - t')$ of the (η-dependent) pseudo-differential operator $\mathrm{Op}_t(a)(\eta)$.

Remark 2.7. Under the assumptions of Theorem 2.6 for every $\beta, \delta \in \mathbb{R}$ there are coefficients $c_k(\beta, \delta)$ (depending on ψ) such that

$$h(\psi)(a)(\eta, \tau + i\beta) \sim \sum_{k=0}^{\infty} c_k(\beta, \delta) D_\tau^k H(\psi)(a)(\eta, \tau + i\delta)$$

in the sense of an asymptotic sum in $S^\mu_{(\mathrm{cl})}(\mathbb{R}^{q+1}; E, \tilde{E})$. If $\psi(0) = 1$, it follows that $c_0(\beta, \delta) = 1$ for every $\beta, \delta \in \mathbb{R}$.

A particularly simple proof of Theorem 2.6 may be found in [13].

In our applications, the complex variable plays the role of a Mellin covariable with imaginary part τ. Therefore, we now slightly change the notation and pass from $\zeta = \tau + i\delta$ to $w = \delta + i\tau$. Instead of Theorem 2.6 we could consider an antiholomorphic variant as well by talking about $\tau - i\delta$ rather than $\tau + i\delta$; the change to w then gives us an analogue of Theorem 2.6 with interchanged real and imaginary parts.

Definition 2.8. Let $S^\mu_{(\mathrm{cl})}(U \times \mathbb{R}^q \times \mathbb{C}; E, \tilde{E})$ denote the space of all $a(y, \eta, w) \in \mathcal{A}(\mathbb{C}_w, C^\infty(U \times \mathbb{R}^q, \mathcal{L}(E, \tilde{E}))$ such that

$$a(y, \eta, \delta + i\tau) \in S^\mu_{(\mathrm{cl})}(U \times \mathbb{R}^{q+1}_{\eta,\tau}; E, \tilde{E})$$

for every $\delta \in \mathbb{R}$, uniformly in $c \le \delta \le c'$ for every $c \le c'$.

The space $S^\mu_{(\mathrm{cl})}(\mathbb{R}^q \times \mathbb{C}; E, \tilde{E})$ is Fréchet in a natural way. In the following we write

$$S^\mu_{(\mathrm{cl})}(U \times \mathbb{R}^q \times \Gamma_\delta; E, \tilde{E}) \qquad (2.39)$$

for symbols of the form $a(y, \eta, \delta + i\tau)$ that belong with respect to (y, η, τ) to the space $S^\mu_{(\mathrm{cl})}(U \times \mathbb{R}^{q+1}; E, \tilde{E})$.

Corollary 2.9. *Setting*

$$H(\psi)(a)(y, \eta, \beta + i\tau) := h(\psi)(\tilde{a})(y, \eta, \tau - i\beta)$$

for $\tilde{a}(y,\eta,\tau) := a(y,\eta,i\tau)$, $\psi \in C_0^\infty(\mathbb{R})$, *where* $h(\psi)$ *refers to the cut-off operator with respect to covariable* τ, *the assertions of Theorem* 2.6 (*combined with a translation in direction of* $\mathrm{Re}\, w$) *can be interpreted in the following way:*

1.
$$H(\psi): S^\mu_{(\mathrm{cl})}(U \times \mathbb{R}^q \times \Gamma_\delta; E, \tilde{E}) \to S^\mu_{(\mathrm{cl})}(U \times \mathbb{R}^q \times \mathbb{C}; E, \tilde{E}) \tag{2.40}$$
is a continuous operator.
2. *The map* $\psi \to H(\psi)a$ *for fixed* $a(y,\eta,w) \in S^\mu_{(\mathrm{cl})}(U \times \mathbb{R}^q \times \Gamma_\delta; E, \tilde{E})$ *defines a continuous operator*
$$C_0^\infty(\mathbb{R}) \to S^\mu_{(\mathrm{cl})}(U \times \mathbb{R}^q \times \mathbb{C}; E, \tilde{E}).$$
3. *For every* $a(y,\eta,w) \in S^\mu_{(\mathrm{cl})}(U \times \mathbb{R}^q \times \Gamma_\delta; E, \tilde{E})$ *there is an* $h(y,\eta,w) \in S^\mu_{(\mathrm{cl})}(U \times \mathbb{R}^q \times \mathbb{C}; E, \tilde{E})$ *such that*
$$a(y,\eta,w) = h(y,\eta,w)\big|_{\mathrm{Re}\, w = \delta} \in S^{-\infty}(U \times \mathbb{R}^q \times \Gamma_\delta; E, \tilde{E}).$$

Remark 2.10. $a(y,\eta,w) \in S^\mu_{(\mathrm{cl})}(U \times \mathbb{R}^q \times \mathbb{C}; E, \tilde{E})$ and $a(y,\eta,w)\big|_{\mathrm{Re}\, w=\delta} \in S^{\mu-1}_{(\mathrm{cl})}(U \times \mathbb{R}^q \times \Gamma_\delta; E, \tilde{E})$ for some $\delta \in \mathbb{R}$ implies $a(y,\eta,w) \in S^{\mu-1}_{(\mathrm{cl})}(U \times \mathbb{R}^q \times \mathbb{C}; E, \tilde{E})$.

We now apply the kernel cut-off operator $H(\psi)$ with respect to the covariable $\tau = \mathrm{Im}\, w, w \in \Gamma_\delta$, to symbols of the space
$$\mathcal{R}^{\mu,d}_{M+G}(U \times \mathbb{R}^q \times \Gamma_\delta, \boldsymbol{g}; \boldsymbol{w})_R \tag{2.41}$$
for $R = (V_0, V_1, \mathcal{P}, \mathcal{Q}) \in \mathbf{As}_{M+G}(X, \boldsymbol{g})$; the meaning of Γ_δ in (2.41) is analogous to (2.39).

There is a specialisation of Definition 2.8 for the symbol space (2.41) which yields a corresponding space of holomorphic symbols in $w \in \mathbb{C}$:
$$\mathcal{R}^{\mu,d}_{M+G}(U \times \mathbb{R}^q \times \mathbb{C}, \boldsymbol{g}; \boldsymbol{w})_R. \tag{2.42}$$
This is studied in detail in De Donno and Schulze [3]. The definition of the subclass $R^{\mu,d}_G(U \times \mathbb{R}^q \times \mathbb{C}, \boldsymbol{g}; \boldsymbol{w})_{\mathcal{P},\mathcal{Q}}$ is straightforward, and we can set
$$\mathcal{R}^{\mu,d}_{M+G}(U \times \mathbb{R}^q \times \mathbb{C}, \boldsymbol{g}; \boldsymbol{w})_R := \{H(\psi)a + g : a(y,\eta,w) \in \mathcal{R}^{\mu,d}_{M+G}(U \times \mathbb{R}^q \times \Gamma_\delta, \boldsymbol{g}; \boldsymbol{w})_R,\ g(y,\eta,w) \in \mathcal{R}^{\mu,d}_G(U \times \mathbb{R}^q \times \mathbb{C}, \boldsymbol{g}; \boldsymbol{w})_{\mathcal{P},\mathcal{Q}}\}$$
for any $\psi \in C_0^\infty(\mathbb{R})$ which is equal to 1 in a neigbourhood of 0.

More generally, we have a map
$$H(\psi): \mathcal{R}^{\mu,d}_{M+G}(U \times \mathbb{R}^q \times \Gamma_\delta, \boldsymbol{g}; \boldsymbol{w})_R \to \mathcal{R}^{\mu,d}_{M+G}(U \times \mathbb{R}^q \times \mathbb{C}, \boldsymbol{g}; \boldsymbol{w})_R \tag{2.43}$$
for every $\psi \in C_0^\infty(\mathbb{R})$, that may be obtained in analogy of the preceding assertions on the abstract context.

Theorem 2.11. *For* $\psi \in C_0^\infty(\mathbb{R})$ *and every element* $a(y,\eta,w)$ *of* (2.41) *we have*
$$h(y,\eta,w) := H(\psi)a(y,\eta,w) \in \mathcal{R}^{\mu,d}_{M+G}(U \times \mathbb{R}^q \times \mathbb{C}, \boldsymbol{g}; \boldsymbol{w})_R,$$

where

$$a(y,\eta,w) = h(y,\eta,w)\big|_{\mathrm{Re}\,w=\delta} \quad \mathrm{mod}\ \mathcal{R}_G^{-\infty,d}(U\times\mathbb{R}^q\times\Gamma_\delta,\boldsymbol{g},\boldsymbol{w})_{\mathcal{P},\mathcal{Q}}$$

when $\psi\equiv 1$ *in a neighbourhood of* 0.

Note that, although the map (2.43) is formally analogous to (2.40), it may appear surprising that operator-valued symbols which contain the τ-variable in the form $\omega_{\eta,w}(r)=\omega(r[\eta,\delta+i\tau])$, for a cut-off function ω, cf. the expression (2.8), can be transformed into holomorphic ones, modulo smoothing symbols, cf. also [3].

To complete the structure of (2.31) it remains to explain (2.32). Starting from arbitrary elements

$$\tilde{p}(r,y,\tilde{\varrho},\tilde{\eta},\tilde{w})\in C^\infty(\overline{\mathbb{R}}_+\times U,\mathcal{B}^{\mu,d}(X;\boldsymbol{v};\mathbb{R}^{1+q}\times\Gamma_\delta)),$$
$$\tilde{h}(r,y,z,\tilde{\eta},\tilde{w})\in C^\infty(\overline{\mathbb{R}}_+\times U,\mathcal{B}^{\mu,d}(X;\boldsymbol{v};\mathbb{C}\times\mathbb{R}^q\times\Gamma_\delta))$$

for any fixed $\delta\in\mathbb{R}$ and setting

$$p(r,y,\varrho,\eta,\delta+i\tau):=\tilde{p}(r,y,r\varrho,r(\delta+i\tau)), \tag{2.44}$$
$$h(r,y,z,\eta,\delta+i\tau):=\tilde{h}(r,y,z,r\eta,r(\delta+i\tau)), \tag{2.45}$$

it suffices to apply a kernel cut-off operator (2.43) with respect to the variable τ, in order to generate the required holomorphic dependence in w. The new element here, compared with the analogous procedure of Remark 1.6, is the extra degeneracy in the variable $r\in\mathbb{R}_+$. Details are elaborated in [18], see also [8]. Concerning the present situation, to generate the space (2.31), it suffices to insert

$$(H(\psi)h)(r,y,z,\eta,w) \quad\text{and}\quad (H(\psi)p)(r,y,\varrho,\eta,w)$$

in (2.32) in place of h and p there, for arbitrary families (2.44) and (2.45), respectively.

This definition is correct; the choice of the cut-off function ψ only affects (2.32) modulo a remainder of the kind (2.33).

2.2. The Edge Algebra

Let W be a compact manifold with edge Y and boundary and $\mathbb{W}$ its stretched manifold. Concerning notation, in particular, for the double $2\mathbb{W}=\tilde{\mathbb{W}}$, we refer to Section 1.1. In the present section we study parameter-dependent edge-boundary value problems. These will be families $\mathcal{A}(\lambda)$ of operators, with parameter $\lambda\in\mathbb{R}^l$, constituting a vector space

$$\mathcal{Y}^{\mu,d}(\mathbb{W},\boldsymbol{g};\boldsymbol{w};\mathbb{R}^l), \tag{2.46}$$

with $\mu\in\mathbb{Z}$ as order and $d\in\mathbb{N}$ as type, and weight data $\boldsymbol{g}=(\gamma,\gamma-\mu,\Theta)$, $\gamma\in\mathbb{R}$, with a (finite or infinite) weight interval Θ, and a tuple $\boldsymbol{w}=(E,F;J_-,J_+;L_-,L_+)$ of vector bundles $E,F\in\mathrm{Vect}(\mathbb{W})$, $J_\pm\in\mathrm{Vect}(\mathbb{W}')$, $L_\pm\in\mathrm{Vect}(Y)$. Here $\mathbb{W}'$ is the stretched manifold associated with the boundary W' of W (in the sense $W'=\partial(W\setminus Y)\cup Y$). Also W' is a manifold with edge Y, now without boundary; the base of the model cone near Y is ∂X.

In local considerations vector bundles $E \in \mathrm{Vect}(\mathbb{W})$ will be restricted to $\mathbb{W}_{\mathrm{reg}}$ or to neighbourhoods of $\mathbb{W}_{\mathrm{sing}}$. To simplify notation we will denote them again by E. Moreover, if

$$\chi : [0,1) \times X \times U'' \to \overline{\mathbb{R}}_+ \times X \times \Omega \tag{2.47}$$

is a 'singular chart', we assume χ to be of the form

$$\chi(r,x,y) = (\chi_0(r), x, \chi''(y))$$

for a diffeomorphism $\chi_0 : [0,1) \to \overline{\mathbb{R}}_+$, $\chi_0(r) = r$ for $0 \le r \le \varepsilon$ for some $\varepsilon > 0$, and a chart $\chi'' : U'' \to \Omega$ on Y. The pull back of E under χ^{-1} will also be denoted by E; also for the restriction of E to $X^\wedge$ for any fixed $y \in \Omega$ we use the same letter. Similar notation will be applied for bundles over $\mathbb{W}'$. Clearly, in pull backs of operators on $\mathbb{R}_+ \times X \times \Omega$ to $(0,1) \times X \times U''$, we take into account transition maps from the bundles, using corresponding invariance properties of our constructions.

Choose an open covering of a neighbourhood $\mathbb{V}$ of $\mathbb{W}_{\mathrm{sing}}$ in $\mathbb{W}$ by charts

$$\chi_k : [0,1) \times X \times U''_k \to \overline{\mathbb{R}}_+ \times X \times \Omega, \tag{2.48}$$

$k = 1, \dots, N$, $\Omega \subseteq \mathbb{R}^q$ open, such that $\{U''_1, \dots, U''_N\}$ is an open covering of Y by induced charts $\chi''_k : U''_k \to \Omega$. Let $\{\varphi_1, \dots, \varphi_N\}$, $\{\psi_1, \dots, \psi_N\}$ be systems of functions $\varphi_k, \psi_k \in C_0^\infty([0,1) \times X \times U''_k)$ such that $\sum \varphi_k = 1$ in a neighbourhood of $\mathbb{W}_{\mathrm{sing}}$, and $\psi_k \equiv 1$ on $\operatorname{supp} \varphi_k$.

Set $\varphi'_k = \varphi_k|_{[0,1)\times(\partial X)\times U''_k}$, $\psi'_k = \psi_k|_{[0,1)\times(\partial X)\times U''_k}$, $\varphi''_k = \varphi_k|_Y$, and $\psi''_k = \psi_k|_Y$. Then, using edge-amplitude functions

$$a_k(y,\eta,\lambda) \in \mathcal{R}^{\mu,d}(\Omega \times \mathbb{R}^{q+l}_{\eta,\lambda}, \boldsymbol{g}; \boldsymbol{w}_k),$$

$\boldsymbol{w}_k = (E, F; J_-, J_+; l_-, l_+)$, we can pass to operator pull backs $\mathcal{A}_k(\lambda)$ of pseudo-differential operators $\mathrm{Op}_y(a_k)(\lambda)$ to $[0,1) \times X \times U''_k$ with respect to (2.46) (which also take into account the transition functions of the involved bundles) and form

$$\mathcal{A}_{\mathrm{edge}}(\lambda) := \sum_{k=1}^{N} \operatorname{diag}(\varphi_k, \varphi'_k, \varphi''_k) \mathcal{A}_k(\lambda) \operatorname{diag}(\psi_k, \psi'_k, \psi''_k). \tag{2.49}$$

By definition, the local amplitude functions a_k contain certain fixed cut-off factors $\sigma(r)$ and $\tilde{\sigma}(r)$, cf. the formula (2.11). For the global construction we assume them to be independent of k. Without loss of generality, let $\tilde{\sigma} \equiv 1$ on $\operatorname{supp} \sigma$. Moreover, choose another cut-off function $\tilde{\tilde{\sigma}}$ such that $\sigma \equiv 1$ on $\operatorname{supp} \tilde{\tilde{\sigma}}$.

The space (2.46) is then defined as the set of all operator families of the form

$$\mathcal{A}(\lambda) = \mathcal{A}_{\mathrm{edge}}(\lambda) + \operatorname{diag}((1-\sigma)\mathcal{B}_{\mathrm{reg}}(\lambda)(1-\tilde{\tilde{\sigma}}), 0) + \mathcal{C}(\lambda) \tag{2.50}$$

where $\mathcal{A}_{\mathrm{edge}}(\lambda)$ has the form (2.49), further $\mathcal{B}_{\mathrm{reg}}(\lambda)$ is an arbitrary element of $\mathcal{B}^{\mu,d}(\mathbb{W}_{\mathrm{reg}}; \boldsymbol{v}; \mathbb{R}^l)$ for $\boldsymbol{v} = (E, F; J_-, J_+)$, and $\mathcal{C}(\lambda)$ belongs to

$$\mathcal{Y}^{-\infty,d}(\mathbb{W}, \boldsymbol{g}; \boldsymbol{w}; \mathbb{R}^l). \tag{2.51}$$

The latter space of smoothing operators is defined as follows:

First recall that in Section 1.1 we introduced the spaces $\mathcal{W}^{s,\gamma}(\tilde{\mathbb{W}})$. For every $\tilde{E} \in \mathrm{Vect}(\tilde{\mathbb{W}})$ there is a straightforward generalisation to spaces $\mathcal{W}^{s,\gamma}(\tilde{\mathbb{W}}, \tilde{E})$ of corresponding distributional sections. In $\tilde{E}$ we fix a Hermitian metric and on $\tilde{\mathbb{W}}$ a Riemannian metric, where $\tilde{\mathbb{W}}$ is treated as a compact C^∞ manifold with boundary. We then have a natural scalar product in the space $L^2(\tilde{\mathbb{W}}, \tilde{E})$.

Let h^γ denote any strictly positive function in $C^\infty(\tilde{\mathbb{W}}_{\mathrm{reg}})$ that is equal to r^γ in a small neighbourhood of $\tilde{\mathbb{W}}_{\mathrm{sing}}$. The multiplication by $h^{-\frac{n}{2}}$ induces a bijection

$$h^{-\frac{n}{2}} : L^2(\tilde{\mathbb{W}}, \tilde{E}) \to \mathcal{W}^{0,0}(\tilde{\mathbb{W}}, \tilde{E})$$

which gives us a scalar product also in $\mathcal{W}^{0,0}(\tilde{\mathbb{W}}, \tilde{E})$. This gives rise to non-degenerate sesquilinear pairings

$$\mathcal{W}^{s,\gamma}(\tilde{\mathbb{W}}, \tilde{E}) \times \mathcal{W}^{-s,-\gamma}(\tilde{\mathbb{W}}, \tilde{E}) \to \mathbb{C}$$

for all $s, \gamma \in \mathbb{R}$.

Also for arbitrary $s, \gamma \in \mathbb{R}$ the space $\mathcal{W}^{s,\gamma}(\tilde{\mathbb{W}}, \tilde{E})$ is a Hilbert space. To define a scalar product, we consider the double $2\tilde{\mathbb{W}}$ of $\tilde{\mathbb{W}}$ which is a closed compact C^∞ manifold. Let $2\tilde{E}$ denote any vector bundle on $2\tilde{\mathbb{W}}$ such that $2\tilde{E}|_{\tilde{\mathbb{W}}} = \tilde{E}$. We then have the standard Sobolev space $H^s(2\tilde{\mathbb{W}}, 2\tilde{E})$ of sections of smoothness $s \in \mathbb{R}$.

Let $\sigma \in C^\infty(\tilde{\mathbb{W}})$ be any function that is equal to 1 on $[0, \frac{1}{3}) \times X \times Y$ and vanishes outside $[0, \frac{1}{2}) \times X \times Y$. Let us identify $(1-\sigma)u$ for $u \in H^s_{\mathrm{loc}}(\tilde{\mathbb{W}}_{\mathrm{reg}}, \tilde{E})$ with an element of $H^s(2\tilde{\mathbb{W}}, 2\tilde{E})$ (vanishing on $2\tilde{\mathbb{W}} \setminus \tilde{\mathbb{W}}_{\mathrm{reg}}$). Then a scalar product in $\mathcal{W}^{s,\gamma}(\tilde{\mathbb{W}}, \tilde{E})$ is

$$\begin{aligned}(u,v)_{\mathcal{W}^{s,\gamma}(\tilde{\mathbb{W}},\tilde{E})} = \sum_{k=1}^{N} \int & \langle\eta\rangle^{2s} (\kappa^{-1}_{\langle\eta\rangle} F(((\chi''_k)^{-1})^* \varphi''_k \sigma u)(\eta), \\ & \kappa^{-1}_{\langle\eta\rangle} F(((\chi''_k)^{-1})^* \varphi''_k \sigma v)(\eta))_{\mathcal{K}^{s,\gamma}(\tilde{X}^\wedge,\tilde{E})} d\eta \\ & + ((1-\sigma)u, (1-\sigma)v)_{H^s(2\tilde{\mathbb{W}},2\tilde{E})}.\end{aligned}$$

Analogously to the constructions of Section 1.1 we have the spaces

$$\mathcal{W}_0^{s,\gamma}(\mathbb{W}_\pm, \tilde{E}) = \{\tilde{u} \in \mathcal{W}^{s,\gamma}(\tilde{\mathbb{W}}, \tilde{E}) : \operatorname{supp} \tilde{u} \subseteq \mathbb{W}_\pm\}$$

and

$$\mathcal{W}^{s,\gamma}(\mathbb{W}, E) = \{\tilde{u}|_{\operatorname{int} \mathbb{W}_{\mathrm{reg}}} : \tilde{u} \in \mathcal{W}^{s,\gamma}(\tilde{\mathbb{W}}, \tilde{E})\} \tag{2.52}$$

for $\tilde{E}|_{\mathbb{W}} = E$, with a natural identification

$$\mathcal{W}^{s,\gamma}(\mathbb{W}, E) = \mathcal{W}^{s,\gamma}(\tilde{\mathbb{W}}, \tilde{E}) / \mathcal{W}_0^{s,\gamma}(\mathbb{W}_-, \tilde{E}); \tag{2.53}$$

this gives us a Hilbert space structure also in $\mathcal{W}^{s,\gamma}(\mathbb{W}, E)$.

Let us now pass to subspaces with asymptotics. First, on $\tilde{X}^\wedge \times \mathbb{R}^q$ we can form the spaces

$$\mathcal{W}^{s,\gamma}_{\tilde{P}}(\tilde{X}^\wedge \times \mathbb{R}^q, \tilde{E}) := \mathcal{W}^{s,\gamma}(\mathbb{R}^q, \mathcal{K}^{s,\gamma}_{\tilde{P}}(\tilde{X}^\wedge, \tilde{E}))$$

for every $\tilde{P} \in \mathrm{As}(\tilde{X}, \boldsymbol{g})$, $\boldsymbol{g} = (\gamma, \Theta)$. Then $\mathcal{W}^{s,\gamma}_{\tilde{P}}(\tilde{\mathbb{W}}, \tilde{E})$ is defined as the subspace of all $u \in H^s_{\mathrm{loc}}(\tilde{\mathbb{W}}_{\mathrm{reg}}, \tilde{E})$ such that $\omega\varphi u = \chi^* v$ for some $v \in \mathcal{W}^{s,\gamma}_{\tilde{P}}(\tilde{X}^\wedge \times \mathbb{R}^q, \tilde{E})$ for every χ of the kind (1.1). Then, by the same scheme as (2.52), (2.53) we can form Fréchet spaces

$$\mathcal{W}_P^{s,\gamma}(\mathbb{W}, E) \tag{2.54}$$

with asymptotics of type $P \in \mathrm{As}(X, \boldsymbol{g})$, where we use the fact that every P can be obtained from some $\tilde{P} = \{(p_j, m_j, \tilde{L}_j)\}$ by restricting the spaces $\tilde{L}_j$ from $\tilde{X}$ to X.

Analogous constructions apply to $\mathbb{W}'$, vector bundles $J \in \mathrm{Vect}(\mathbb{W}')$ and asymptotic types $P' \in \mathrm{As}(\partial X, \boldsymbol{g}')$ for weight data $\boldsymbol{g}' = (\gamma', \Theta)$, i.e., we have the spaces

$$\mathcal{W}^{s,\gamma'}(\mathbb{W}', J) \quad \text{and} \quad \mathcal{W}_{P'}^{s,\gamma'}(\mathbb{W}', J), \tag{2.55}$$

respectively, $s, \gamma' \in \mathbb{R}$.

For every $\gamma, \varrho \in \mathbb{R}$ we have non-degenerate sesquilinear pairings

$$\mathcal{W}^{s,\gamma-\varrho}(\mathbb{W}', J) \times \mathcal{W}^{-s,-\gamma-\varrho}(\mathbb{W}', J) \to \mathbb{C},$$

$s \in \mathbb{R}$, induced by the $\mathcal{W}^{0,-\varrho}(\mathbb{W}', J)$-scalar product. We will apply this to $\varrho = \frac{1}{2}$. In the following E, F, J_-, J_+ are bundles as before, and we assume $L_-, L_+ \in \mathrm{Vect}(Y)$. Given an operator

$$\mathcal{A} : \begin{array}{c} \mathcal{W}^{s,\gamma}(\mathbb{W}, E) \\ \oplus \\ \mathcal{W}^{s-\frac{1}{2},\gamma-\frac{1}{2}}(\mathbb{W}', J_-) \\ \oplus \\ H^{s-\frac{n+1}{2}}(Y, L_-) \end{array} \to \begin{array}{c} \mathcal{W}^{s-\mu,\gamma-\mu}(\mathbb{W}, F) \\ \oplus \\ \mathcal{W}^{s-\frac{1}{2}-\mu,\gamma-\mu-\frac{1}{2}}(\mathbb{W}', J_+) \\ \oplus \\ H^{s-\frac{n+1}{2}-\mu}(Y, L_-) \end{array} \tag{2.56}$$

that is continuous for every $s > -\frac{1}{2}$ we define its formal adjoint $\mathcal{A}^*$ in the sense

$$(\mathcal{A}u, v)_{\mathfrak{W}^0_+} = (u, \mathcal{A}^* v)_{\mathfrak{W}^0_-}$$

for $\mathfrak{W}^0_+ = \mathcal{W}^{0,0}(\mathbb{W}, F) \oplus \mathcal{W}^{0,-\frac{1}{2}}(\mathbb{W}', J_+) \oplus H^0(Y, L_+)$, $\mathfrak{W}^0_- = \mathcal{W}^{0,0}(\mathbb{W}, E) \oplus \mathcal{W}^{0,-\frac{1}{2}}(\mathbb{W}', J_-) \oplus H^0(Y, L_-)$, for all $u \in C_0^\infty(\mathrm{int}\,\mathbb{W}_{\mathrm{reg}}, E) \oplus C_0^\infty(\mathbb{W}'_{\mathrm{reg}}, J_-) \oplus C^\infty(Y, L_-)$, $v \in C_0^\infty(\mathrm{int}\,\mathbb{W}_{\mathrm{reg}}, F) \oplus C_0^\infty(\mathbb{W}'_{\mathrm{reg}}, J_+) \oplus C^\infty(Y, L_+)$. An operator (2.50) is said to belong $\mathcal{Y}^{-\infty,0}(\mathbb{W}, \boldsymbol{g}; \boldsymbol{w})$ for $\boldsymbol{g} = (\gamma, \gamma-\mu, \Theta)$, $\boldsymbol{w} = (E, F; J_-, J_+; L_-, L_+)$, if there are asymptotic types $(P, P') \in \mathrm{As}(X, (\gamma-\mu, \Theta)) \times \mathrm{As}(\partial X, (\gamma-\mu-\frac{1}{2}, \Theta))$ and $(Q, Q') \in \mathrm{As}(X, (-\gamma, \Theta)) \times \mathrm{As}(\partial X, (-\gamma-\frac{1}{2}, \Theta))$ such that $\mathcal{A}$ and $\mathcal{A}^*$ induce continuous operators

$$\mathcal{A} : \begin{array}{c} \mathcal{W}^{s,\gamma}(\mathbb{W}, E) \\ \oplus \\ \mathcal{W}^{s',\gamma-\frac{1}{2}}(\mathbb{W}', J_-) \\ \oplus \\ H^{s''}(Y, L_-) \end{array} \to \begin{array}{c} \mathcal{W}_P^{\infty,\gamma-\mu}(\mathbb{W}, F) \\ \oplus \\ \mathcal{W}_{P'}^{\infty,\gamma-\mu-\frac{1}{2}}(\mathbb{W}', J_+) \\ \oplus \\ C^\infty(Y, L_+) \end{array} \tag{2.57}$$

and

$$\mathcal{A}^* : \begin{array}{c} \mathcal{W}^{s,-\gamma+\mu}(\mathbb{W},F) \\ \oplus \\ \mathcal{W}^{s',-\gamma+\mu-\frac{1}{2}}(\mathbb{W}',J_+) \\ \oplus \\ H^{s''}(Y,L_+) \end{array} \to \begin{array}{c} \mathcal{W}_Q^{\infty,-\gamma}(\mathbb{W},E) \\ \oplus \\ \mathcal{W}_{Q'}^{\infty,-\gamma-\frac{1}{2}}(\mathbb{W}',J_-), \\ \oplus \\ C^\infty(Y,L_-) \end{array} \tag{2.58}$$

respectively, for all $s > -\frac{1}{2}$, $s', s'' \in \mathbb{R}$. Moreover, $\mathcal{Y}^{-\infty,d}(\mathbb{W},\boldsymbol{g};\boldsymbol{w})$ denotes the space of all operators

$$\mathcal{C} = \mathcal{C}_0 + \sum_{j=1}^{d} \mathcal{C}_j \operatorname{diag}(T^j, 0, 0) \tag{2.59}$$

for arbitrary $\mathcal{C}_j \in \mathcal{Y}^{-\infty,0}(\mathbb{W},\boldsymbol{g};\boldsymbol{w})$ and differential operators T^j of order j on $\mathbb{W}$, $T^j = \tilde{T}^j|_{\mathbb{W}}$ for differential operators $\tilde{T}^j$ on $\tilde{\mathbb{W}}$ with smooth coefficients up to $\partial\tilde{\mathbb{W}} = \tilde{\mathbb{W}}_{\text{sing}}$ which are in the splitting of variables (r,x,y) near $\tilde{\mathbb{W}}_{\text{sing}}$ of the form $\frac{\partial^j}{\partial x_n^j} \otimes \mathrm{id}_{\tilde{E}}$ for all j. Elements of $\mathcal{Y}^{-\infty,d}(\mathbb{W},\boldsymbol{g};\boldsymbol{w})$ are said to be of type d.

Let $\mathcal{Y}^{-\infty,d}(\mathbb{W},\boldsymbol{g};\boldsymbol{w})_{\mathcal{P},\mathcal{Q}}$ denote the set of all $\mathcal{C} \in \mathcal{Y}^{-\infty,d}(\mathbb{W},\boldsymbol{g};\boldsymbol{w})$ such that the summands $\mathcal{C}_j$ in (2.59) satisfy the continuity properties (2.57) and (2.58) for fixed pairs of asymptotic types $\mathcal{P}$, $\mathcal{Q}$, $j = 0,\dots,d$. This space is Fréchet in a natural way, and we set

$$\mathcal{Y}^{-\infty,d}(\mathbb{W},\boldsymbol{g};\boldsymbol{w},\mathbb{R}^l)_{\mathcal{P},\mathcal{Q}} = \mathcal{S}(\mathbb{R}^l, \mathcal{Y}^{-\infty,d}(\mathbb{W},\boldsymbol{g};\boldsymbol{w})_{\mathcal{P},\mathcal{Q}}). \tag{2.60}$$

Moreover, let $\mathcal{Y}^{-\infty,d}(\mathbb{W},\boldsymbol{g};\boldsymbol{w};\mathbb{R}^l)$ be the union of all spaces (2.60) over $\mathcal{P}$, $\mathcal{Q}$, and set

$$\mathcal{Y}^{-\infty}(\mathbb{W},\boldsymbol{g};\boldsymbol{w};\mathbb{R}^l) = \bigcup_{d\in\mathbb{N}} \mathcal{Y}^{-\infty,d}(\mathbb{W},\boldsymbol{g};\boldsymbol{w};\mathbb{R}^l).$$

Thus the class of parameter-dependent operators (2.46) is completely defined.

Parameter-dependent operators will also be employed in the variant

$$\mathcal{Y}^\mu(\mathbb{W},\boldsymbol{g};\boldsymbol{w};\Gamma_\delta \times \mathbb{R}^l) \tag{2.61}$$

which is defined as the space of all operator families $\mathcal{A}(w,\lambda)$, parametrised by $(w,\lambda) \in \Gamma_\delta \times \mathbb{R}^l$, such that $\mathcal{A}(\delta + i\tau,\lambda)$ belongs to $\mathcal{Y}^\mu(\mathbb{W},\boldsymbol{g};\boldsymbol{w};\mathbb{R}^{1+l}_{\tau,\lambda})$.

Given a pair of vector bundles $\mathfrak{E} := (E,J)$ for $E \in \operatorname{Vect}(\mathbb{W})$, $J \in \operatorname{Vect}(\mathbb{W}')$, we set

$$\mathcal{W}^{s,\gamma}(\mathbb{W},\mathfrak{E}) := \mathcal{W}^{s,\gamma}(\mathbb{W},E) \oplus \mathcal{W}^{s-\frac{1}{2},\gamma-\frac{1}{2}}(\mathbb{W}',J), \tag{2.62}$$

$$\mathcal{W}^{s,\gamma}_{\mathcal{P}}(\mathbb{W},\mathfrak{E}) := \mathcal{W}^{s,\gamma}_P(\mathbb{W},E) \oplus \mathcal{W}^{s-\frac{1}{2},\gamma-\frac{1}{2}}_{P'}(\mathbb{W}',J) \tag{2.63}$$

for asymptotic types $\mathcal{P} = (P,P') \in \operatorname{As}(X,(\gamma,\Theta)) \times \operatorname{As}(\partial X,(\gamma - \frac{1}{2},\Theta))$.

In the following theorem we set $\boldsymbol{g} = (\gamma,\gamma-\mu,\Theta)$, $\Theta = (-(k+1),0]$, $k \in \mathbb{N} \cup \{\infty\}$, and $\boldsymbol{w} = (E,F;J_-,J_+;L_-,l_+)$, $\mathfrak{E}_- = (E,J_-)$, $\mathfrak{E}_+ = (F,J_+)$.

Theorem 2.12. *Every $\mathcal{A} \in \mathcal{Y}^\mu(\mathbb{W}, \boldsymbol{g}; \boldsymbol{w}; \mathbb{R}^l)$ induces families of continuous operators*

$$\mathcal{A}(\lambda) : \begin{array}{c} \mathcal{W}^{s,\gamma}(\mathbb{W}, \mathfrak{E}_-) \\ \oplus \\ H^{s-\frac{n+1}{2}}(Y, L_-) \end{array} \longrightarrow \begin{array}{c} \mathcal{W}^{s-\mu,\gamma-\mu}(\mathbb{W}, \mathfrak{E}_+) \\ \oplus \\ H^{s-\mu-\frac{n+1}{2}}(Y, L_+) \end{array} \tag{2.64}$$

and

$$\mathcal{A}(\lambda) : \begin{array}{c} \mathcal{W}^{s,\gamma}_{\mathcal{P}}(\mathbb{W}, \mathfrak{E}_-) \\ \oplus \\ H^{s-\frac{n+1}{2}}(Y, L_-) \end{array} \longrightarrow \begin{array}{c} \mathcal{W}^{s-\mu,\gamma-\mu}_{\mathcal{Q}}(\mathbb{W}, \mathfrak{E}_+) \\ \oplus \\ H^{s-\frac{n+1}{2}}(Y, L_+) \end{array} \tag{2.65}$$

for $s > d - \frac{1}{2}$, $d = d_{\mathcal{A}}$ (the type of $\mathcal{A}$) and every pair $\mathcal{P}$ of asymptotic types, with some resulting pair $\mathcal{Q}$, depending on $\mathcal{A}$ and $\mathcal{P}$ (not on s).

The proof of this result can be obtained by considering the summands of (2.50) separately. $\mathcal{C}(\lambda)$ is clear by definition, and the second summand is essentially a variant of (1.14). The first term on the right of (2.50) is a finite sum of local operators with amplitude functions as in Theorem 2.5. It is now sufficient to apply the Theorem 1.4 to our specific situation, taking into account the definition of the edge Sobolev spaces and their subspaces with asymptotics.

Parameter-dependent edge boundary value problems $\mathcal{A} \in \mathcal{Y}^\mu(\mathbb{W}, \boldsymbol{g}; \boldsymbol{w}; \mathbb{R}^l)$ have a parameter-dependent principal symbolic structure

$$\sigma(\mathcal{A}) = (\sigma_\psi(\mathcal{A}), \sigma_\partial(\mathcal{A}), \sigma_\wedge(\mathcal{A})). \tag{2.66}$$

To give a definition we write $\mathcal{A} = (\mathcal{A}_{ij})_{i,j=1,2,3}$ and observe that

$$\mathcal{A}_{11} \in L^\mu_{\mathrm{cl}}(\operatorname{int} \mathbb{W}_{\mathrm{reg}}; E, F; \mathbb{R}^l), \tag{2.67}$$

$$\mathcal{A}' := (\mathcal{A}_{ij})_{i,j=1,2} \in \mathcal{B}^{\mu,d}(\mathbb{W}_{\mathrm{reg}}, \boldsymbol{v}; \mathbb{R}^l) \tag{2.68}$$

for $\boldsymbol{v} = (E, F; J_-, J_+)$. We then set $\sigma_\psi(\mathcal{A}) := \sigma_\psi(\mathcal{A}_{11})$ which is the standard homogeneous principal symbol of order μ with parameter $\lambda \in \mathbb{R}^l$; moreover $\sigma_\partial(\mathcal{A}) := \sigma_\partial(\mathcal{A}')$ is the principal boundary symbol of $\mathcal{A}'$ with parameter $\lambda \in \mathbb{R}^l$, cf. Section 1.2, in particular, the generalisation of (1.22) to the non-compact manifold $\mathbb{W}_{\mathrm{reg}}$ with boundary $\mathbb{W}'_{\mathrm{reg}}$ (and with the corresponding vector bundles).

Because of the edge-degeneracy in the splitting of variables $(r, x, y) \in \overline{\mathbb{R}}_+ \times X \times Y$ we can write

$$\sigma_\psi(\mathcal{A})(r, x, y, \varrho, \xi, \eta, \lambda) = r^{-\mu} \tilde{\sigma}_\psi(\mathcal{A})(r, x, y, r\varrho, \xi, r\eta, r\lambda)$$

where

$$\tilde{\sigma}_\psi(\mathcal{A})(r, x, y, \tilde{\varrho}, \xi, \tilde{\eta}, \tilde{\lambda}), \quad (\tilde{\varrho}, \xi, \tilde{\eta}, \tilde{\lambda}) \neq 0,$$

is smooth in r up to 0. Analogously for $(r, x', y) \in \overline{\mathbb{R}}_+ \times \partial X \times Y$ we have

$$\sigma_\partial(\mathcal{A})(r, x', y, \varrho, \xi', \eta, \lambda) = r^{-\mu} \tilde{\sigma}_\partial(\mathcal{A})(r, x', y, r\varrho, \xi', r\eta, r\lambda)$$

with

$$\tilde{\sigma}_\partial(\mathcal{A})(r, x', y, \tilde{\varrho}, \xi', \tilde{\eta}, \tilde{\lambda}), \quad (\tilde{\varrho}, \xi', \tilde{\eta}, \tilde{\lambda}) \neq 0$$

being smooth in r up to 0.

Finally, $\sigma_\wedge(\mathcal{A})(y,\eta,\lambda)$ denotes the parameter-dependent homogeneous principal edge symbol which is an operator family that can be expressed in terms of the local amplitude functions

$$a(y,\eta,\lambda) \in R^{\mu,d}(U \times \mathbb{R}^q \times \mathbb{R}^l, \boldsymbol{g}; \boldsymbol{w}),$$

cf. the notation in the formula (2.2). Here $U \subseteq \mathbb{R}^q$ is open and corresponds to a chart on the edge Y. Applying (2.4) in the variant with (η,λ) in place of η we set

$$\sigma_\wedge(\mathcal{A})(y,\eta,\lambda) := \sigma_\wedge(a)(y,\eta,\lambda)$$

for $\sigma_\wedge(a)(y,\eta,\lambda) := \operatorname{diag}(\sigma_\wedge(b)(y,\eta,\lambda),0) + \sigma_\wedge(g)(y,\eta,\lambda)$. Here $\sigma_\wedge(g)$ is the homogeneous principal part of g as a classical operator-valued symbols, while

$$\begin{aligned} \sigma_\wedge(b)(y,\eta,\lambda) \quad &:= \quad r^{-\mu}\{\omega(r|\eta,\lambda|)\operatorname{op}_M^{\gamma-\frac{n}{2}}(h_0)(y,\eta,\lambda)\tilde{\omega}(r|\eta,\lambda|) \\ &+ \quad (1-\omega(r|\eta,\lambda|))\operatorname{op}_r(p_0)(y,\eta,\lambda)(1-\tilde{\tilde{\omega}}(r|\eta,\lambda|)) \\ &+ \quad \sigma_\wedge(m)(y,\eta,\lambda) \end{aligned}$$

with $\sigma_\wedge(m)$ being the homogeneous principal part of m as a classical operator-valued symbol, and

$$h_0(r,y,z,\eta,\lambda) := \tilde{h}(0,y,z,r\eta,r\lambda), \quad p_0(r,\varrho,\eta,\lambda) := \tilde{p}(0,r\varrho,r\eta,r\lambda),$$

cf. the λ-dependent analogue of (2.6) and (2.7).

Theorem 2.13. *$\mathcal{A} \in \mathcal{Y}^\mu(\mathbb{W},\boldsymbol{g};\boldsymbol{v};\mathbb{R}^l)$ for*

$$\boldsymbol{g} = (\gamma-\nu,\gamma-(\mu+\nu),\Theta), \quad \boldsymbol{v} = (\widetilde{E},F;\widetilde{J},J_+;\widetilde{L},L_+)$$

and $\mathcal{B} \in \mathcal{Y}^\nu(\mathbb{W},\boldsymbol{h};\boldsymbol{w};\mathbb{R}^l)$ for

$$\boldsymbol{h} = (\gamma,\gamma-\nu,\Theta), \quad \boldsymbol{w} = (E,\widetilde{E};J_-,\widetilde{J};L_-,\widetilde{L})$$

implies

$$\mathcal{AB} \in \mathcal{Y}^{\mu+\nu}(\mathbb{W},\boldsymbol{g}\circ\boldsymbol{h};\boldsymbol{v}\circ\boldsymbol{w};\mathbb{R}^l)$$

for $\boldsymbol{g}\circ\boldsymbol{h} = (\gamma,\gamma-(\mu+\nu),\Theta)$, $\boldsymbol{v}\circ\boldsymbol{w} = (E,F;J_-,J_+;L_-,L_+)$, and $d_{\mathcal{AB}} = \max(\nu+d_{\mathcal{A}},d_{\mathcal{B}})$, and we have

$$\sigma(\mathcal{AB}) = \sigma(\mathcal{A})\sigma(\mathcal{B})$$

with componentwise composition (cf. the formula (2.66)).

Theorem 2.13 is a parameter-dependent analogue of a corresponding composition result on edge boundary values problems, cf. [9, Section 4.5.2]. The proof in the parameter-dependent case does not contain additional difficulties.

2.3. Ellipticity of Edge-Boundary Value Problems

This section studies parameter-dependent ellipticity of edge-degenerate boundary value problems on a (compact) manifold W with edge Y and boundary. As usual we formulate results on the associated stretched manifold $\mathbb{W}$.

Let $\boldsymbol{w} = (E, F; J_-, J_+; L_-, L_+)$, $\boldsymbol{g} = (\gamma, \gamma - \mu, \Theta)$, $\Theta = (-(k+1), 0]$, $k \in \mathbb{N} \cup \{\infty\}$.

Definition 2.14. An element $\mathcal{A}(\lambda) \in \mathcal{Y}^\mu(\mathbb{W}, \boldsymbol{g}; \boldsymbol{w}; \mathbb{R}^l)$ is called parameter-dependent elliptic of order μ, if

1.
$$\sigma_\psi(\mathcal{A}) : \pi^*_{\mathbb{W}_{\text{reg}}} E \to \pi^*_{\mathbb{W}_{\text{reg}}} F \tag{2.69}$$
for $\pi_{\mathbb{W}_{\text{reg}}} : (T^*\mathbb{W}_{\text{reg}} \times \mathbb{R}^l) \setminus 0 \to \mathbb{W}_{\text{reg}}$ is an isomorphism and, near $\mathbb{W}_{\text{sing}}$ in the splitting of variables $(r, x, y) \in \overline{\mathbb{R}}_+ \times X \times Y$,
$$\tilde{\sigma}_\psi(\mathcal{A})(r, x, y, \tilde{\varrho}, \xi, \tilde{\eta}, \tilde{\lambda})$$
are isomorphisms between the respective fibres of E and F over all (r, x, y), including $r = 0$, for all $(\tilde{\varrho}, \xi, \tilde{\eta}, \tilde{\lambda}) \neq 0$.
2.
$$\sigma_\partial(\mathcal{A}) : \pi^*_{\mathbb{W}'_{\text{reg}}} \begin{pmatrix} H^s(\mathbb{R}_+, E') \\ \oplus \\ J_- \end{pmatrix} \to \pi^*_{\mathbb{W}'_{\text{reg}}} \begin{pmatrix} H^{s-\mu}(\mathbb{R}_+, F') \\ \oplus \\ J_+ \end{pmatrix} \tag{2.70}$$
for $\pi_{\mathbb{W}'_{\text{reg}}} : (T^*\mathbb{W}'_{\text{reg}} \times \mathbb{R}^l) \setminus 0 \to \mathbb{W}'_{\text{reg}}$ is an isomorphism for any $s > \max(\mu, d) - \frac{1}{2}$, $d = d_{\mathcal{A}}$, and near $\mathbb{W}'_{\text{sing}}$, in the splitting of variables $(r, x', y) \in \overline{\mathbb{R}}_+ \times (\partial X) \times Y$,
$$\tilde{\sigma}_\partial(\mathcal{A})(r, x', y, \tilde{\varrho}, \xi', \tilde{\eta}, \tilde{\lambda})$$
are isomorphisms between the respective fibres of $H^s(\mathbb{R}_+, E') \oplus J_-$ and $H^{s-\mu}(\mathbb{R}_+, F') \oplus J_+$ over all (r, x', y), including $r = 0$, for all $(\tilde{\varrho}, \xi', \tilde{\eta}, \tilde{\lambda}) \neq 0$.
3.
$$\sigma_\wedge(\mathcal{A}) : \pi^*_Y \begin{pmatrix} \mathcal{K}^{s,\gamma}(X^\wedge, \mathfrak{E}_-) \\ \oplus \\ L_- \end{pmatrix} \to \pi^*_Y \begin{pmatrix} \mathcal{K}^{s-\mu,\gamma-\mu}(X^\wedge, \mathfrak{E}_+) \\ \oplus \\ L_+ \end{pmatrix} \tag{2.71}$$
for $\pi_Y : (T^*Y \times \mathbb{R}^l) \setminus 0 \to Y$ is an isomorphism for any $s > \max(\mu, d) - \frac{1}{2}$, $d = d_{\mathcal{A}}$.

Conditions (ii) and (iii) are independent of the choice of s. Note that in (ii) the spaces may equivalently be replaced by $\mathcal{S}(\overline{\mathbb{R}}_+, E) \oplus J_-$ and $\mathcal{S}(\overline{\mathbb{R}}_+, F) \oplus J_+$, and in (iii) by $\mathcal{S}^\gamma(X^\wedge, \mathfrak{E}_-) \oplus L_-$ and $\mathcal{S}^{\gamma-\mu}(X^\wedge, \mathfrak{E}_+) \oplus L_+$, respectively. Here, for instance, for $\mathfrak{E}_- = (E, J_-)$,
$$\mathcal{S}^\gamma(X^\wedge, \mathfrak{E}_-) := \mathcal{S}^\gamma(X^\wedge, E) \oplus \mathcal{S}^{\gamma - \frac{1}{2}}((\partial X), J_-),$$
cf. also the formula (2.25).

In the sequel we also talk about ellipticity rather than parameter-dependent ellipticity of order μ.

Theorem 2.15. *Let $\mathcal{A} \in \mathcal{Y}^\mu(\mathbb{W}, \boldsymbol{g}; \boldsymbol{w}; \mathbb{R}^l)$ be elliptic. Then $\mathcal{A}$ has a parametrix $\mathcal{P} \in \mathcal{Y}^{-\mu}(\mathbb{W}, \boldsymbol{g}^{-1}; \boldsymbol{w}^{-1}; \mathbb{R}^l)$ for $\boldsymbol{g}^{-1} = (\gamma - \mu, \gamma, \Theta)$, $\boldsymbol{w}^{-1} = (E, F; J_+, J_-; L_+, L_-)$ of type $d_\mathcal{P} = \max(d_\mathcal{A} - \mu, 0)$ where*

$$\mathcal{I} - \mathcal{P}\mathcal{A} = \mathcal{C}_\mathrm{l} \in \mathcal{Y}^{-\infty}(\mathbb{W}, \boldsymbol{g}_\mathrm{l}; \boldsymbol{w}_\mathrm{l}; \mathbb{R}^l),$$

for $\boldsymbol{g}_\mathrm{l} = (\gamma, \gamma, \Theta)$, $\boldsymbol{w}_\mathrm{l} = (E, E; J_-, J_-; L_-, L_-)$,

$$\mathcal{I} - \mathcal{A}\mathcal{P} = \mathcal{C}_\mathrm{r} \in \mathcal{Y}^{-\infty}(\mathbb{W}, \boldsymbol{g}_\mathrm{r}; \boldsymbol{w}_\mathrm{r}; \mathbb{R}^l),$$

and $\boldsymbol{g}_\mathrm{r} = (\gamma - \mu, \gamma - \mu, \Theta)$, $\boldsymbol{w}_\mathrm{r} = (F, F; J_+, J_+; L_+, L_+)$ and we have $d_{\mathcal{C}_\mathrm{l}} = \max(\mu, d_\mathcal{A})$, $d_{\mathcal{C}_\mathrm{r}} = \max(d_\mathcal{A} - \mu, 0)$.

Theorem 2.16. *Let $\mathcal{A} \in \mathcal{Y}^\mu(\mathbb{W}, \boldsymbol{g}; \boldsymbol{w}; \mathbb{R}^l)$ be elliptic. Then $\mathcal{A}$ induces a family of Fredholm operators (2.64), $s > \max(\mu, d_\mathcal{A}) - \frac{1}{2}$, of index zero. There exists a constant $c > 0$ such that (2.64) is invertible for all $|\lambda| \geq c$.*

Theorem 2.17. [14] *For every $\mu \in \mathbb{Z}$, $\gamma \in \mathbb{R}$, $E \in \mathrm{Vect}(\mathbb{W})$, there exists an elliptic operator*

$$R^\mu \in \mathcal{Y}^\mu(\mathbb{W}, \boldsymbol{g}; (E, E); \mathbb{R}^l)$$

of type 0, $\boldsymbol{g} = (\gamma, \gamma - \mu, (-\infty, 0])$, such that

$$R^\mu(\lambda) : \mathcal{W}^{s,\gamma}(\mathbb{W}, E) \to \mathcal{W}^{s-\mu,\gamma-\mu}(\mathbb{W}, E)$$

is a family of isomorphisms for all $s > \max(\mu, 0) - \frac{1}{2}$ and all $\lambda \in \mathbb{R}^l$, and we have

$$(R^\mu(\lambda))^{-1} \in \mathcal{Y}^{-\mu}(\mathbb{W}, \boldsymbol{g}; (E, E); \mathbb{R}^l),$$

$\boldsymbol{g}^{-1} = (\gamma - \mu, \gamma, (-\infty, 0])$, also of being type 0.

Remark 2.18. Theorems 2.15 and 2.16 generalise analogous results for a manifold W' with edge Y and without boundary, cf. [2] and [4]. Also Theorem 2.17 has a corresponding analogue for the boundaryless case. Moreover, in a refined version, order reducing results also hold for boundary value problems without the transmission property, see [31].

3. Corner Conormal Symbols

3.1. Meromorphic Families

We now introduce families $\mathcal{A}(w, \lambda)$ of elements in $\mathcal{Y}^{\mu,d}(\mathbb{W}, \boldsymbol{g}; \boldsymbol{w}; \mathbb{R}^l)$ holomorphically depending on a parameter $w \in \mathbb{C}$.

The definition of the corresponding space

$$\mathcal{Y}^{\mu,d}(\mathbb{W}, \boldsymbol{g}; \boldsymbol{w}; \mathbb{C}_w \times \mathbb{R}^l) \tag{3.1}$$

will be given along the lines of the expression (2.50). The space (3.1) is defined as the set of all operator families

$$\mathcal{A}(w, \lambda) = \mathcal{A}_\mathrm{edge}(w, \lambda) + \mathrm{diag}\big((1 - \sigma)\mathcal{B}_\mathrm{reg}(w, \lambda)(1 - \tilde{\sigma}), 0\big) + \mathcal{C}(w, \lambda), \tag{3.2}$$

where the summands are defined as follows: We begin with the smoothing part $\mathcal{C}(w,\lambda) \in \mathcal{Y}^{-\infty,d}(\mathbb{W}, \boldsymbol{g}; \boldsymbol{w}; \mathbb{C} \times \mathbb{R}^l)$. The latter space is defined to be the set of all

$$\mathcal{C}(w,\lambda) \in \mathcal{A}(\mathbb{C}, \mathcal{Y}^{-\infty,d}(\mathbb{W}, \boldsymbol{g}; \boldsymbol{w}; \mathbb{R}^l)_{\mathcal{P},\mathcal{Q}}) \tag{3.3}$$

for any $d \in \mathbb{N}$ and asymptotics $\mathcal{P}$ and $\mathcal{Q}$ as in (2.60), such that

$$\mathcal{C}(\delta + i\tau, \lambda) \in \mathcal{Y}^{-\infty,d}(\mathbb{W}, \boldsymbol{g}; \boldsymbol{w}; \mathbb{R}^{1+l}_{\tau,\lambda})_{\mathcal{P},\mathcal{Q}}$$

for every $\delta \in \mathbb{R}$, uniformly in $c \leq \delta \leq c'$ for arbitrary $c \leq c'$. The space $\mathcal{Y}^{-\infty}(\mathbb{W}, \boldsymbol{g}; \boldsymbol{w}; \mathbb{C}\times\mathbb{R}^l)$ is then defined as the union of all spaces $\mathcal{Y}^{-\infty,d}(\mathbb{W}, \boldsymbol{g}; \boldsymbol{w}; \mathbb{C}\times \mathbb{R}^l)_{\mathcal{P},\mathcal{Q}}$ over all $d \in \mathbb{N}$ and $\mathcal{P}, \mathcal{Q}$. The operator family $\mathcal{C}(w,\lambda)$ in (3.2) is assumed to belong to that space.

Furthermore, by (1.24) (in the version of a non-compact C^∞ manifold with boundary) we have the space

$$\mathcal{B}^{\mu,d}(\mathbb{W}_{\text{reg}}; \boldsymbol{v}; \mathbb{C} \times \mathbb{R}^l). \tag{3.4}$$

In (3.2) we assume that $\mathcal{B}_{\text{reg}}(w,\lambda)$ is an element of (3.4).

The operator family $\mathcal{A}_{\text{edge}}(w,\lambda)$ is defined in terms of summands $\mathcal{A}_k(w,\lambda)$, similarly as in (2.49). Locally, these operator functions have the form $\operatorname{Op}_y(a_k)(w,\lambda)$ with $a_k(y,\eta,w,\lambda)$ belonging to the space of edge amplitude functions

$$\mathcal{R}^{\mu,d}(\Omega \times \mathbb{R}^q_\eta \times \mathbb{C}_w \times \mathbb{R}^l_\lambda, \boldsymbol{g}; \boldsymbol{w}) \tag{3.5}$$

for $\boldsymbol{w} = (E, F; J_-, J_+; l_-, l_+)$ and some type $d = d_{\mathcal{A}}$, cf. Section 2.1 (recall that the local amplitude functions a_k contain cut-off factors $\sigma(r)$ and $\tilde{\sigma}(r)$ with $\tilde{\sigma} \equiv 1$ on $\operatorname{supp}\sigma$, similarly as in (2.11)). This completes the definition of (3.1).

Remark 3.1. $\mathcal{A}(w,\lambda) \in \mathcal{Y}^{\mu,d}(\mathbb{W}, \boldsymbol{g}; \boldsymbol{w}; \mathbb{C} \times \mathbb{R}^l)$ and $\mathcal{A}(\delta + i\tau, \lambda) \in \mathcal{Y}^{\mu-1,d}(\mathbb{W}, \boldsymbol{g}; \boldsymbol{w};\ \mathbb{R}^{1+l}_{\tau,\lambda})$ for any fixed $\delta \in \mathbb{R}$ implies $\mathcal{A}(w,\lambda) \in \mathcal{Y}^{\mu-1,d}(\mathbb{W}, \boldsymbol{g}; \boldsymbol{w}; \mathbb{C} \times \mathbb{R}^l)$.

Let us set

$$\mathcal{Y}^{\mu}(\mathbb{W}, \boldsymbol{g}; \boldsymbol{w}; \mathbb{C} \times \mathbb{R}^l) = \bigcup_{d\in\mathbb{N}} \mathcal{Y}^{\mu,d}(\mathbb{W}, \boldsymbol{g}; \boldsymbol{w}; \mathbb{C} \times \mathbb{R}^l).$$

Let us fix an element $R \in \mathbf{As}_{M+G}(X, \boldsymbol{g})$, $R = (V_0, V_1, \mathcal{P}, Q)$. Then

$$\mathcal{Y}^{\mu,d}(\mathbb{W}, \boldsymbol{g}; \boldsymbol{w}; \mathbb{C} \times \mathbb{R}^l)_R$$

denotes the subspace of all elements $\mathcal{A}(w,\lambda) \in \mathcal{Y}^{\mu,d}(\mathbb{W}, \boldsymbol{g}; \boldsymbol{w}; \mathbb{C} \times \mathbb{R}^l)$ such that the smoothing family $\mathcal{C}(w,\lambda)$ is as in (3.3), and the local amplitude functions a_k belong to

$$\mathcal{R}^{\mu,d}_{M+G}(\Omega \times \mathbb{R}^q \times \mathbb{C} \times \mathbb{R}^l; \boldsymbol{g}; \boldsymbol{w})_R$$

for all k. A similar notation will be used for spaces of (w,λ)-dependent families with w varying on Γ_δ instead of $\mathbb{C}$.

Remark 3.2. Every $\mathcal{A}(w,\lambda) \in \mathcal{Y}^{\mu}(\mathbb{W}, \boldsymbol{g}; \boldsymbol{w}; \mathbb{C} \times \mathbb{R}^l)$ induces a family of continuous operators (2.64), $s > d_{\mathcal{A}} - \frac{1}{2}$, which is holomorphic in $w \in \mathbb{C}$.

Theorem 3.3. *For every*

$$\mathcal{A}(w,\lambda) \in \mathcal{Y}^{\mu,d}(\mathbb{W},\boldsymbol{g};\boldsymbol{w};\Gamma_\delta \times \mathbb{R}^l)_R, \quad \delta \in \mathbb{R} \ , R \in \mathrm{As}_{M+G}(X,\boldsymbol{g}),$$

there exists an $\mathcal{H}(w,\lambda) \in \mathcal{Y}^{\mu,d}(\mathbb{W},\boldsymbol{g};\boldsymbol{w};\mathbb{C} \times \mathbb{R}^l)_R$ *such that*

$$\mathcal{H}(w,\lambda)\big|_{\Gamma_\delta \times \mathbb{R}^l} - \mathcal{A}(w,\lambda) \in \mathcal{Y}^{-\infty,d}(\mathbb{W},\boldsymbol{g};\boldsymbol{w};\Gamma_\delta \times \mathbb{R}^l).$$

Proof. Comparing (2.50) in the version with parameters $(w,\lambda) \in \Gamma_\delta \times \mathbb{R}^l$ and the expression (3.2) we see that we may consider the summands separately. The smoothing summand can obviously be ignored. To treat $\mathcal{B}_{\mathrm{reg}}(w,\lambda)$ in the middle we can apply kernel cut-off to the involved local (interior and boundary) amplitude functions with respect to $w \in \Gamma_\delta$, cf. Corollary 2.9. The first summand can also be modified by kernel cut-off that we apply to the amplitude functions, (cf. also Theorem 2.11) as well as the other amplitude functions coming from the non-smoothing contributions, cf. also [18] and [3]. □

Proposition 3.4. $\mathcal{A} \in \mathcal{Y}^{\mu,d}(\mathbb{W},\boldsymbol{g};\boldsymbol{w};\mathbb{C} \times \mathbb{R}^l)_R$ *and* $\mathcal{A}(\delta + i\tau,\lambda) \in \mathcal{Y}^{\mu-1,d}(\mathbb{W},\boldsymbol{g};\boldsymbol{w};\Gamma_\delta \times \mathbb{R}^l)_R$ *for some fixed* $\delta \in \mathbb{R}$ *implies* $\mathcal{A} \in \mathcal{Y}^{\mu-1,d}(\mathbb{W},\boldsymbol{g};\boldsymbol{w};\mathbb{C} \times \mathbb{R}^l)_R$.

Proof. The proof follows by applying Remark 2.10 to the various local amplitude functions which played a role in the proof Theorem 3.3. □

Definition 3.5.
1. Let $R \in \mathbf{As}_{M+G}(X,\boldsymbol{g})$, and let $\mathbf{As}(\mathbb{W})^\bullet_R$ denote the system of all sequences
$$T := \{(p_j, m_j, L_j)\}_{j\in\mathbb{Z}}, \tag{3.6}$$
with $p_j \in \mathbb{C}$, $m_j \in \mathbb{N}$, such that $\pi_{\mathbb{C}}T := \{p_j\}_{j\in\mathbb{N}}$ intersects the strip $\{w : c \leq \mathrm{Re}\, w \leq c'\}$ in a finite set for every $c \leq c'$, and $L_j \subset \mathcal{Y}^{-\infty,d}(\mathbb{W},\boldsymbol{g};\boldsymbol{w})_R$ is a finite-dimensional subspace of operators of finite rank, for some $d \in \mathbb{N}$.
2. Let
$$\mathcal{M}^{\mu,d}_{R,T}(\mathbb{W},\boldsymbol{g};\boldsymbol{w}) \tag{3.7}$$
for $R = (V_0, V_1, \mathcal{P}, \mathcal{Q}) \in \mathbf{As}_{M+G}(X,\boldsymbol{g})$, and $T = \{(p_j, m_j, L_j)\}_{j\in\mathbb{Z}} \in \mathbf{As}(\mathbb{W})^\bullet_R$ denote the space of all operator functions
$$\mathcal{A}(w) = \mathcal{H}(w) + \mathcal{C}(w) \tag{3.8}$$
for arbitrary $\mathcal{H}(w) \in \mathcal{Y}^{\mu,d}(\mathbb{W},\boldsymbol{g};\boldsymbol{w};\mathbb{C})_R$ and
$$\mathcal{C}(w) \in \mathcal{A}(\mathbb{C} \setminus \pi_{\mathbb{C}}T,\ \mathcal{Y}^{-\infty,d}(\mathbb{W},\boldsymbol{g};\boldsymbol{w})_{\mathcal{P},\mathcal{Q}})$$
such that $\mathcal{C}(w)$ is meromorphic with poles at p_j of multiplicity $m_j + 1$ and Laurent coefficients at $(w - p_j)^{-(k+1)}$ belonging to L_j for all $0 \leq k \leq m_j$, $j \in \mathbb{Z}$, and
$$\chi(\delta + i\tau)\mathcal{C}(\delta + i\tau) \in \mathcal{Y}^{-\infty,d}(\mathbb{W},\boldsymbol{g};\boldsymbol{w};\mathbb{R}_\tau)_{\mathcal{P},\mathcal{Q}}$$
for every $\pi_{\mathbb{C}}T$-excision function $\chi(w)$, and every $\delta \in \mathbb{R}$, uniformly in $c \leq \delta \leq c'$ for every $c \leq c'$.

Let us write $\mathcal{M}^{\mu,d}_{R,\mathcal{O}}(\mathbb{W}, \boldsymbol{g}; \boldsymbol{w})$ for the space (3.7) when $\pi_{\mathbb{C}} T = \emptyset$ (this coincides with $\mathcal{Y}^{\mu,d}(\mathbb{W}, \boldsymbol{g}; \boldsymbol{w}; \mathbb{C})_R$). Moreover, set

$$\mathcal{M}^{\mu}_{R,T}(\mathbb{W}, \boldsymbol{g}; \boldsymbol{w}) = \bigcup \mathcal{M}^{\mu,d}_{R,T}(\mathbb{W}, \boldsymbol{g}; \boldsymbol{w}), \tag{3.9}$$

where the union is taken over all $d \in \mathbb{N}$. We finally write $\mathcal{M}^{\mu}(\mathbb{W}, \boldsymbol{g}; \boldsymbol{w})$ for the union of the latter spaces over all $R \in \mathbf{As}_{M+G}(X, \boldsymbol{g})$, $T \in \mathbf{As}(\mathbb{W})^{\bullet}_{R}$.

By definition there is a decomposition as a non-direct sum

$$\mathcal{M}^{\mu}_{R,T}(\mathbb{W}, \boldsymbol{g}; \boldsymbol{w}) = \mathcal{M}^{\mu}_{R,\mathcal{O}}(\mathbb{W}, \boldsymbol{g}; \boldsymbol{w}) + \mathcal{M}^{-\infty}_{R,T}(\mathbb{W}, \boldsymbol{g}; \boldsymbol{w}), \tag{3.10}$$

cf. analogously, for the case without boundary [15, formula (2.5.2)].

Theorem 3.6. *Let $\mathcal{A} \in \mathcal{M}^{\mu}(\mathbb{W}, \boldsymbol{g}; \boldsymbol{v})$ for*

$$\boldsymbol{g} = (\gamma - \nu, \gamma - (\mu + \nu), \Theta), \ \ \boldsymbol{v} = (\tilde{E}, F; \tilde{J}, J_+; \tilde{L}, L_+),$$

and $\mathcal{B} \in \mathcal{M}^{\nu}(\mathbb{W}, \boldsymbol{h}; \boldsymbol{w})$ for

$$\boldsymbol{h} = (\gamma, \gamma - \nu, \Theta), \ \ \boldsymbol{w} = (E, \tilde{E}; J_-, \tilde{J}; L_-, \tilde{L}).$$

Then we have

$$\mathcal{AB} \in \mathcal{M}^{\mu+\nu}(\mathbb{W}, \boldsymbol{g} \circ \boldsymbol{h}; \boldsymbol{v} \circ \boldsymbol{w})$$

for $\boldsymbol{g} \circ \boldsymbol{h} = (\gamma, \gamma - (\mu + \nu); \Theta))$, $\boldsymbol{v} \circ \boldsymbol{w} = (E, F; J_-, J_+; L_-, L_+)$, and $d_{\mathcal{AB}} = \max(\nu + d_{\mathcal{A}}, d_{\mathcal{B}})$.

This theorem can be regarded as an application of Theorem 2.13 to meromorphic parameter-dependent families. If both families are holomorphic this is fairly straightforward. In the general case we can apply the decomposition (3.10) which leads to the case that one factor is meromorphic and of order $-\infty$ with finite rank Laurent coefficients. The desired composition behaviour also this case can easily be obtained.

3.2. Ellipticity and Meromorphic Inverses

Definition 3.7. An element $\mathcal{A} \in \mathcal{M}^{\mu}_{R,T}(\mathbb{W}, \boldsymbol{g}; \boldsymbol{w})$ for $\boldsymbol{g} = (\gamma, \gamma - \mu, \Theta)$ is called elliptic of order μ, if for some $\beta \in \mathbb{R}$ with $\pi_{\mathbb{C}} T \cap \Gamma_{\beta} = \emptyset$ the family $\mathcal{A}(\beta + i\tau) \in \mathcal{Y}^{\mu}(\mathbb{W}, \boldsymbol{g}; \boldsymbol{w}; \mathbb{R}_{\tau})_R$ is parameter-dependent elliptic in the sense of Definition 2.14.

Remark 3.8. Definition 3.7 is independent of the choice of β.

In fact, writing $\mathcal{A}$ in the form (3.8), the ellipticity of $\mathcal{A}(\beta + i\tau)$ only depends on $\mathcal{H}(\beta + i\tau)$, since $\mathcal{C}(\beta + i\tau)$ is of order $-\infty$ in τ. Proposition 3.4 then shows that the principal symbols which determine the ellipticity are independent of β.

Theorem 3.9. *Let $\mathcal{A} \in \mathcal{M}^{\mu}_{R,T}(\mathbb{W}, \boldsymbol{g}; \boldsymbol{w})$ be elliptic. Then there exists a countable set $D \subset \mathbb{C}$, such that $D \cap \{w \in \mathbb{C} : c \leq \operatorname{Re} w \leq c'\}$ is finite for every $c \leq c'$, and*

$$\mathcal{A}(w) : \begin{array}{c} \mathcal{W}^{s,\gamma}(\mathbb{W}, \boldsymbol{E}) \\ \oplus \\ H^s(Y, L_-) \end{array} \longrightarrow \begin{array}{c} \mathcal{W}^{s-\mu,\gamma-\mu}(\mathbb{W}, \boldsymbol{F}) \\ \oplus \\ H^{s-\mu}(Y, L_+) \end{array} \tag{3.11}$$

is invertible for all $w \in \mathbb{C} \setminus D$ and for all $s > \max(\mu, d_{\mathcal{A}}) - \frac{1}{2}$.

Proof. Because of Remark 3.2 and Definition 3.5 the family of operators (3.11) is holomorphic in $U := \mathbb{C} \setminus \pi_{\mathbb{C}}T$. At the same time it is elliptic in the space $\mathcal{Y}^{\mu,d}(\mathbb{W}, \boldsymbol{g}; \boldsymbol{w})$ for every $w \in U$. Thus, (3.11) is a holomorphic Fredholm family in U. Let us write $\mathcal{A}(w)$ in the form (3.8). Then, by virtue of the parameter-dependent ellipticity of $\mathcal{H}(w)\big|_{\Gamma_\beta}$ for every $\beta \in \mathbb{R}$, there is a $C > 0$ such that $\mathcal{H}(w)$ is invertible for $|\operatorname{Im} w| > C$ and $\operatorname{Re} w = \beta$ for any fixed $\beta \in \mathbb{R}$. Since β is involved as a parameter that only affects lower order terms and because of the continuity in β, for every $c < c'$ we can choose a constant $C = C(c, c')$ such that $\mathcal{A}(w)$ is invertible for all $|\operatorname{Im} w| > C(c, c')$ and $c < \beta < c'$. We are now in a well known situation of a holomorphic Fredholm family $\mathcal{A}(w)$ in an open set $U \subset \mathbb{C}$ which is invertible in at least one point $w_1 \in U$; it follows then that $\mathcal{A}(w)$ is invertible for all $w \in U \setminus D$, except for a countable subset of U that intersects every $K \Subset U$ in a finite set. It remains to note that the points of $\pi_{\mathbb{C}}T$ are not accumulation points of D. The technique is the same as in [22, Lemma 4.3.13]. □

Theorem 3.10. *Every elliptic element $\mathcal{A} \in \mathcal{M}^{\mu}_{R,T}(\mathbb{W}, \boldsymbol{g}; \boldsymbol{w})$ has an inverse*

$$\mathcal{A}^{-1} \in \mathcal{M}^{-\mu}_{P,Q}(\mathbb{W}, \boldsymbol{g}^{-1}; \boldsymbol{w}^{-1}), d_{\mathcal{A}^{-1}} = \max(d_{\mathcal{A}} - \mu, 0),$$

for suitable asymptotic types

$$P \in \mathbf{As}_{M+G}(X, \boldsymbol{g}^{-1}) \quad \text{and} \quad Q \in \mathbf{As}(\mathbb{W})^{\bullet}_{P}.$$

Proof. We invert $\mathcal{A}$ with respect to the multiplication of Theorem 3.6. Let us first apply Theorem 2.15 to $\mathcal{A}\big|_{\Gamma_\beta} \in \mathcal{Y}^{\mu}(\mathbb{W}, \boldsymbol{g}; \boldsymbol{w}; \Gamma_\beta)_R$ for any β such that $\pi_{\mathbb{C}}T \cap \Gamma_\beta = \emptyset$. Then there is a parametrix $\mathcal{P}_1 \in \mathcal{Y}^{-\mu}(\mathbb{W}, \boldsymbol{g}^{-1}; \boldsymbol{w}^{-1}; \Gamma_\beta)_{P_1}$ for some $P_1 \in \mathbf{As}_{M+G}(X, \boldsymbol{g}^{-1})$. Theorem 3.3 applied to $\mathcal{P}_1$ gives us an element $\mathcal{P}_0 \in \mathcal{Y}^{-\mu}(\mathbb{W}, \boldsymbol{g}^{-1}; \boldsymbol{w}^{-1}; \mathbb{C})_{P_1}$ such that $\mathcal{P}_0\big|_{\Gamma_\beta}$ coincides with $\mathcal{P}_1$ modulo a family of order $-\infty$. It follows that $\mathcal{P}_0\big|_{\Gamma_\beta}$ is also a parametrix of $\mathcal{A}\big|_{\Gamma_\beta}$. We now apply Theorem 3.6 and obtain $\mathcal{P}_0\mathcal{A} \in \mathcal{M}^0(\mathbb{W}, \boldsymbol{g}_l; \boldsymbol{w}_l)$ (cf. the notation of Theorem 2.15). Moreover, we have $\mathcal{I} - \mathcal{P}_0\mathcal{A}\big|_{\Gamma_\beta} \in \mathcal{Y}^{-\infty}(\mathbb{W}, \boldsymbol{g}_l; \boldsymbol{w}_l; \Gamma_\beta)$. From Remark 3.1 we obtain $\mathcal{N} := \mathcal{I} - \mathcal{P}_0\mathcal{A} \in \mathcal{Y}^{-\infty}(\mathbb{W}, \boldsymbol{g}_l; \boldsymbol{w}_l; \mathbb{C})_{P_2} = \mathcal{M}^{-\infty}_{P_2,\mathcal{O}}(\mathbb{W}, \boldsymbol{g}_l; \boldsymbol{w}_l)$ for some $P_2 \in \mathbf{As}_{M+G}(X, \boldsymbol{g}_l)$. It remains to observe that there is an $\mathcal{L} \in \mathcal{M}^{-\infty}_{P_3,S}(\mathbb{W}, \boldsymbol{g}_l; \boldsymbol{w}_l)$ such that $(\mathcal{I} - \mathcal{N})^{-1} = \mathcal{I} - \mathcal{L}$. Then it follows that $\mathcal{A}^{-1} = (\mathcal{I} - \mathcal{L})\mathcal{P}_0$ which belongs to the space $\mathcal{M}^{-\mu}_{P,Q}(\mathbb{W}, \boldsymbol{g}^{-1}; \boldsymbol{v}^{-1})$ by Theorem 3.6. □

References

[1] M.S. Agranovich and M.I. Vishik, Elliptic problems with parameter and parabolic problems of general type, *Uspekhi Mat. Nauk.* **19** (1964), 53–161.

[2] S. Behm, *Pseudo-Differential Operators with Parameters on Manifolds with Edges*, Ph.D. Thesis, University of Potsdam, 1995.

[3] G. De Donno and B.-W. Schulze, Meromorphic symbolic structures for boundary value problems on manifolds with edges, Preprint 2003/10, Institut für Mathematik, Potsdam, 2003.

[4] Ch. Dorschfeldt, *Algebras of Pseudo-Differential Operators near Edge and Corner Singularities*, Akademie Verlag, Berlin, 1998.

[5] B.V. Fedosov, B.-W. Schulze and N.N. Tarkhanov, The index of elliptic operators on manifolds with conical points *Selecta mathematica, New Ser.* **5** (1999), 467–506.

[6] B.V. Fedosov, B.-W. Schulze and N.N. Tarkhanov, Analytic index formulas for elliptic corner operators, Ann. *Inst. Fourier* **52** (2002), 899–982.

[7] J.B. Gil, B.-W. Schulze and J. Seiler, Cone pseudo-differential operators in the edge symbolic calculus, *Osaka J. Math.* **37** (2000), 219–258.

[8] G. Harutjunjan and B.-W Schulze, Asymptotics and relative index on a cylinder with conical cross section, Preprint 2002/27, Institut für Mathematik, Potsdam, 2002.

[9] D. Kapanadze and B.-W. Schulze, *Crack Theory and Edge Singularities*, Kluwer Academic Publishers, Dordrecht, 2003.

[10] V.A. Kondratyev, Boundary value problems for elliptic equations in domains with conical points, *Trudy Mosk. Mat. Obshch.* **16** (1967), 209–292.

[11] T. Krainer, On the inverse of parabolic boundary value problems for large times, *Japanese J. Math.*, to appear.

[12] T. Krainer and B.-W. Schulze, Long-time asymptotics with geometric singularities in the spatial variables, in *Israel Math. Conference Proceedings*, Vol. 16, 2003.

[13] T. Krainer, Volterra families of pseudo-differential operators, in Advances in Partial Differential Equations (*Parabolicity, Volterra Calculus, and Conical Singularities*), Birkhäuser, 2002, 1–45.

[14] X. Liu and B.-W. Schulze, Ellipticity on manifolds with edges and boundary, in preparation.

[15] L. Maniccia and B.-W. Schulze, An algebra of meromorphic corner symbols, *Bull. des Sci. Math.* **127** (2003), 55–99.

[16] V. Nazaikinskij, A. Savin, B.-W. Schulze and B. Ju. Sternin, Elliptic theory on manifolds with nonisolated singularities: II. Products in elliptic theory on manifolds with edges, Preprint 2002/15, Institut für Mathematik, Potsdam, 2002.

[17] V. Nazaikinskij, A. Savin, B.-W. Schulze and B. Ju. Sternin, Elliptic theory on manifolds with nonisolated singularities: III. The spectral flow of families of conormal symbols, Preprint 2002/20, Institut für Mathematik, Potsdam, 2002.

[18] A. Oliaro and B.-W. Schulze, Parameter-dependent boundary value problems on manifolds with edges, Preprint 2002/25, Institut für Mathematik, Potsdam, 2002.

[19] S. Rempel and B.-W. Schulze, *Index Theory of Elliptic Boundary Problems*, Akademie-Verlag, Berlin, 1982.

[20] E. Schrohe and B.-W. Schulze, Boundary value problems in Boutet de Monvel's calculus for manifolds with conical singularities II, Advances in Partial Differential Equations (*Boundary Value Problems, Schrödinger Operators, Deformation Quantization*), Akademie Verlag, Berlin, 1995, 70–205.

[21] E. Schrohe, Fréchet algebra techniques for boundary value problems on noncompact manifolds: Fredholm criteria and functional calculus via spectral invariance, *Math. Nachr.* **199** (1999), 145–185.

[22] E. Schrohe and B.-W. Schulze, Boundary value problems in Boutet de Monvel's calculus for manifolds with conical singularities I, Advances in Partial Differential Equations (*Pseudo-Differential Calculus and Mathematical Physics*), Akademie Verlag, Berlin, 1994, 97–209.

[23] E. Schrohe and J. Seiler, Ellipticity and invertibility in the cone algebra on L_p-Sobolev spaces, *Integral Equations Operator Theory* **41** (2001), 93–114.

[24] B.-W. Schulze, Corner Mellin operators and reduction of orders with parameters, *Ann. Sc. Norm. Sup. Pisa Cl. Sci.* **16** (1989), 1–81.

[25] B.-W. Schulze, Pseudo-differential operators on manifolds with edges, *Symposium "Partial Differential Equations", Holzhau* 1988, Teubner-Texte zur Mathematik, Vol. 112, Teubner, Leipzig, 1989, 259–287.

[26] B.-W. Schulze, *Pseudo-Differential Operators on Manifolds with Singularities*, North-Holland, Amsterdam, 1991.

[27] B.-W. Schulze, The Mellin pseudo-differential calculus on manifolds with corners, pseudo-differential operators on manifolds with edges, in *Symposium "Analysis in Domains and on Manifolds with Singularities", Breitenbrunn* 1990, Teubner-Texte zur Mathematik, Vol. 131, Teubner, Leipzig, 1992, 208–289.

[28] B.-W. Schulze, Mellin representations of pseudo-differential operators on manifolds with corners, *Ann. Glob. Anal. Geom.* **8** (1990), 261–297.

[29] B.-W. Schulze, *Boundary Value Problems and Singular Pseudo-Differential Operators*, J. Wiley, Chichester, 1998.

[30] B.-W. Schulze, Operators with symbol hierarchies and iterated asymptotics, *Publ. Res. Inst. Math. Sci.* **38** (2002), 735–802.

[31] B.-W. Schulze and J. Seiler, The edge algebra structure of boundary value problems, *Ann. Global Anal. Geom.* **22** (2002), 197–265.

[32] B.-W. Schulze, B. Ju. Sternin and V. Shatalov, On the index of differential operators on manifolds with conical singularities, *Ann. Global Anal. Geom.* **16** (1998), 141–172.

T. Krainer and B.-W. Schulze
Institut für Mathematik
Universität Potsdam
An Neuen Palais 10
14469 Potsdam
Deutschland
e-mail: `krainer@math.uni-potsdam.de`
`schulze@math.uni-potsdam.de`

Operator Theory:
Advances and Applications, Vol. 155, 65–75

A New Proof of Global Smoothing Estimates for Dispersive Equations

Michael Ruzhansky and Mitsuru Sugimoto

Abstract. The aim of this article is to provide a new method to prove global smoothing estimates for dispersive equations such as Schrödinger equations. For the purpose, the Egorov-type theorem via canonical transformation in the form of a class of Fourier integral operators is established, and their weighted L^2-boundedness is also proved. The boundedness result is not covered by previous one such as Asada and Fujiwara [1]. By using them, a different proof for the result obtained by Ben-Artzi & Klainerman [2] is provided. This new idea gives a clear understanding of smoothing effects of dispersive equations, and further developments are also expected. In fact, some extended results based on the same idea are also announced.

Mathematics Subject Classification (2000). Primary 35Q40; Secondary 35B65.

Keywords. Dispersive equation, smoothing effect, canonical transformation.

1. Introduction

We consider Fourier integral operators, which can be globally written in the form

$$Tu(x) = (2\pi)^{-n} \int_{\mathbf{R}^n} \int_{\mathbf{R}^n} e^{i\phi(x,y,\xi)} p(x,y,\xi) u(y)\, dy d\xi \quad (x \in \mathbf{R}^n), \tag{1.1}$$

where $p(x,y,\xi)$ is an amplitude function and $\phi(x,y,\xi)$ is a real phase function. Especially, if $p(x,y,\xi) = 1$ and $\phi(x,y,\xi)$ satisfies the graph condition

$$\begin{aligned} \Lambda &= \{(x,\phi_x,y,-\phi_y);\ \phi_\xi = 0\} \\ &= \{(x,\xi),\chi(x,\xi)\} \subset T^*\mathbf{R}^n \times T^*\mathbf{R}^n, \end{aligned}$$

This work was completed with the aid of "UK-Japan Joint Project Grant" by "The Royal Society" and "Japan Society for the Promotion of Science".

we have the relation

$$T \cdot A(X,D) = B(X,D) \cdot T + (lower),$$
$$B(x,\xi) = (A \circ \chi)(x,\xi).$$

In this way, Fourier integral operators are recognized as a tool of the realization of the canonical transformation. This fact is known as Egorov's theory, and by taking phase function appropriately, properties of the operator $B(X,D)$ can be extracted from those of the well known operator $A(X,D)$.

If we take

$$\phi(x,y,\xi) = x \cdot \xi - y \cdot \psi(\xi) \tag{1.2}$$

as a special case, we have

$$Tu(x) = F^{-1}[(Fu)(\psi(\xi))](x),$$

where F (F^{-1} *resp.*) denotes the (inverse *resp.*) Fourier transformation. Hence we have the relation

$$T \cdot \sigma(D) = (\sigma \circ \psi)(D) \cdot T, \qquad a(D) = (\sigma \circ \psi)(D), \tag{1.3}$$

for constant variable operators $\sigma(D)$ and $a(D)$. For example, for a positive and positively homogeneous function $a(\xi)$ of order 2, we have

$$T \cdot (-\triangle) = a(D) \cdot T,$$

if we take

$$\sigma(\eta) = |\eta|^2, \quad \psi(\xi) = \sqrt{a(\xi)}\frac{\nabla a(\xi)}{|\nabla a(\xi)|}.$$

Since the property of the Laplacian is well known in various situations, we can expect the same for $a(D)$.

By using Egorov's theorem, many qualitative properties of solutions of partial differential equations (propagation of singularity for example) has been investigated. Our main interest is to have quantitative (L^2-property for example) by the same idea. As an example, we present a new proof of global smoothing estimates for Schrödinger equations in this article.

2. Smoothing Effect of Schrödinger Equations

We consider the following Schrödinger equation:

$$\begin{cases} (i\partial_t + \triangle_x)\, u(t,x) = 0, \\ \qquad\qquad u(0,x) = \varphi(x) \in L^2(\mathbf{R}^n). \end{cases} \tag{2.1}$$

By Plancherel's theorem, we know that the solution operator $e^{it\triangle}$ preserves the L^2-norm, that is, we have

$$\|u(t,\cdot)\|_{L^2(\mathbf{R}^n_x)} = \|\varphi\|_{L^2(\mathbf{R}^n)},$$

for fixed $t \in \mathbf{R}$. On the other hand, the extra gain of regularity of order $1/2$ in x can be observed if we integrate the solution in t. For example, in the case $n = 1$, we have

$$\left\| |D_x|^{1/2} u(\cdot, x) \right\|_{L^2(\mathbf{R}_t)} \le C \|\varphi\|_{L^2(\mathbf{R})},$$

for any fixed $x \in \mathbf{R}$. This result was given by Kenig, Ponce and Vega [5], and will play an important role later in Section 4 (Proposition 4.1).

In the higher dimensional case $n \ge 2$, similar smoothing properties were given:

$$\|Au\|_{L^2(\mathbf{R}_t \times \mathbf{R}_x^n)} \le C \|\varphi\|_{L^2(\mathbf{R}_x^n)}, \tag{2.2}$$

where A is one of the following:

$$\begin{aligned}
&(1) \quad A = \langle x \rangle^{-s} |D|^{1/2};\ s > 1/2,\\
&(2) \quad A = \langle x \rangle^{-s} \langle D \rangle^{1/2};\ s \ge 1 \quad (s > 1 \text{ if } n = 2),\\
&(3) \quad A = |x|^{\alpha - 1} |D|^{\alpha};\ 1 - n/2 < \alpha < 1/2.
\end{aligned}$$

The type (1) was given by Ben-Artzi and Klainerman [2] ($n \ge 3$), and Chihara [3] ($n \ge 2$). The type (2) was given by Kato and Yajima [4] ($n \ge 3$), and Walther [13] ($n \ge 2$) which also showed that it is not true for $s < 1$ ($s \le 1$ if $n = 2$). The type (3) was given by Kato and Yajima [4] ($n \ge 3$, $0 \le \alpha < 1/2$ or $n = 2$, $0 < \alpha < 1/2$), and Sugimoto [12] ($n \ge 2$). Watanabe [15] showed that it is not true for $\alpha = 1/2$.

Each proof was carried out by proving one of the following estimates (or their variants):

$$\left\| \widehat{A^* f}_{|\rho S^{n-1}} \right\|_{L^2(\rho S^{n-1})} \le C \sqrt{\rho} \|f\|_{L^2(\mathbf{R}^n)} \qquad \text{(Restriction theorem)}, \tag{2.3}$$

where, $\rho S^{n-1} = \{\xi;\ |\xi| = \rho\}$, $(\rho > 0)$.

$$\sup_{\operatorname{Im} \zeta > 0} |(R(\zeta) A^* f, A^* f)| \le C \|f\|^2_{L^2(\mathbf{R}^n)} \qquad \text{(Resolvent estimate)}, \tag{2.4}$$

where $R(\zeta) = (-\triangle - \zeta)^{-1}$.

Estimate (2.3) implies the dual one of estimate (2.2). Estimate (2.4) implies (2.2) since the resolvent $R(\zeta)$ is the Laplace transform of the solution operator $e^{it\triangle}$ of equation (2.1):

$$R(\zeta) = \frac{1}{i} \int_0^\infty e^{it\triangle} e^{i\zeta t}\, dt \quad (\operatorname{Im} \zeta > 0).$$

The fact that (2.4) implies (2.3) due to the formula

$$\operatorname{Im} \left(R(\rho^2 + i0) f, f \right) = \frac{1}{4(2\pi)^{n-1} \rho} \left\| \hat{f}_{|\rho S^{n-1}} \right\|^2_{L^2(\rho S^{n-1})}$$

was also used.

Following the idea in Introduction, we present a completely different idea to prove them. In this note, we give another proof of the result given by Ben-Artzi and

Klainerman [2] which treated the type (1) above. In order to treat more general equations, we consider the following dispersive equation:

$$\begin{cases} (i\partial_t - a(D))\, u(t,x) = 0, \\ \qquad\qquad u(0,x) = \varphi(x) \in L^2(\mathbf{R}^n). \end{cases} \tag{2.5}$$

where $a(\xi) \in C^\infty(\mathbf{R}^n \setminus 0)$ is real-valued and positively homogeneous of order 2. Equation (2.5) with $a(\xi) = |\xi|^2$ is the Schrödinger equation (2.1).

Theorem 2.1. *Suppose $n \geq 3$ and $s > 1/2$. Assume that the real-valued function $a(\xi) \in C^\infty(\mathbf{R}^n \setminus 0)$ satisfies $a(\lambda\xi) = \lambda^2 a(\xi)$ and $\nabla a(\xi) \neq 0$ for all $\lambda > 0$ and $\xi \in \mathbf{R}^n \setminus 0$. Then the solution u to (2.5) satisfies*

$$\langle x\rangle^{-s} |D|^{1/2} u(t,x) \in L^2(\mathbf{R}_t \times \mathbf{R}^n_x).$$

Remark 2.2. Recently Chihara [3] proved Theorem 2.1 in a general setting by proving the restriction theorem (2.3) or the resolvent estimates (2.4). But we can prove more general result by using our idea. See Theorem 5.1 and Remark 5.2 in Section 5.

3. Main Tool

Based on the argument in Introduction, we introduce the main tool for the proof of Theorem 2.1.

Let $\psi, \psi^{-1} : \mathbf{R}^n \setminus 0 \to \mathbf{R}^n \setminus 0$ be C^∞-maps satisfying $(\psi \circ \psi^{-1})(\xi) = (\psi^{-1} \circ \psi)(\xi) = \xi$, $\psi(\lambda\xi) = \lambda\psi(\xi)$, and $\psi^{-1}(\lambda\xi) = \lambda\psi^{-1}(\xi)$ for all $\lambda > 0$ and $\xi \in \mathbf{R}^n \setminus 0$. Then we have

$$C^{-1} \leq |\det \partial\psi(\xi)| \leq C, \quad C^{-1} \leq \left|\det \partial\psi^{-1}(\xi)\right| \leq C$$

for some $C > 0$. We set

$$\begin{aligned} Iu(x) &= (2\pi)^{-n} \int_{\mathbf{R}^n}\int_{\mathbf{R}^n} e^{i(x\cdot\xi - y\cdot\psi(\xi))} u(y)\, dy d\xi, \\ I^{-1}u(x) &= (2\pi)^{-n} \int_{\mathbf{R}^n}\int_{\mathbf{R}^n} e^{i\left(x\cdot\xi - y\cdot\psi^{-1}(\xi)\right)} u(y)\, dy d\xi. \end{aligned} \tag{3.1}$$

We remark that we have

$$Iu(x) = F^{-1}[(Fu)(\psi(\xi))](x), \qquad I^{-1}u(x) = F^{-1}\left[(Fu)(\psi^{-1}(\xi))\right](x).$$

Since $I^{-1} \cdot I = I \cdot I^{-1} = id$, we have the formula

$$a(D) = I \cdot \sigma(D) \cdot I^{-1}, \quad a(\xi) = (\sigma \circ \psi)(\xi) \tag{3.2}$$

by (1.3). By Plancherel's theorem, the operators I and I^{-1} are L^2-bounded. Furthermore, we have the boundedness on weighted spaces. For $k \in \mathbf{R}$, let $L^2_k(\mathbf{R}^n)$ be the set of functions f such that the norm

$$\|f\|_{L^2_k(\mathbf{R}^n)} = \left(\int_{\mathbf{R}^n} \left|\langle x\rangle^k f(x)\right|^2 dx\right)^{1/2}; \quad \langle x\rangle^k = \left(1 + |x|^2\right)^{k/2}$$

is finite.

Proposition 3.1. *Suppose $n \geq 3$ and $0 \leq k \leq 1$. Then the operators I and I^{-1} defined by (3.1) are $L^2_{-k}(\mathbf{R}^n)$-bounded.*

Proof. We prove the boundedness of I. The boundedness of I^{-1} can be given by replacing ψ by ψ^{-1}. If we use the equality

$$e^{-iy\cdot\xi} = \frac{1 + iy\cdot\partial_\xi}{\langle y\rangle^2} e^{-iy\cdot\xi},$$

change of variables, and integration by parts, we can justify the following:

$$\begin{aligned}
&Iu(x) \\
&= (2\pi)^{-n} \int\int e^{i(x\cdot\psi^{-1}(\xi) - y\cdot\xi)} d(\xi) u(y) dy d\xi \\
&= (2\pi)^{-n} \int\int e^{i(x\cdot\psi^{-1}(\xi) - y\cdot\xi)} \left(\frac{d(\xi)(1 + x\partial\psi^{-1}(\xi)^t y) - i(\partial d)(\xi)\cdot y}{\langle y\rangle^2} \right) u(y) dy d\xi \\
&= (2\pi)^{-n} \int\int e^{i(x\cdot\xi - y\cdot\psi(\xi))} \left(\frac{1 + xA(\xi)^t y + \mathbf{a}(\xi)\cdot y}{\langle y\rangle^2} \right) u(y) dy d\xi,
\end{aligned}$$

where

$$\begin{aligned}
A(\xi) &= (\partial\psi)^{-1}(\xi), \quad a(\xi) = -i(\partial d)(\psi(\xi))\tilde{d}(\xi), \\
d(\xi) &= \left|\det \partial\psi^{-1}(\xi)\right|, \quad \tilde{d}(\xi) = |\det \partial\psi(\xi)|.
\end{aligned}$$

Hence we have

$$I = I\frac{1}{\langle x\rangle^2} + xA(D)\frac{{}^t x}{\langle x\rangle^2} + \mathbf{a}(D)\cdot I\frac{x}{\langle x\rangle^2}. \tag{3.3}$$

We remember here that I is L^2-bounded by Plancherel's theorem. We remark that all entries of $|D|\mathbf{a}(D)$ and $A(D)$ are L^2-bounded, and $|D|^{-1}$ is bounded from L^2 to L^2_{-1}. Using them, we obtain the L^2_{-1}-boundedness of I from (3.3). Then, by the interpolation, we have the desired result. To justify these boundedness, use the following results of Kurtz and Wheeden [6], Stein and Weiss [11]:

Lemma 3.2 ([6]). *Suppose $-n/2 < \delta < n/2$. Then we have*

$$\left\||x|^\delta m(D)u\right\|_{L^2(\mathbf{R}^n)} \leq C \sum_{|\gamma|\leq n} \sup_{\xi\in\mathbf{R}^n} \left||\xi|^{|\gamma|}\partial^\gamma m(\xi)\right| \left\||x|^\delta u\right\|_{L^2(\mathbf{R}^n)}.$$

Lemma 3.3 ([11]). *Suppose $\gamma < n/2$, $\delta < n/2$, $m < n$, and $\gamma + \delta + m = n$. Then we have*

$$\left\||x|^{-\gamma}|D|^{m-n}u\right\|_{L^2(\mathbf{R}^n)} \leq C\left\||x|^\delta u\right\|_{L^2(\mathbf{R}^n)}.$$

□

4. Proof of Theorem 2.1

By the formula (3.2), we may show the result by replacing $a(D)$ by $\sigma(D)$. In fact, applying I^{-1} to equation (2.5) we have

$$\begin{cases} (i\partial_t - \sigma(D))\, v(t,x) = 0 \\ \qquad\qquad v(0,x) = g(x), \end{cases} \tag{4.1}$$

where

$$v = I^{-1}u, \quad g = I^{-1}\varphi.$$

Suppose that we can show

$$\left\| \langle x \rangle^{-s} |D|^{1/2} v(t,x) \right\|_{L^2(\mathbf{R}_t \times \mathbf{R}^n_x)} \leq C \|g\|_{L^2(\mathbf{R}^n_x)}. \tag{4.2}$$

Substituting $v = I^{-1}u$ and $g = I^{-1}\varphi$ and noticing $|\psi(D)|^{1/2} = I \cdot |D|^{1/2} \cdot I^{-1}$ by (3.2) we have

$$\left\| \langle x \rangle^{-s} I^{-1} |\psi(D)|^{1/2} u(t,x) \right\|_{L^2(\mathbf{R}_t \times \mathbf{R}^n_x)} \leq C \left\| I^{-1} \varphi \right\|_{L^2(\mathbf{R}^n_x)}.$$

Furthermore, for $0 \leq s \leq 1$, $|\psi(D)|^{1/2} |D|^{-1/2}$ is L^2_{-s}-bounded by Lemma 3.2, and I, I^{-1} by Proposition 3.1. Hence we have

$$\left\| \langle x \rangle^{-s} |D|^{1/2} u(t,x) \right\|_{L^2(\mathbf{R}_t \times \mathbf{R}^n_x)} \leq C \|\varphi\|_{L^2(\mathbf{R}^n_x)},$$

which means that we have the same estimates for equation (2.5).

By the microlocalization and the rotation, we may assume that the initial data φ in equation (2.5) satisfies $\operatorname{supp} \hat{\varphi} \subset \Gamma$, where $\Gamma \subset \mathbf{R}^n \setminus 0$ is a sufficiently small conic neighborhood of $e_n = (0, \ldots 0, 1)$. We remark that the argument so far can be justified for microlocal C^∞-maps $\psi : \Gamma \to \tilde{\Gamma}$ and $\psi^{-1} : \tilde{\Gamma} \to \Gamma$, where $\Gamma, \tilde{\Gamma} \subset \mathbf{R}^n \setminus 0$ are cones. All our task is to find $\psi : \Gamma \to \tilde{\Gamma}$ and $\sigma(\eta)$ such that $\det \partial\psi(e_n) \neq 0$, $a(\xi) = (\sigma \circ \psi)(\xi)$ $(\xi \in \Gamma)$, and prove the estimate (4.2) for equation (4.1). We also remark that we have $\operatorname{supp} \hat{g} \subset \tilde{\Gamma}$ if we notice $g = I^{-1}\varphi$ and (3.2).

The assumption $\nabla a(e_n) \neq 0$ implies the following two cases:

(i) $\partial_n a(e_n) \neq 0$. Then, by Euler's identity $a(\xi) = (\xi/2) \cdot \nabla a(\xi)$, we have $a(e_n) \neq 0$. Hence, in this case, we may assume

$$a(\xi) > 0 \ (\xi \in \Gamma), \quad \partial_n a(e_n) \neq 0.$$

(ii) $\partial_n a(e_n) = 0$. Then, by assumption $\nabla a(e_n) \neq 0$, there exits $j \neq n$ such that $\partial_j a(e_n) \neq 0$. Hence, in this case, we may assume

$$\partial_1 a(e_n) \neq 0.$$

Case (i)

We take

$$\sigma(\eta) = \eta_n^2, \quad \psi(\xi) = (\xi_1, \dots, \xi_{n-1}, \sqrt{a(\xi)}).$$

Then we have $a(\xi) = (\sigma \circ \psi)(\xi)$ and

$$\det \partial\psi(e_n) = \begin{vmatrix} E_{n-1} & 0 \\ * & \partial_n a(e_n)/\left(2\sqrt{a(e_n)}\right) \end{vmatrix} \neq 0,$$

where E_{n-1} is the identity matrix of order $n-1$. The estimate for $\sigma(D) = D_n^2$ was given by the following (Kenig, Ponce and Vega [5, p.56]):

Proposition 4.1. *In the case $n = 1$, we have*

$$\sup_{x\in\mathbf{R}} \left\| |D_x|^{1/2} e^{itD_x^2} f(x) \right\|_{L^2(\mathbf{R}_t)} \leq C\|f\|_{L^2(\mathbf{R}_x)}.$$

Hence we have

$$\left\| \langle x\rangle^{-s} |D_n|^{1/2} e^{it\sigma(D)} g(x) \right\|_{L^2(\mathbf{R}_t\times\mathbf{R}_x^n)} \leq C\|g\|_{L^2(\mathbf{R}_x^n)}$$

for $s > 1/2$. Here we have used the trivial inequality $\langle x\rangle^{-s} \leq \langle x_n\rangle^{-s}$, Schwarz's inequality, and Plancherel's theorem. We remark that ψ maps Γ to another small conic neighborhood $\tilde{\Gamma}$ of e_n. Since $\operatorname{supp}\hat{g} \subset \tilde{\Gamma}$ and $|\xi_n|$ is equivalent to $|\xi|$ on $\tilde{\Gamma}$, we have the estimate

$$\left\| \langle x\rangle^{-s} |D|^{1/2} e^{it\sigma(D)} g(x) \right\|_{L^2(\mathbf{R}_t\times\mathbf{R}_x^n)} \leq C\|g\|_{L^2(\mathbf{R}_x^n)},$$

that is, estimate (4.2).

Case (ii)

We take

$$\sigma(\eta) = \eta_1\eta_n, \quad \psi(\xi) = \left(\frac{a(\xi)}{\xi_n}, \xi_2, \dots, \xi_n\right)$$

Then we have $a(\xi) = (\sigma \circ \psi)(\xi)$ and

$$\det \partial\psi(e_n) = \begin{vmatrix} \partial_1 a(e_n) & * \\ 0 & E_{n-1} \end{vmatrix} \neq 0.$$

The estimate for $\sigma(D) = D_1 D_n$ was given by the following (Linares and Ponce [7, p.528]):

Proposition 4.2. *In the case $n = 2$, we have*

$$\sup_{y\in\mathbf{R}} \left\| |D_x|^{1/2} e^{itD_xD_y} f(x,y) \right\|_{L^2(\mathbf{R}_t\times\mathbf{R}_x)} \leq C\|f\|_{L^2(\mathbf{R}_{x,y}^2)}.$$

Hence, similarly in the case (i), we have

$$\left\| \langle x\rangle^{-s} |D|^{1/2} e^{it\sigma(D)} g(x) \right\|_{L^2(\mathbf{R}_t\times\mathbf{R}_x^n)} \leq C\|g\|_{L^2(\mathbf{R}_x^n)}$$

for $s > 1/2$, that is, estimate (4.2).

5. Announcement of Further Results

Finally, we mention further results. By developing the argument in Section 4, we can have more generalized results on smoothing effects for dispersive equations. We exhibit here some of them without the proofs which will appear in [9] and [10].

Let consider equation (2.5), but we assume that real-valued function $a(\xi)$ is of order m. Let $a_m(\xi) \in C^\infty(\mathbf{R}^n \setminus 0)$ be a positively homogeneous function of order m satisfying $\nabla a_m(\xi) \neq 0$ $(\xi \neq 0)$. We assume one of the following:

(H) $a(\xi) = a_m(\xi)$.

(L) $a(\xi) \in C^\infty(\mathbf{R}^n)$, $\nabla a(\xi) \neq 0$, and $|\partial^\alpha(a(\xi) - a_m(\xi))| \leq C\langle\xi\rangle^{m-1-|\alpha|}$ for large $|\xi|$.

For example $a(\xi_1, \xi_2) = \xi_1^3 + \xi_2^3 + \xi_1$ satisfies (L).

Under the assumption above, we have the following results corresponding to the types (1), (2), and (3) respectively in Section 2:

Theorem 5.1 (Type (1)). *Suppose $m > 0$ and $n \geq 2$. Assume* (H) *or* (L). *Then the solution u to* (2.5) *satisfies*

$$\langle x\rangle^{-s}|D|^{(m-1)/2}u(t,x) \in L^2(\mathbf{R}_t \times \mathbf{R}_x^n)$$

for $s > 1/2$.

Remark 5.2. Chihara [3] proved Theorem 5.1 with $m > 1$ under the assumption (H).

Theorem 5.3 (Type (2)). *Suppose $m > 1$ and $n > m + 1$, or $n > m > 1$ in the case $a(\xi) > 0$. Assume* (H) *or* (L). *Then the solution u to* (2.5) *satisfies*

$$\langle x\rangle^{-m/2}\langle D\rangle^{(m-1)/2}u(t,x) \in L^2(\mathbf{R}_t \times \mathbf{R}_x^n).$$

Remark 5.4. Walther [14] proved Theorem 5.3 with $a(\xi) = |\xi|^m$ and $n > m > 1$. The optimality of the orders $-m/2$ and $(m-1)/2$ are also mentioned there.

For the proofs of Theorems 5.1 and 5.3, we use the following result which is a generalized version of Proposition 3.1:

Proposition 5.5. *Suppose $n \geq 2$ and $0 \leq |k| \leq n/2$. Then the operators I and I^{-1} defined by* (3.1) *are $L_k^2(\mathbf{R}^n)$-bounded.*

In order to state one more result, we introduce some notations. Let

$$\{(x(t), \xi(t));\, t \in \mathbf{R}\}$$

be the classical orbit, that is, the solutions of the ordinary differential equation

$$\begin{cases} \dot{x}(t) = (\nabla a)(\xi(t)), & \dot{\xi}(t) = 0 \\ x(0) = 0, & \xi(0) = \omega, \end{cases}$$

and consider the set of the path of all classical orbits

$$\begin{aligned} \Gamma_a &= \left\{(x(t), \xi(t)); t \in \mathbf{R}, \omega \in S^{n-1}\right\} \\ &= \{(\lambda\nabla a(\xi), \xi); \xi \in \mathbf{R}^n \setminus 0, \lambda \in \mathbf{R}\}. \end{aligned}$$

In the case $a(\xi) = |\xi|^2$ for example, we have

$$\Gamma_a = \{(x,\xi) \in T^*\mathbf{R}^n \setminus 0;\ |x \wedge \xi| = 0\}.$$

For the symbol $\sigma(x,\xi)$ of the pseudo-differential operator $\sigma(X,D)$, we use the notation $\sigma(x,\xi) \sim |x|^a|\xi|^b$ if the symbol $\sigma(x,\xi)$ is smooth in $x \neq 0$, $\xi \neq 0$ and satisfies

$$\sigma(\lambda x,\xi) = \lambda^a \sigma(x,\xi), \quad \sigma(x,\lambda\xi) = \lambda^b \sigma(x,\xi) \qquad (\lambda > 0).$$

Theorem 5.6 (Type (3)). *Suppose $m \in \mathbf{N}$ and $n \geq 2D$ Assume that* (H), $a(\xi) > 0$, *and the Gaussian curvature of $\Sigma_a = \{\xi; a(\xi) = 1\}$ never vanishes. Also assume that*

$$\tau(x,\xi) \sim |x|^{-1/2}|\xi|^{(m-1)/2}$$

and

$$\tau(x,\xi) = 0 \quad \text{if } (x,\xi) \in \Gamma_a \text{ and } x \neq 0. \tag{5.1}$$

Then the solution u to (2.5) *satisfies*

$$\tau(X,D)u(t,x) \in L^2(\mathbf{R}_t \times \mathbf{R}^n_x).$$

Remark 5.7. In the case $m = 2$, Theorem 5.6 corresponds to the critical case $\alpha = 1/2$ of Kato and Yajima [4] which cannot be allowed without the additional structure condition (5.1).

For the proof of Theorem 5.6, we need more refined boundedness theorem of Fourier integral operators. We should mention here that the global L^2-boundedness with the phase function (1.2) is not covered by previous results, for example, Asada and Fujiwara [1], The result of [1] is motivated by the construction of fundamental solution of Schrödinger equation in the way of Feynman's path integral, and it requires the boundedness of all the derivatives of entries of the matrix

$$D(\phi) = \begin{pmatrix} \partial_x\partial_y\phi & \partial_x\partial_\xi\phi \\ \partial_\xi\partial_y\phi & \partial_\xi\partial_\xi\phi \end{pmatrix}.$$

- (Asada and Fujiwara [1]) Assume that $|\det D(\phi)| \geq C > 0$ and and all the derivatives of $p(x,y,\xi)$ and those of each entry of $D(\phi)$ are bounded. Then T defined by (1.1) is bounded on $L^2(\mathbf{R}^n)$.

With our phase function (1.2), the boundedness of the entries of $\partial_\xi\partial_\xi\phi$ fails. But the following result will appear in [8]:

Theorem 5.8. *Let the operator T be defined by* (1.1) *with $\phi(x,y,\xi) = x\cdot\xi + \varphi(y,\xi)$. Assume that*

$$|\det \partial_y\partial_\xi\varphi(y,\xi)| \geq C > 0,$$

and all the derivatives of entries of $\partial_y\partial_\xi\varphi$ are bounded. Also assume that

$$\left|\partial_\xi^\alpha\varphi(y,\xi)\right| \leq C_\alpha\langle y\rangle \quad (\forall|\alpha| \geq 1),$$
$$\left|\partial_x^\alpha\partial_y^\beta\partial_\xi^\gamma p(x,y,\xi)\right| \leq C_{\alpha\beta\gamma}\langle x\rangle^{-|\alpha|} \quad (\forall\alpha,\beta,\gamma)$$

or

$$\left|\partial_y^\alpha \partial_\xi^\beta \varphi(y,\xi)\right| \le C_\alpha \langle y\rangle^{1-|\alpha|} \quad (\forall \alpha, |\beta| \ge 1),$$

$$\left|\partial_x^\alpha \partial_y^\beta \partial_\xi^\gamma p(x,y,\xi)\right| \le C_{\alpha\beta\gamma} \langle y\rangle^{-|\beta|} \quad (\forall \alpha, \beta, \gamma)$$

Then T is bounded on $L^2_k(\mathbf{R}^n)$ for any $k \in \mathbf{R}$.

This theorem says that, if amplitude functions $p(x,y,\xi)$ have some decaying properties with respect to x or y, we do not need the boundedness of $\partial_\xi \partial_\xi \phi$ for the L^2-boundedness, as required in [1], and we can have weighted estimates, as well.

References

[1] K. Asada and D. Fujiwara, On some oscillatory integral transformations in $L^2(\mathbf{R}^n)$, *Japanese J. Math. (N.S.)* **4** (1978), 299–361.

[2] M. Ben-Artzi and S. Klainerman, Decay and regularity for the Schrödinger equation, *J. Analyse Math.* **58** (1992), 25–37.

[3] H. Chihara, Smoothing effects of dispersive pseudodifferential equations, *Comm. Partial Differential Equations* **27** (2002), 1953–2005.

[4] T. Kato and K. Yajima, Some examples of smooth operators and the associated smoothing effect, *Rev. Math. Phys.* **1** (1989), 481–496.

[5] C.E. Kenig, G. Ponce and L. Vega, Oscillatory integrals and regularity of dispersive equations, *Indiana Univ. Math. J.* **40** (1991), 33–69.

[6] D.S. Kurtz and R.L. Wheeden, Results on weighted norm inequalities for multipliers, *Trans. Amer. Math. Soc.* **255** (1979), 343–362.

[7] F. Linares and G. Ponce, On the Davey-Stewartson systems, *Ann. Inst. H. Poincaré Anal. Non Linéaire* **10** (1993), 523–548.

[8] M. Ruzhansky and M. Sugimoto, Global L^2-boundedness theorems for a class of Fourier integral operators, preprint.

[9] M. Ruzhansky and M. Sugimoto, A smoothing property of Schrödinger equations, preprint.

[10] M. Ruzhansky and M. Sugimoto, Smoothing effects for dispersive equations via canonical transformations, in preparation.

[11] E.M. Stein and G. Weiss, Fractional integrals on n-dimensional Euclidean space, *J. Math. Mech.* **7** (1958), 503–514.

[12] M. Sugimoto, Global smoothing properties of generalized Schrödinger equations, *J. Analyse Math.* **76** (1998), 191–204.

[13] B.G. Walther, A sharp weighted L^2-estimate for the solution to the time-dependent Schrödinger equation, *Ark. Mat.* **37** (1999), 381–393.

[14] B.G. Walther, Regularity, decay, and best constants for dispersive equations, *J. Funct. Anal.* **189** (2002), 325–335.

[15] K. Watanabe, Smooth perturbations of the selfadjoint operator $|\Delta|^{\alpha/2}$, *Tokyo J. Math.* **14** (1991), 239–250.

Michael Ruzhansky
Department of Mathematics
Imperial College
180 Queen's Gate
London SW7 2BZ
UK
e-mail: ruzh@ic.ac.uk

Mitsuru Sugimoto
Department of Mathematics
Graduate School of Science
Osaka University
Toyonaka
Osaka 560-0043
Japan
e-mail: sugimoto@math.wani.osaka-u.ac.jp

Operator Theory:
Advances and Applications, Vol. 155, 77–90

Gevrey Hypoellipticity of p-Powers of Non-Hypoelliptic Operators

Giuseppe De Donno

Abstract. We characterize the hypoellipticity in C^∞ and Gevrey G^λ classes of 2-variable PDO's containing powers of anisotropic principal terms. We use an approach based on methods from microlocal analysis. Conditions are imposed on the coefficients of lower order terms. Also a semilinear version is proposed considering C^∞ nonlinear perturbations. See Theorem 1.1 and Theorem 1.6.

Mathematics Subject Classification (2000). Primary 35S05.

Keywords. Partial differential equations, hypoellipticity, Gevrey classes.

1. Introduction

This paper deals with the properties of hypoellipticity and Gevrey hypoellipticity of a class of partial differential operators with multiple characteristics containing p-powers of anisotropic principal terms, i.e., the derivatives with highest anisotropic order; in two variables $z = (x, y)$, belonging to the set Ω, neighborhood of a point $z_0 = (x_0, y_0) \in \mathbb{R}^2$. In particular we consider equations of the form:

$$P(x, y, D_x, D_y)u + F(x, y, D_x^l D_y^j u)\big|_{\frac{m}{d}l+j<k^*} = 0, \tag{1.1}$$

where the linear part is given by:

$$P(x, y, D_x, D_y) = Q^p + \sum_{k^* \le \frac{m}{d}l+j<pm} b_{lj}(x, y)D_x^l D_y^j \tag{1.2}$$

and the operator Q is of the form:

$$Q(x, y, D_x, D_y) = D_y^m - a(x, y)D_x^d, \tag{1.3}$$

with $m, d, j, l, p \in \mathbb{Z}_+$, $p \ge 2$, $0 < k^* < pm$, $D_x = -i\frac{\partial}{\partial x}$, $D_y = -i\frac{\partial}{\partial y}$; the anisotropic principal terms are represented by the operator Q. We shall also say that $\frac{m}{d}l + j$ is the anisotropic order of $D_x^l D_y^j$, so the nonlinearity involves derivatives of anisotropic order less than k^*. For related results in the case $p = 1$, see De Donno-Oliaro [5], where assuming $\operatorname{Re} a(x,y) \neq 0$ and looking for a couple (l^*, j^*) with $k^* := \frac{m}{d} l^* + j^* > m - \frac{1}{2}$, such that $\operatorname{Im} b_{l^* j^*}(x,y) \neq 0$, a sufficient condition for hypoellipticity with loss of the smoothness of solutions equal to $m - k^*$ has been established for (1.1) and (1.2). On the coefficients $\operatorname{Im} b_{lj}(x,y)$ of the terms having anisotropic order larger than k^*, we assumed to be of the same sign of $\operatorname{Im} b_{l^* j^*}(x,y)$ (allowing they can vanish); the same assumption on $-\operatorname{Im} a(x,y)$. Also local solvability was showed for (1.1) and (1.2); and G^λ hypoellipticity, $\lambda \geq \frac{1}{k^* - (m-1)}$, for (1.1). As standard, the Gevrey space $G^\lambda(\Omega)$, is defined by the estimates:

$$f \in C^\infty(\Omega), \quad \sup_K |\partial_z^\alpha f(z)| \leq C_K^{|\alpha|+1} (\alpha!)^\lambda, \text{ for every compact } K \subset\subset \Omega,$$

with $\lambda \geq 1$.

In the case $\operatorname{Im} b_{l^* j^*}(x,y)$ vanishes at some points (but not identically), Gevrey solvability for small Gevrey index was proved, too.

In De Donno-Oliaro [5] in spite of the conditions imposed on the lower order terms are optimal (the results are sharp in the particular case of constant coefficients), we just obtained a small hypoelliptic interval $(m - \frac{1}{2}, m)$, in which we look for a couple (l^*, j^*) in order to impose our hypotheses (the interval is $(m-1, m)$ in the case of constant coefficients). For operators (constant coefficients) containing p-powers of anisotropic principal terms of the form

$$(D_y^2 - D_x)^p + i D_x^{l^*} D_y^{j^*} \quad , \tag{1.4}$$

k^* can runs in the interval $(p, 2p)$, which is $p-1$ anisotropic orders in width, see De Donno-Oliaro-Rodino [6].

In order to enlarge the hypoelliptic interval of the variable coefficients case, according with the model (1.4) of constant coefficients, we consider general operators in the form (1.2) having p-powers of the operator Q. Unfortunately the influence of higher order terms of Q^p, $p \geq 4$, limit k^* in the interval $(pm - 1 - \frac{1}{2}, pm)$, that anyway contains the anisotropic order $pm - \frac{1}{2}$ not accepted in the case $p = 1$ ($k^* > m - \frac{1}{2}$). For $p \leq 3$, the influence of (the higher order terms of) Q^p is irrelevant, since $pm - 1 - \frac{1}{2} \leq p(m - \frac{1}{2})$, and $p(m - \frac{1}{2})$ is the best limitation from the below we can obtain with our techniques. Let us also observe that in the worst case we get $k^* \in (pm - \frac{1}{2} - \frac{1}{d}, pm) \supset (pm - \frac{1}{2}, pm)$, see Remark 1.3.

On the contrary, in the generic differential polynomial symbols

$$(\eta^m - a(x,y)\xi^d)^p + \sum_{(l,j)\in I} c_{lj}(x,y)\xi^l \eta^j \ , \tag{1.5}$$

where $(\xi,\eta) \in \mathbb{R}^2$ are the dual variables of (x,y); containing as particular case the symbol of the operators (1.2), k^* belonging in the interval $\left(p(m - \frac{1}{2}), pm\right)$ should be accepted for any integer p; see Theorem 2.1. The subset of indices I corresponds to lower order terms.

In this paper we propose a complete analysis of the influence of the lower-order terms of (1.5) on the hypoellipticity of (1.1) and (1.2) in the C^∞ category and in the Gevrey spaces G^λ beyond the critical index $m/(m-1)$. The arguments in our proofs are based mainly on microlocal tools: pseudo-differential operators, wave front sets, allowing relevant simplifications in the study, and $S^m_{\rho,\delta}$ techniques. In the present paper, for simplicity, we consider real valued functions $a(x,y)$ ($\operatorname{Im} a(x,y) \equiv 0$) in (1.3). We suppose C^∞ coefficients in (1.2), (1.3), and in the follow we always ask that:

$$d < m, \quad a(x_0, y_0) \neq 0, \tag{1.6}$$

$$F(z;t) = \sum_{r \in \mathbb{Z}^M_+} C_r(z) t^r, \ C_r \in C^\infty(\Omega), \ t \in \mathbb{Z}^M, \tag{1.7}$$

where, for every compact $K \subset\subset \Omega$, $\sup_{z \in K} |D^\alpha C_r(z)| \leq C_{\alpha,K} \lambda_r$ and moreover $\tilde{F}(t) = \sum_r \lambda_r t^r$ is entire analytic.

The nonzero requirement on $a(x,y)$ in (1.6) is an invariant nondegeneracy condition, usually required in the study of the hypoellipticity (and local solvability) of the linear operator (1.2) with $p = 1$, in the C^∞ and Gevrey G^λ, $\lambda > \frac{m}{m-1}$, see for example, Liess-Rodino[16], De Donno-Rodino[7], where Gevrey hypoellipticity for PDE's with high multiplicity is proved in the isotropic case. For $p = 1$ see also Gramchev-Popivanov[13], where hypotheses are given on the roots of the full symbol of the operator, and Corli[3], about results of non-hypoellipticity. Let us also observe that if $\operatorname{Im} a(x,y) \neq 0$ then the operator is quasi-elliptic; the results on hypoellipticity (and local solvability) are well known in this case.

A new important element with respect to the case $p = 1$, for formulating our results in this paper, is the so called k^{max} order, i.e., the highest anisotropic order $\frac{m}{d} l + j$, $k^* < \frac{m}{d} l + j < pm$ of $D^l_x D^j_y$ such that the corresponding coefficient is not identically null. When $k^* \geq \frac{m}{d} l + j$ for all couples (l,j) we say that does not exists k^{max}.

We split the symbol of (1.2) in the following way:

$$\begin{aligned} p(x,y,\xi,\eta) = (\eta^m - a(x,y)\xi^d)^p &+ \sum_{k^* \leq \frac{m}{d} l + j \leq k^{max}} \tilde{b}_{lj}(x,y) \xi^l \eta^j \\ &+ \sum_{\frac{m}{d} l + j < k^*} \tilde{b}_{lj}(x,y) D^l_x D^j_y \end{aligned} \tag{1.8}$$

where $\tilde{b}_{lj}(x,y) = b_{lj}(x,y)$, for $\frac{m}{d} l + j > pm - 1$ since the operator Q^p captures derivatives of anisotropic order less or equal than $pm-1$, and $\tilde{b}_{lj}(x,y) = a^{rs}_{xy}(x,y) + \operatorname{Re} b_{lj}(x,y) + i \operatorname{Im} b_{lj}(x,y)$, for $m \leq k^* \leq \frac{m}{d} l + j \leq pm - 1$, with r, s representing suitable derivative orders respectively with respect x and y variables. We are not interested in the coefficients of mixed derivatives having anisotropic order less than k^*.

Moreover we define the anisotropic characteristic manifold

$$\Sigma := \{(x,y,\xi,\eta) \in \Omega \times \mathbb{R}^2 \setminus \{0\}, \ (\eta^m - a(x,y)\xi^d)^p = 0\}. \tag{1.9}$$

Let us state the main results.

Theorem 1.1. *Let* $(l^*, j^*) \in \mathbb{Z}_+^2$ *be the unique couple having anisotropic order* $k^* := \frac{m}{d}l^* + j^*$, *with* $\max\{p(m-\frac{1}{2}), k^{max} - \frac{1}{2}\} < k^* < pm$. *We suppose* $a(x,y)$, $b_{lj}(x,y) \in C^\infty(\Omega)$ *and assume that for* $(x,y,\xi,\eta) \in \Sigma$ *the following conditions hold:*

i) $\mathrm{Im}\, b_{l^*j^*}(x,y) \neq 0$

ii) $\mathrm{Im}\, b_{l^*j^*}(x,y)\mathrm{Im}\, b_{lj}(x,y)\xi^{l+l^*}\eta^{j+j^*} \geq 0$ *for all* (l,j) *such that* $k^* < \frac{m}{d}l + j \leq k^{max}$

iii) $\mathrm{Re}\, \tilde{b}_{lj}(x,y)\xi^l\eta^j \geq 0$ *for all* (l,j) *such that* $k^* \leq \frac{m}{d}l + j \leq k^{max}$

iii′) *for odd p:* $\mathrm{Re}\, \tilde{b}_{lj}(x,y) \equiv 0$ *for all* (l,j) *such that* $k^* \leq \frac{m}{d}l + j \leq k^{max}$.

Assume moreover that (1.6) *holds. Then the operator* P *in* (1.2) *is* C^∞*-hypoelliptic with loss of regularity equal to* $pm - k^*$ *derivatives. Moreover, taking analytic coefficients* a, b_{lj}, (1.2) *is* G^λ*-hypoelliptic for*

$$\lambda \geq \begin{cases} \frac{p}{k^*-p(m-1)} & , \quad k^{max} \leq \frac{p-1}{p}k^* + m \\ \frac{1}{k^*-(k^{max}-1)} & , \quad k^{max} > \frac{p-1}{p}k^* + m \end{cases} . \tag{1.10}$$

Remark 1.2. It is always possible to rephrase the previous assumptions in Theorem 1.1 directly on the coefficients of P. For example, if $a(x,y) > 0$ and m, d are odd, the conditions i), ii), iii) are respectively equivalent to: i′) $\mathrm{Im}\, b_{l^*j^*}(x,y) > 0\ (<0)$; ii′) for all (l,j) such that $\frac{d}{m}l + j > k^*$, $\mathrm{Im}\, b_{lj}(x,y) \geq 0\ (\leq 0)$ for $j + j^*$ and $l + l^*$ both even or both odd and $\mathrm{Im}\, b_{lj}(x,y) \equiv 0$ otherwise; iii′) $\mathrm{Re}\, \tilde{b}_{lj}(x,y) \geq 0)$ for j and l both even or both odd and $\mathrm{Re}\, \tilde{b}_{lj}(x,y) \equiv 0$ otherwise.

Remark 1.3. In the operators (1.2), the order k^{max} is always larger than or equal to $pm-1$. Assuming that it is exactly $pm-1$ (just the influence of Q^p), the condition $k^* > \max\{p(m-\frac{1}{2}), k^{max} - \frac{1}{2}\}$ gives us the sharp limitation from the below $p(m-\frac{1}{2})$ whenever $pm - 1 - \frac{1}{2} \leq p(m-\frac{1}{2})$, that is $p \leq 3$. Otherwise, allowing k^{max} to be the largest order less than pm, $k^{max} = pm - \frac{1}{d}$, we again obtain the best limitation $p(m-\frac{1}{2})$ if and only if $p \leq 1 + \frac{2}{d}$, that represents the following conditions: $\{d = 1$ and $p = 1, 2\}$ or $\{d = 2$ and $p = 2\}$. Observe that $k^* > pm - \frac{1}{d} - \frac{1}{2}$ for $k^* > k^{max} - \frac{1}{2}$, in this case. In general for $k^{max} = pm - \frac{h}{d}$, for $1 \leq h \leq d$, the best limitation is obtained for $p \leq 1 + \frac{2h}{d}$. Let us also observe that the step from an anisotropic order and the next one in intervals of the type $(pm-1, pm)$ is exactly $\frac{1}{d}$, since we have $dpm - d < ml + dj < dpm$.

Remark 1.4. If $a(x,y)$ in the operator Q in (1.3) is a constant function, in Theorem 1.1 we can consider $k^* > p(m-1)$ if k^{max} there is not, $k^* > \max\{\frac{p}{p+1}(k^{max} + m - 1), k^{max} - \frac{1}{2}\}$ otherwise. It follows directly by the proof of the Theorem 2.1 in the next section, see Remark 2.11.

Remark 1.5. The hypoellipticity results in G^λ for P in Theorem1.1 imply local Gevrey G^λ solvability of the adjoint operator P^*. This follows from the main theorem in Albanese-Corli-Rodino [1]

As example of operator with odd p we consider the following:

Example.

$$(D_y^3 - (y^2+1)D_x^2)^3 + (18y+i)D_x^2 D_y^5 \tag{1.11}$$

where we have not k^{max}, since there are not anisotropic orders strictly larger than $k^* = \frac{3}{2}2 + 5 = 8 > 3(3-\frac{1}{2})$. Of course hypothesis $i)$ in 1.1 is satisfied. The symbol of (1.11) is

$$(\eta^3 - (y^2+1)\xi^2)^3 - (18y)D_x^2 D_y^5 + (18y+i)D_x^2 D_y^5 + \text{lower o. ts.}$$

so $\operatorname{Re} \tilde{b}_{l^*j^*}(x,y) \equiv 0$. We have hypoellipticity and G^λ hypoellipticity for $\lambda \geq \frac{3}{8-3(3-1)} = \frac{3}{2}$. More general we can consider operators of the form

$$(D_y^m - (y^2+1)D_x^d)^3 + (6my+i)D_x^d D_y^{2m-1} \quad ,$$

hypoelliptic and G^λ hypoelliptic for $\lambda \geq \frac{3}{2}$.

As a relevant example with constant function $a(x,y)$ we can consider p-powers of the Schrödinger operator:

Example. for $p = 2$

$$(D_y^2 - D_x)^2 + i(y^2+1)D_x D_y \quad ,$$

is hypoelliptic and G^λ hypoelliptic for $\lambda \geq 2$, see also [6] for the isotropic frame. Instead for the operator

$$(D_y^2 - D_x)^3 + i(y^2+1)D_x D_y^2 + D_x D_y^3 \quad ,$$

nothing we can conclude about its hypoellipticity since the term $k^{max} = 5$ is too high. In general operators of the form

$$(D_y^2 - D_x)^p + i(y^2+1)D_x^l D_y^{p+1-2l} \quad ,$$

are hypoelliptic and G^λ hypoelliptic for $\lambda \geq 2$.

As full example of operator satisfying Theorem 1.1 we consider:

Example.

$$(D_y^6 - (y^2+1)D_x^5)^2 + (y^4+i)D_x^6 D_y^4 + x^2 D_x^8 D_y^2 \quad ,$$

where $k^{max} = \frac{58}{5}$, $k^* = \frac{56}{5} > k^{max} - \frac{1}{2} > 2(6-\frac{1}{2})$; i), ii), iii) in Theorem 1.1 hold, so we have hypoellipticity and G^λ hypoellipticity for $\lambda \geq \frac{5}{3}$, since $k^{max} = \frac{p-1}{p}k^* + m$, see (1.10).

In the following the Sobolev anisotropic space $H^s_{(\frac{d}{m},1)}(\mathbb{R}^2)$ will be used, defined by:

$$\|f\|^2_{H^s_{(\frac{d}{m},1)}} := \int (1 + |\xi|^{2\frac{d}{m}} + |\eta|^2)^s \, |\widehat{f}(\xi,\eta)|^2 \, d\xi \, d\eta \ < \infty, \tag{1.12}$$

$\widehat{f}(\xi,\eta)$ being the Fourier transform of $f(x,y)$. For $s > \frac{1+\frac{m}{d}}{2}$, $H^s_{(\frac{d}{m},1)}$ is an algebra, cf. the inhomogeneous Schauder estimates in Garello [9, Proposition 2.5].

Theorem 1.6. *Under the assumptions of Theorem* 1.1 *on* P *and the condition* (1.7) *on* F, *let* u *be a solution of* (1.1) *which belongs to* $H^s_{(\frac{d}{m},1),\mathrm{loc}}(\Omega)$, *for* $s \geq s_0$, *where* s_0 *is a sufficiently large fixed real number. Then* $u \in C^\infty(\Omega)$.

Let us observe that several results have been written for microlocal regularity for nonlinear elliptic PDEs with analytic nonlinearities as in (1.7), cf., Lascar[10], see also Popivanov-Iordanov[20].

About the critical bound from below for s, we recall that the regularity statement of Theorem1.6 is not true if the solution u belongs to H^s with $s > 0$ but small enough. For explicit constructions of strongly singular solutions in H^s_{loc} $s < \frac{n}{2}$ of semilinear elliptic PDE, cf. Biagioni-Gramchev [2] and Gramchev [11].

In the next Section 2 we prove Theorem 1.1 using $S^m_{\rho,\delta}$ estimates; Theorem 1.6 is proved in Section 3.

2. Hypoellipticity for the Class (1.5) of Differential Polynomials

In this section we prove $S^m_{\rho,\delta}$ estimates of the following class of differential polynomial in suitable symplectic co-ordinates:

$$p(x,y,\xi,\eta) = (\eta^m - a(x,y)\xi^d)^p + \sum_{(l,j)\in I} c_{lj}(x,y)\xi^l\eta^j \tag{2.1}$$

Where $m,d,j,l,p \in \mathbb{Z}_+$, $d < m$, $p \geq 2$ even, and $a : \Omega \to \mathbb{R}$, $c_{lj} : \Omega \to \mathbb{C}$ smooth functions. As before we denote by $z = (x,y)$ the real variables in Ω, open subset in $\mathbb{R}^2$, neighborhood of a point z_0; $\zeta = (\xi,\eta)$ the dual variables of z. We define the following sets for $k \in \mathbb{Q}_+$, $0 < k < pm$:

$$I_k = \{(l,j) \in \mathbb{Z}_+ \times \mathbb{Z}_+ : \frac{m}{d}l + j = k\}$$

and fix $k = k^*$ such that $p(m - \frac{1}{2}) < k^* < pm$. We use the notations k^- for all $k < k^*$, k^+ for all $k > k^*$ and k^{max} for the biggest value k^+. We define $I = I_- \cup I_{k^*} \cup I_+$, with $I_- = \cup I_{k^-}$, $I_+ = \cup I_{k^+}$.

We consider the anisotropic norm $\lambda(\zeta) = |\xi|^{\frac{d}{m}} + |\eta|$, cf. the expression of the anisotropic Sobolev spaces $H^s_{(\frac{d}{m},1)}(\mathbb{R}^2)$, in (1.12). We recall that $\Sigma := \{(z,\zeta) \in \Omega \times \mathbb{R}^2 \setminus \{0\} : (\eta^m - a(x,y)\xi^d)^p = 0\}$ is the anisotropic characteristic manifold of $p(x,y,\xi,\eta)$ in (2.1); let Λ be a neighborhood of Σ, we denote with Γ the set $\Omega \times \Lambda$, and we state the following:

Theorem 2.1. *Assume that* I_{k^*} *consists of just one couple* (l^*,j^*), $k^* = \frac{m}{d}l^* + j^*$ *with* $\max\{p(m-\frac{1}{2}), k^{max} - \frac{1}{2}\} < k^* < pm$ *such that:*

i) $\operatorname{Im} c_{l^*j^*}(x,y) \neq 0$ *for all* $(x,y) \in \Omega$;
ii) *for all* (l,j) *such that* $k^* < \frac{m}{d}l + j \leq k^{max}$,
$\operatorname{Im} c_{l^*j^*}(x,y)\operatorname{Im} c_{lj}(x,y)\xi^{l+l^*}\eta^{j+j^*} \geq 0$, $(z,\zeta) \in \Gamma$;
iii) $\operatorname{Re} c_{lj}(x,y)\xi^l\eta^j \geq 0$, *for all* (l,j) *such that* $k^* \leq \frac{m}{d}l + j \leq k^{max}$;
iv) $a(x,y) \neq 0$ *for all* $(x,y) \in \Omega$.

Then

$$|p(x,y,\xi,\eta)| \geq b\lambda(\zeta)^{k^*} \quad in\ \Omega \times \mathbb{R}^2,$$

for a suitable constant $b > 0$*; and for all* $(\alpha,\beta) \in \mathbb{Z}_+^2$, $(\gamma,\theta) \in \mathbb{Z}_+^2$ *and for all* $K \subset\subset \Omega$ *we have with suitable constants* $L_{\alpha,\beta,\gamma,\delta}$ *and* B *that:*

$$\frac{|D_x^\alpha D_y^\beta D_\xi^\gamma D_\eta^\theta p(x,y,\xi,\eta)|\,\lambda(\zeta)^{\rho(\gamma\frac{m}{d}+\theta)\,-\,\delta(\alpha\frac{m}{d}+\beta)}}{|p(x,y,\xi,\eta)|} \leq L_{\alpha,\beta,\gamma,\theta}\ ,\ |\xi|+|\eta| > B, \tag{2.2}$$

with

$$\rho = \min\{\frac{k^* - p(m-1)}{p}, k^* - k^{max} + 1\}\ ,\ \delta = \max\{\frac{pm - k^*}{p}, k^{max} - k^*\}. \tag{2.3}$$

Remark 2.2. Observe that it is always $\delta < \rho$ since

$$\rho = \begin{cases} \frac{k^*-p(m-1)}{p} & , \quad k^{max} \leq \frac{p-1}{p}k^* + m \\ k^* - k^{max} + 1 & , \quad k^{max} > \frac{p-1}{p}k^* + m \end{cases},$$

$$\delta = \begin{cases} \frac{pm-k^*}{p} & , \quad k^{max} \leq \frac{p-1}{p}k^* + m \\ k^{max} - k^* & , \quad k^{max} > \frac{p-1}{p}k^* + m \end{cases},$$

and we have assumed $k^* > \max\{p(m-\frac{1}{2}), k^{max} - \frac{1}{2}\}$.

Remark 2.3. By formula (2.2) and by Mascarello-Rodino([18], Theorem 3.3.6), we have that an operator $P(x,y,D_x,D_y)$, associated to the symbol $p(x,y,\xi,\eta)$ in (2.1), is C^∞-microlocally hypoelliptic in Γ; i.e. $\Gamma \cap WF\,Pu = \Gamma \cap WF\,u$, for all $u \in \mathcal{D}'(\Omega)$, where $WF\,u$ is the Hörmander wave front set. A microhypoelliptic operator is hypoelliptic too.

Remark 2.4. If the coefficients are analytic, formula (2.2) holds for $L_{\alpha\beta\gamma\theta} = L^{\alpha+\beta+\gamma+\theta+1}\alpha!\beta!\gamma!\theta!$, so by Kajitani-Wakabayashi([17], Theorem 1.9), we have that an operator $P(z,D)$, associated to the symbol $p(z,\zeta)$ in (2.1) is G^λ-microlocally hypoelliptic in Γ for $\lambda \geq \max\left\{\frac{1}{\rho}, \frac{1}{1-\delta}\right\} = \frac{1}{\rho}$ (or equivalently $\frac{1}{1-\delta}$), that is $\lambda \geq \frac{p}{k^*-p(m-1)}$ for $k^{max} \leq \frac{p-1}{p}k^*+m$ and $\lambda \geq \frac{1}{k^*-(k^{max}-1)}$ for $k^{max} > \frac{p-1}{p}k^*+m$.

Remark 2.5. When $\rho < 1$, and $\delta > 0$, one can prove by means of interpolation theory as in Wakabayashi([23] Theorem 2.6) that (2.2) is valid for any $(\alpha,\beta,\gamma,\theta) \in \mathbb{Z}_+^4$, if (2.2) holds for $\alpha+\beta+\gamma+\theta = 1$. Hence it is sufficient to verify (2.2) for $\alpha+\beta+\gamma+\theta = 1$ since the relations $k^* < k^{max} < pm$ give the inequalities required.

Remark 2.6. For the proof of the Theorem 1.1 it will be sufficient to apply Theorem 2.1. We recall that in the operators of the form (1.2) always is $k^{max} \geq pm - 1$.

Proof of Theorem 2.1. First we estimate the numerator of (2.2) and then we give some lemmas to estimate the denominator, see Lemma 2.7, Lemma 2.9, Lemma 2.10. If $\alpha = 1$, we get

$$\begin{aligned} &|D_x p(x,y,\zeta)|\,\lambda(\zeta)^{-\delta\frac{m}{d}} = \\ &\quad |p(\eta^m - a(x,y)\xi^d)^{p-1} D_x a(x,y)\xi^d + \sum_{(l,j)\,\in\, I} D_x c_{lj}(x,y)\,\xi^l\eta^j|\lambda(\zeta)^{-\delta\frac{m}{d}} \\ &\quad \le L_1(|\eta^m - a(x,y)\xi^d|^{p-1}|\xi|^d + \sum_{(l,j)\,\in\, I} |\xi|^l|\eta|^j|\,)\lambda(\zeta)^{-\delta\frac{m}{d}}; \end{aligned}$$

and similarly for $\beta = 1$

$$|D_y p(x,y,\zeta)|\,\lambda(\zeta)^{-\delta} \le L_2(|\eta^m - a(x,y)\xi^d|^{p-1}|\xi|^d + \sum_{(l,j)\,\in\, I} |\xi|^l|\eta|^j|\,)\lambda(\zeta)^{-\delta},$$

for suitable constants $L_1,\ L_2$. If $\gamma = 1$,

$$|D_\xi\, p(z,\xi,\eta)|\lambda(\zeta)^{\rho\frac{m}{d}} \le L_3(|\eta^m - a(x,y)\xi^d|^{p-1}|\xi|^{d-1} + \sum_{(l,j)\,\in\, I} |\xi|^{l-1}|\eta|^j\,)\lambda(\zeta)^{\rho\frac{m}{d}};$$

and for $\theta = 1$

$$|D_\eta\, p(z,\xi,\eta)|\lambda(\zeta)^{\rho} \le L_4(|\eta^m - a(x,y)\xi^d|^{p-1}|\eta|^{m-1} + \sum_{(l,j)\,\in\, I} |\xi|^{l}|\eta|^{j-1}\,)\lambda(\zeta)^{\rho},$$

with suitable constants L_3, L_4.

To prove (2.2), it will be then sufficient to show the boundedness, for $|\zeta| > B$, of the functions

$$Q_1(z,\zeta) = \frac{\left(|\eta^m - a(x,y)\xi^d|^{p-1}|\xi|^d + \sum_{(l,j)\,\in\, I} |\xi|^l|\eta|^j|\right)\lambda(\zeta)^{-\delta}}{|p(z,\zeta)|},$$

$$Q_2(z,\zeta) = \frac{\left(|\eta^m - a(x,y)\xi^d|^{p-1}|\eta|^{m-1} + \sum_{(l,j)\,\in\, I} |\xi|^l|\eta|^{j-1}\right)\lambda(\zeta)^{\rho}}{|p(z,\zeta)|},$$

$$Q_3(z,\zeta) = \frac{\left(|\eta^m - a(x,y)\xi^d|^{p-1}|\xi|^{d-1} + \sum_{(l,j)\,\in\, I} |\xi|^{l-1}|\eta|^j\right)\lambda(\zeta)^{\rho\frac{m}{d}}}{|p(z,\zeta)|}.$$

First we introduce three regions:

$$\begin{array}{ll} R_1: & c\,|\xi|^d \le |\eta|^m \le C\,|\xi|^d \\ R_2: & \qquad |\eta|^m \ge C\,|\xi|^d \\ R_3: & \qquad |\eta|^m \le c\,|\xi|^d \end{array}, \tag{2.4}$$

for suitable constants c, C satisfying for every compact set K in Ω the inequalities $c < \frac{1}{2}\min_{(x,y)\in K} |a(x,y)|$, and $C > 2\max_{(x,y)\in K} |a(x,y)|$, cf. [4], [5], [7], [8]. We understand the neighborhood Λ to be the region R_1.

The following estimates then hold:

$$\lambda(\zeta)^{-\delta} \leq \begin{cases} const.\,|\eta|^{-\delta} & , \quad \zeta \in R_1 \quad (I) \\ const.|\eta|^{-\delta} & , \quad \zeta \in R_2 \quad (II) \\ const.|\xi|^{-\delta\frac{d}{m}} & , \quad \zeta \in R_3\,; \quad (III) \end{cases} \tag{2.5}$$

and

$$\lambda(\zeta)^{\rho} \leq \begin{cases} const.\,|\eta|^{\rho} & , \quad \zeta \in R_1 \\ const.\,|\eta|^{\rho} & , \quad \zeta \in R_2 \\ const.\,|\xi|^{\rho\frac{d}{m}} & , \quad \zeta \in R_3\,; \end{cases} \tag{2.6}$$

note that (II) and (III) in (2.5) hold for all $\zeta \in \mathbb{R}^2$ but for our aim we may limit ourselves to consider them respectively in R_2 and in R_3. By abuse of notation, in the following we shall also denote by R_1, R_2, R_3 the sets $\Omega \times R_1, \Omega \times R_2, \Omega \times R_3$; recall that $\Gamma = \Omega \times \Lambda$.

The following three lemmas give us some relevant estimates from the below of $|p(x, y, \xi, \eta)|$ in (2.1).

Lemma 2.7. *Let $p(z, \zeta)$ be the function* (2.1), *such that* i), ii) *and* iii) *in Theorem* 2.1 *hold. Then there are positive constants $K_1 < 1$, B, such that, for $(z, \zeta) \in R_1$, $|\zeta| > B$:*

$$|p(z,\zeta)| \geq K_1 \left((\eta^m - a(x,y)\xi^d)^{2p} + |\mathrm{Im}\, c_{l^*j^*}(x,y)|^2 |\xi|^{2l^*} |\eta|^{2j^*} \right)^{\frac{1}{2}}. \tag{2.7}$$

Proof. We have that

$$\begin{aligned} |p(z,\zeta)|^2 &= \left((\eta^m - a(x,y)\,\xi^d)^p + \textstyle\sum_{(l,j)\in I} \mathrm{Re}\, c_{lj}(x,y)\,\xi^l \eta^j \right)^2 \\ &+ \left(\mathrm{Im}\, c_{l^*j^*}(x,y) \xi^{l^*} \eta^{j^*} + \textstyle\sum_{(l,j)\in I_+} \mathrm{Im}\, c_{lj}(x,y)\,\xi^l \eta^j \right. \\ &+ \left. \textstyle\sum_{(l,j)\in I_-} \mathrm{Im}\, c_{lj}(x,y)\,\xi^l \eta^j \right)^2 . \end{aligned} \tag{2.8}$$

By developing (2.8) and removing the terms

$$(\sum_{(l,j)\in I} \mathrm{Re}\, c_{lj}(x,y)\xi^l\eta^j)^2\,, \quad (\sum_{(l,j)\in I_+} \mathrm{Im}\, c_{lj}(x,y)\xi^l\eta^j + \sum_{(l,j)\in I_-} \mathrm{Im}\, c_{lj}(x,y)\xi^l\eta^j)^2$$

respectively from the real and imaginary part of $p(z, \zeta)$, we can write

$$|p(z,\zeta)|^2 \geq (\eta^m - a(x,y)\,\xi^d)^{2p} + \mathrm{Im}\, c_{l^*j^*}(x,y)^2\, \xi^{2l^*} \eta^{2j^*} + \sum_{i=1}^{4} J_i(z,\zeta)$$

where

$$J_1(z,\zeta) = 2\mathrm{Im}\, c_{l^*j^*}(x,y) \sum_{(l,j)\in I_+} \mathrm{Im}\, c_{lj}(x,y)\,\xi^{l^*+l}\eta^{j^*+j}\,, \tag{2.9}$$

$$J_2(z,\zeta) = 2(\eta^m - a(x,y)\,\xi^d)^p \sum_{(l,j)\in I_{k^*}\cup I_+} \mathrm{Re}\, c_{lj}(x,y)\,\xi^l\eta^j \tag{2.10}$$

$$J_3(z,\zeta) = 2\mathrm{Im}\, c_{l^*j^*}(x,y) \sum_{(l,j)\in I_-} \mathrm{Im}\, c_{lj}(x,y)\,\xi^{l^*+l}\eta^{j^*+j}\,, \tag{2.11}$$

$$J_4(z,\zeta) = 2(\eta^m - a(x,y)\,\xi^d)^p \sum_{(l,j)\in I_-} \operatorname{Re} c_{lj}(x,y)\,\xi^l\eta^j \tag{2.12}$$

The functions are non-negative, (2.9) and (2.10) are also non-negative respectively by hypotheses ii) and iii), since p is even. Let us fix attention on $J_3(z,\zeta)$ and $J_4(z,\zeta)$ defined respectively by (2.11) and (2.12). We have for all $\epsilon > 0$

$$\frac{1}{2}(\operatorname{Im} c_{l^*j^*}(x,y))^2\,\eta^{2j^*}\xi^{2l^*} + J_3(z,\zeta) \geq (\frac{1}{2} - \epsilon)(\operatorname{Im} c_{l^*j^*}(x,y))^2\,\eta^{2j^*}\xi^{2l^*},$$

in R_1, $|\zeta| > B$. In fact by (2.4) in R_1 and hypothesis (i) in Theorem 2.1, for all $\epsilon > 0$ we get for B sufficiently large

$$\frac{|J_3(z,\zeta)|}{(\operatorname{Im} c_{l^*j^*}(x,y))^2\,\eta^{2j^*}\xi^{2l^*}} \leq const \sum_{(l,j)\in I_-} \frac{|\xi|^{l^*+l}|\eta|^{j^*+j}}{\xi^{2l^*}\eta^{2j^*}} \leq$$

$$\leq const \sum_{(l,j)\in I_-} \frac{|\eta|^{j^*+j+(l^*+l)\frac{m}{d}}}{\eta^{2j^*+2l^*\frac{m}{d}}} < \epsilon, \quad |\zeta| > B.$$

We remark that $k^* = \frac{m}{d}l^* + j^* > k^- = \frac{m}{d}l + j$. Concerning (2.12) we also have that:

$$\frac{1}{2}(\eta^m - a(x,y)\,\xi^d)^{2p} + \frac{1}{2}\operatorname{Im} c_{l^*j^*}(x,y)^2\,\xi^{2l^*}\eta^{2j^*} + J_4(z,\zeta)$$
$$\geq (\frac{1}{2} - \varepsilon)[(\eta^m - a(x,y)\,\xi^d)^{2p} + \operatorname{Im} c_{l^*j^*}(x,y)^2\,\xi^{2l^*}\eta^{2j^*}]$$

since

$$\frac{|J_4(z,\zeta)|}{(\eta^m - a(x,y)\,\xi^d)^{2p} + \operatorname{Im} c_{l^*j^*}(x,y)^2\,\xi^{2l^*}\eta^{2j^*}} < \varepsilon, \ |\zeta| > B. \tag{2.13}$$

In fact, considering the region $\xi^d = (\frac{1}{a(x,y)} - r)\eta^m$ in the plane (ξ,η), with $(x,y) \in K \subset\subset \Omega$, $|r| < 1$; we obtain by (2.4) in R_1 that the left part of (2.13) is estimated by

$$\sum_{(l,j)\in I_-} \frac{r^p|\eta|^{mp+\frac{m}{d}l+j}}{r^{2p}|\eta|^{2mp} + |\eta|^{2k^*}} = \sum_{(l,j)\in I_-} \frac{t^p|\eta|^{mp+\frac{m}{d}l+j-2k^*}}{|\eta|^{2mp-2k^*} + t^{2p}} < \varepsilon, \ \ |t| + |\eta| > B, \tag{2.14}$$

where $t = \frac{1}{r}$, by dividing for η^{2k^*}; since $k^- < k^*$. Recall that:

for all $\varepsilon > 0$ there exists $B_\varepsilon > 0$ such that for $|x| + |y| > B$,

$$\frac{x^\alpha y^\beta}{x^{2\gamma} + x^{2\nu}} < \varepsilon \text{ if and only if } (2\gamma - \alpha)(2\nu - \beta) > \alpha\beta.$$

For $|r| \geq 1$, in R_1, we have

$$\sum_{(l,j)\in I_-} \frac{t^p|\eta|^{mp+\frac{d}{m}l+j-2k^*}}{|\eta|^{2mp-2k^*} + t^{2p}} \leq |\eta|^{-mp+\frac{m}{d}l+j} < \varepsilon, \quad \text{since } k^- < mp. \tag{2.15}$$

Then

$$|p(z,\zeta)| \geq K_1 \left((\eta^m - a(x,y)\xi^d)^{2p} + |\operatorname{Im} c_{l^*j^*}(x,y)|^2|\xi|^{2l^*}|\eta|^{2j^*}\right)^{\frac{1}{2}}, \ |\zeta| > B. \tag{2.16}$$

for a suitable positive constant K_1. □

Remark 2.8. From the previous estimate (2.16) follows easily that $|p(z,\zeta)| \geq |\zeta|^{k^*}$ in R_1.

Lemma 2.9. *Let* $p(z,\zeta)$ *be the function* (2.1). *Then there are positive constants* $K_2 < 1$, B, *such that:*

$$|p(z,\zeta)| \geq K_2 |\eta|^{pm}, \quad (z,\zeta) \in R_2, \ |\zeta| > B. \tag{2.17}$$

Lemma 2.10. *Let* $p(z,\zeta)$ *be the function* (2.1), *such that* iv) *in* (2.1) *holds. Then there are positive constants* $K_3 < 1$, B, *such that:*

$$|p(z,\zeta)| \geq K_3 |\xi|^{pd}, \quad (z,\zeta) \in R_3, \ |\zeta| > B. \tag{2.18}$$

For the proofs of Lemma 2.9 and Lemma 2.10, see De Donno-Rodino [7], concerning the case $p = 1$, the general case $p \geq 2$ actually does not involve more complications.

By the relations (2.4), (2.5), (2.6), and the inequality $|\eta^m - a(x,y)\xi^d| \leq |\eta|^m + C|\xi|^d$, we easily estimate the numerators N_i, $i = 1,2,3$, of Q_i, in the regions R_2 and R_3 in the following way:

$$N_1(\zeta) \leq \begin{cases} const.\, |\eta|^{pm-\delta} & , \quad \zeta \in R_2 \\ const.\, |\xi|^{pd-\delta\frac{d}{m}} & , \quad \zeta \in R_3; \end{cases} \tag{2.19}$$

$$N_2(\zeta) \leq \begin{cases} const.\, |\eta|^{pm-1+\rho} & , \quad \zeta \in R_2 \\ const.\, |\xi|^{pd-\frac{d}{m}(1-\rho)} & , \quad \zeta \in R_3; \end{cases} \tag{2.20}$$

$$N_3(\zeta) \leq \begin{cases} const.\, |\eta|^{pm-\frac{m}{d}(1-\rho)} & , \quad \zeta \in R_2 \\ const.\, |\xi|^{pd-1+\rho} & , \quad \zeta \in R_3. \end{cases} \tag{2.21}$$

We have $N_2(z,\zeta) \leq N_3(z,\zeta)$ in R_2 and R_3 since $d < m$ and $\rho < 1$; so we can just consider the functions $Q_1(z,\zeta)$ and $Q_2(z,\zeta)$ in those regions. Now Lemma 2.9, Lemma 2.10 and the estimates (2.19), (2.20) show the boundedness of $Q_1(z,\zeta)$ and $Q_2(z,\zeta)$ in the regions R_2 and R_3, so for $Q_3(z,\zeta)$, too.

Regarding the region R_1, we observe that $|\eta^m - a(x,y)\xi^d|$ vanishes in it, so estimates of the previous type are not optimal.

In R_1 by (2.5),(2.7) we get easily:

$$Q_1(z,\zeta) \leq const \frac{|\eta^m - a(x,y)\xi^d|^{p-1}|\eta|^{m-\delta}}{((\eta^m - a(x,y)\xi^d)^{2p} + \operatorname{Im} c_{l^*j^*}(x,y)^2\xi^{2l^*}\eta^{2j^*})^{\frac{1}{2}}} + \sum_{(l,j)\in I_{max}} \frac{|\xi|^{\frac{m}{d}l+j-\delta}}{|\eta|^{k^*}} \leq L, \ (x,y) \in K, \ |\zeta| > B, \tag{2.22}$$

where $I_{max} \subset I_+$, s.t. $\frac{m}{d}l + j = k^{max}$. The second term in right-hand side is bounded since $\delta \geq k^{max} - k^*$ in (2.3). About the first term we can argue in the same way we have done in Lemma 2.7, see formulas (2.13), (2.14), (2.15); since also $\delta \geq \frac{pm-k^*}{p}$ in (2.3).

Remark 2.11. If $a(x,y)$ is a constant function, the first part of the right-hand side in (2.22) completely vanishes (first derivatives with respect z), so we just consider $\delta \geq k^{max} - k^*$ for the boundedness of Q_1 in R_1. Imposing the fundamental inequality in the $S^m_{\rho,\delta}$ framework, $\delta < \rho$, with ρ in (2.3) we find the new conditions for k^*: $k^* > \max\{p(m-1), \frac{p}{p+1}(k^{max} + m - 1), k^{max} - \frac{1}{2}\}$.

The study of the boundedness of the functions $Q_i(z,\zeta)$, $i = 2,3$ in the region R_1 actually does not involve further complicated statements, so arguing like the previous step and using the estimates (2.3) on ρ, i.e. $\rho = \min\{\frac{k^*-p(m-1)}{p}, k^* - k^{max} + 1\}$ we have proved that $Q_i(z,\zeta)$, $i = 1,2,3$ is also bounded in R_1.

Now the following remark ends the proof:

Remark 2.12. By formulas (2.7), (2.17), (2.18), we obtain that $|p(z,\zeta)| \geq a\,|\zeta|^{\frac{d}{m}k^*}$, $a > 0$, $|\zeta| > B$. In fact we obtain that $|p(z,\zeta)| \geq const\,|\eta|^{k^*}$ in R_1, then $|\eta|^{k^*} = \frac{1}{2}|\eta|^{k^*} + \frac{1}{2}|\eta|^{k^*} \geq const\,(|\eta|^{k^*} + |\xi|^{\frac{d}{m}k^*}) \geq const\,(|\eta|^{\frac{d}{m}k^*} + |\xi|^{\frac{d}{m}k^*}) \sim const\,|\zeta|^{\frac{d}{m}k^*}$, so $|p(x,\xi)| \geq a\,|\zeta|^{k^*\frac{d}{m}}$. In the same way we get $|p(x,\xi)| \geq a\,|\zeta|^{dp}$ in R_2 and R_3, the result follows . since we have $k^* < pm$. If we refer to the anisotropic weight function $\lambda(\zeta) = |\xi|^{\frac{d}{m}} + |\eta| \sim (|\xi|^d + |\eta|^m)^{\frac{1}{m}}$, we find, in the same way, that $|p(z,\zeta)| \geq b\lambda(\zeta)^{k^*}$ *in* Γ, for $|\xi| > B$, for a suitable positive constant $b > 0$.

□

The next section is devoted to a proof of Theorem 1.6.

3. Sobolev Estimates and Semilinear Version of the Result

Remark 3.1. If we refer to the anisotropic weight function $\lambda(\zeta) = |\xi|^{\frac{d}{m}} + |\eta| \sim (|\xi|^d + |\eta|^m)^{\frac{1}{m}}$, we find, in the same way we have done in Remark 2.12, that $|p(z,\zeta)| \geq b\lambda(\zeta)^{k^*}$ *in* Γ, for $|\xi| > B$, for a suitable positive constant $b > 0$.

Let us assume now $p(z,\zeta)$ defined in all $\Omega \times \mathbb{R}^2$ and denote by P the corresponding operator. Theorem 1.1 and Remark 3.1 show respectively that the loss of regularity of P is $\varepsilon = pm - \frac{d}{m}k^*$ in the isotropic case and $\varepsilon^* = pm - k^*$ in the anisotropic case. Since $\varepsilon^* < \varepsilon$, it is natural to allow a semilinear version of the result of regularity of Theorem 1.1, that is expressed by Theorem 1.6 and that we prove in the sequel.

Let us before consider again the semilinear equation

$$P(x,y,D_x,D_y)u + F(x,y,D^l_xD^j_yu)_{|\frac{m}{d}l+j<k^*} = 0 \tag{3.1}$$

where $p(x, y, D_x, D_y)$ is the model considered in (2.1), such that the hypotheses of the Theorem 2.1 hold. Moreover, let us recall that F is of the type (1.7), i.e. $F(z;t) = \sum_{r\in\mathbb{Z}_+^M} C_r(z)t^r$, $C_r \in C^\infty(\Omega)$, $t \in \mathbb{Z}^M$. Now we prove Theorem1.6:

Proof of Theorem 1.6. Observe that $D_x^l D_y^j u \in H^{s-(\frac{m}{d}l+j)}_{(\frac{d}{m},1),loc}(\Omega)$. Note that $\frac{m}{d}l+j$ actually implies $\frac{m}{d}l + j \leq k^* - \gamma$, with $\gamma > 0$. Then

$$p(x, y, D_x, D_y)u = -F(x, y, D_x^l D_y^j u)_{\big|\frac{m}{d}l+j<k^*} \in H^{s-k^*+\gamma}_{(\frac{d}{m},1),loc}(\Omega),$$

using here the assumption $s \geq s_0$, see Garello ([9], Remark 2.4). By the last part of Remark 2.12 we have that $u \in H^{s+\gamma}_{(\frac{d}{m},1),loc}(\Omega)$.

Using again Garello ([9], Remark 2.4) we get that $p(x, D)u \in H^{s-\frac{k^*}{q}+2\gamma}_{(\frac{d}{m},1),loc}(\Omega)$ and in its own turn $u \in H^{s+2\gamma}_{(\frac{d}{m},1),loc}(\Omega)$. Repeating now the preceding argument we obtain $u \in \bigcap_{t\in\mathbb{R}^+} H^t_{(\frac{d}{m},1),loc}(\Omega)$, that is $u \in C^\infty(\Omega)$. □

Remark 3.2. It is possible to propose a generalization of Theorem 2.1 considering the complete influence of lower order terms of (2.1) in the general and more complicated case when $a(x, y)$ is a complex valued function, or also to operators with involutive characteristics of high multiplicity in more than two space variables. Details will appear in a future paper.

Acknowledgments. The author is grateful to Professors Ryuichi Ashino, Paolo Boggiatto and M. W. Wong for their invitation to give a talk in the special session on pseudo-differential operators at the Fourth ISAAC Congress held at York University. The author also thanks the referee for his useful advice on the first version of this paper.

References

[1] A. Albanese, A. Corli and L. Rodino, Hypoellipticity and local solvability in Gevrey classes, *Math. Nachr.* **252** (2002), 5–16.

[2] H.A. Biagioni and T. Gramchev, Fractional derivative estimates in Gevrey spaces, global regularity and decay for solution to semilinear equations in $\mathbb{R}^n$, *J. Diff. Equations* **194** (2003), 140–165.

[3] A. Corli, On local solvability of linear partial differential operators with multiple characteristics, *J. Diff. Equations* **81** (1989), 275–293.

[4] G. De Donno and A. Oliaro, Hypoellipticity and local solvability of anisotropic PDE's with Gevrey non-linearity, preprint.

[5] G. De Donno and A. Oliaro, Local solvability and hypoellipticity for semilinear anisotropic partial differential equations, *Trans. Amer. Math. Soc.* **335** (2003), 3405–3432.

[6] G. De Donno, A. Oliaro and L. Rodino, Analytic and Gevrey solutions of nonlinear partial differential equations, preprint.

[7] G. De Donno and L. Rodino, Gevrey hypoellipticity for partial differential equations with characteristics of higher multiplicity, *Rend. Sem. Mat. Univ. Pol. Torino* **58** (2000), 435–448.

[8] G. De Donno and L. Rodino, Gevrey hypoellipticity for equations with involutive characteristics of higher multiplicity, *C. R. Acad. Bulg. Sci.* **53** (2000), 25–30.

[9] G. Garello, Inhomogeneuos paramultiplication and microlocal singularities for semilinear equations, *Boll. Un. Mat. Ital.* (**7**) 10-B (1996), 885–902.

[10] B. Lascar, Propagation des singularitees des solutions d'equations aux derivees partielles non lineaire, in *Math. Anal. Appl.*, Part B, Advances in Math, Vol. 7 B, Academic press, 1981, 455–482.

[11] T. Gramchev, Perturbative methods in scales of Banach spaces: applications for Gevrey regularity of solutions to semilinear partial differential equations, *Rend. Sem. Univ. Pol. Torino* **61** (2003), 101–134.

[12] T. Gramchev and P. Popivanov, Local solvability of semilinear partial differential equations, *Ann. Univ. Ferrara Sez. VII - Sc. Mat.* **35** (1989), 147–154.

[13] T. Gramchev and P. Popivanov, *Partial Differential Equations: Approximate Solutions in Scales of Functional Spaces*, Wiley-VCH Verlag, Berlin, 2000.

[14] T. Gramchev and L. Rodino, Gevrey solvability for semilinear partial differential equations with multiple characteristics, *Boll. Un. Mat. Ital. B* **2** (1999), 65–120.

[15] C. Hunt and A. Piriou, Opérateurs pseudo-différentiels anisotropes d'ordre variable, *C. R. Acad. Sci. Paris* **268** (1969), 28–31.

[16] O. Liess and L. Rodino, Linear partial differential equations with multiple involutive characteristics, in *Microlocal Analysis and Spectral Theory*, Editor: L. Rodino, Kluwer Academic Publishers, 1997, 1–38.

[17] K. Kajitani and S. Wakabayashi, Hypoelliptic operators in Gevrey classes, in *Recent Developments in Hyperbolic Equations*: L.Cattabriga, F.Colombini, M.K.V. Murthy, Editor: S. Spagnolo, Longman, 1988, 115–134.

[18] M. Mascarello and L. Rodino, *Partial Differential Equations with Multiple Characteristics*, Wiley-VCH, Berlin, 1997.

[19] P. Popivanov, Local solvability of some classes of linear differential operators with multiple characteristics, *Ann. Univ. Ferrara, VII, Sc. Mat.* **45** (1999), 263–274.

[20] P. Popivanov and I. Iordanov, *Paradifferential operators and propagation of singularities for nonlinear PDE*, Akad. der Wissenschaften der DDR, Institut für Mathematik, 1983.

[21] P. Popivanov and G.S. Popov, Microlocal properties of a class of pseudo-differential operators with multiple characteristics, *Serdica* **6** (1980), 169–183.

[22] L. Rodino, *Linear Partial Differential Operators in Gevrey Spaces*, World Scientific, Singapore, 1993.

[23] S. Wakabayashi, Singularities of solution of the Cauchy problem for hyperbolic system in Gevrey classes, *Japanese J. Math.* **11** (1985), 157–201.

Giuseppe De Donno
Dipartimento di Matematica, Università di Torino, Via Carlo Alberto 10, 10123 Torino, Italy
e-mail: `dedonno@dm.unito.it`

Operator Theory:
Advances and Applications, Vol. 155, 91–106

Continuity in Weighted Sobolev Spaces of L^p Type for Pseudo-Differential Operators with Completely Nonsmooth Symbols

Gianluca Garello and Alessandro Morando

Abstract. We consider symbols $a(x,\xi)$ with a finite number of bounded derivatives with respect to ξ and of weighted Sobolev type in x. Their continuity in weighted Sobolev spaces of sufficiently large order is studied.

Mathematics Subject Classification (2000). Primary 35S05, 35S50.

Keywords. Weighted Sobolev spaces, Besov spaces, pseudo-differential operators.

1. Introduction

Pseudo-differential operators with nonsmooth symbols play an important role in the study of the local solvability and the regularity of solutions for general nonlinear partial differential equations, particularly from the microlocal point of view. As a starting point we consider the paper of M. Beals and M.C. Reeds, [1], 1984, where the symbols $a(x,\xi)$ were considered smooth with respect to the ξ variable and they belong to a Sobolev space of L^2 type, with respect to x. The techniques of pseudo-differential operators with nonsmooth symbols were introduced there as alternative to the arising paradifferential calculus of Bony [2], in the study of the general theory of nonlinear partial differential equations.
About the L^p continuity of the pseudo-differential operator with non-regular symbols, the works of J. Marschall [8], [9] are of great interest. He essentially considers symbol classes characterized by the estimate

$$\|\partial_\xi^\alpha a(\cdot,\xi)\|_{H^{r,p}} < C_\alpha(1+|\xi|)^{m-\rho|\alpha|+\delta r}, \quad \alpha\in\mathbb{Z}_+^n,\ \xi\in\mathbb{R}^n. \tag{1.1}$$

The respective pseudo-differential operators realize to be L^p bounded for $m=0$, when $0\le\delta<\rho=1$ and $r>0$ is suitably large. Once ρ becomes strictly less

The author is supported by F.I.R.B. grant of Italian Government.

than 1, new conditions must be added on the symbol $a(x,\xi)$, but at any rate the L^p continuity is assured only for operators of suitable negative order m.
Such a result is not surprising if we recall that also the operators in the classical Hörmander classes $S^m_{\rho,\delta}$ cannot be L^p bounded, $p \neq 2$, if $0 < \rho < 1$, whatever is $0 \leq \delta < \rho$, see the counterexample in C. Fefferman [4].
Arguing in the framework of the smooth classes $S^0_{\rho,0}$ let us notice that the L^p continuity may be proved under the following condition:

$$|\xi^\gamma \partial_\xi^{\alpha+\gamma} \partial_x^\beta a(x,\xi)| < C_{\alpha,\beta}(1+|\xi|)^{-\rho|\alpha|}, \quad \alpha,\beta \in \mathbb{Z}^n_+, \gamma \in \{0,1\}^n,$$

which essentially arise from the Marcinckiewicz-Lizorkin Lemma that reads:
The Fourier multiplier $p(D)$ is bounded from $L^p(\mathbb{R}^n)$ to itself, $1 < p < \infty$, provided that

$$|\xi^\gamma \partial^\gamma p(\xi)| < C, \quad \gamma \in \{0,1\}^n, \ \xi \in \mathbb{R}^n;$$

see M. Taylor, [[10], Ch XI, Prop. 4.5]. Arguing in this framework, in [7] we show the L^p continuity for a subclass of the pseudodifferential operators whose symbols satisfy (1.1).
In the present paper the emphasis is set on the symbols which are not smooth also in the ξ variable; namely, we consider $a(x,\xi)$ which satisfies a weighted Sobolev estimate with respect to the first variable and admits only a finite number of derivatives with respect to ξ.
The symbol classes are introduced in the next section, together with their expansions in elementary symbols obtained by means of a particular partition of unity due to Triebel [11]; some examples are also offered.
§3 is devoted to the characterization of the weighted Sobolev spaces by means of the weighted Besov and Triebel spaces as defined in [11], [12].
The main result is stated and proved in §4.

2. Symbol Classes

Definition 2.1. $\Lambda(\xi) \in C^\infty(\mathbb{R}^n)$ is a *weight function* if for some positive constants $C = C(\Lambda) > 0$, $\mu_0 \geq 1$, the following assumptions are satisfied, with standard multi-index notation:

1. $\Lambda(\xi) \geq \frac{1}{C}(1+|\xi|)^{\mu_0}, \qquad \xi \in \mathbb{R}^n;$
2. for every $\gamma \in \mathbb{Z}^n_+$:
$$\prod_{j=1}^n (1+\xi_j^2)^{\frac{\gamma_j}{2}} |\partial^\gamma \Lambda(\xi)| \leq C_\gamma \Lambda(\xi), \qquad \xi \in \mathbb{R}^n, \quad C_\gamma > 0;$$
3. $\Lambda(t\xi) \leq C\Lambda(\xi), \qquad t,\xi \in \mathbb{R}^n, \quad \max_{1\leq j\leq n} |t_j| \leq 1, \quad t\xi := (t_1\xi_1, ..., t_n\xi_n);$
4. *(δ-condition)* for some $0 < \delta < 1$:
$$\Lambda(\xi) \leq C\left(\Lambda(\eta) + \Lambda(\xi-\eta) + \Lambda(\eta)^\delta \Lambda(\xi-\eta)^\delta\right), \quad \xi,\eta \in \mathbb{R}^n. \qquad (2.1)$$

Remark 2.2. As shown in Triebel [[11], Lemma 2.1/2] we can always find $\mu_1 \geq \mu_0$ such that $\Lambda(\xi) < C(1+|\xi|)^{\mu_1}$.

Example. The basic examples of weight functions are the *elliptic weights* $\Lambda_m(\xi) := \sqrt{1+\sum_{j=1}^n \xi_j^{2m}}$. Of greater interest are the *multi-quasi-elliptic* weights defined as $\Lambda_{\mathcal{P}}(\xi) := \sqrt{1+\sum_{\alpha\in\mathcal{V}(\mathcal{P})} \xi^{2\alpha}}$ where $\mathcal{V}(\mathcal{P})$ are the vertices of a *complete Newton polyhedron*, see for example [5]. Other examples as $\langle\xi\rangle^s[\log(2+\langle\xi\rangle)]^t$, where $s,t>1$, $\langle\xi\rangle = \sqrt{1+|\xi|^2}$, are given in Triebel [11].

In the whole paper $\Lambda(\xi)$ will be a weight function and X a generic Banach space with norm $\|.\|$.

Definition 2.3. For any nonnegative integer N and $m \in \mathbb{R}$, we define $XM_\Lambda^m(N)$ as the class of measurable functions $a(x,\xi)$ on $\mathbb{R}^{2n}$ such that

$$\begin{aligned} &\textstyle\prod_{j=1}^n (1+\xi_j^2)^{\frac{\gamma_j}{2}} |\partial_\xi^\gamma a(x,\xi)| \leq C_N \Lambda(\xi)^m, \qquad |\gamma| \leq N, \qquad x,\xi \in \mathbb{R}^n;\\ &\textstyle\prod_{j=1}^n (1+\xi_j^2)^{\frac{\gamma_j}{2}} \|\partial_\xi^\gamma a(\cdot,\xi)\| \leq C_N \Lambda(\xi)^m, \qquad |\gamma| \leq N, \qquad \xi \in \mathbb{R}^n. \end{aligned} \tag{2.2}$$

For a fixed $H>1$ let us consider the decomposition of $\mathbb{R}^n$, given by the sequence of n-intervals $\{P_{h,\lambda}^{(H)}\}$, $h \in \mathbb{Z}_+^n$, $\lambda \in \mathbb{E} = \{-1,1\}^n$, described by

$$P_{h,\lambda}^{(H)} := \left\{ \xi \in \mathbb{R}^n \quad ; \quad \frac{1}{H} 2^{h_j} \eta_{h_j} \leq \lambda_j \xi_j \leq H 2^{h_j+1} \quad ; \quad j=1,\ldots,n \right\},$$

where $\eta_h = -1$ if $h=0$ and $\eta_h = 1$ if $h>0$.

Remark 2.4. For a suitable $N_0 = N_0(H)$ it is not very difficult to verify that, for any $\lambda,\epsilon \in \mathbb{E}$, $P_{h,\lambda}^{(H)} \cap P_{k,\varepsilon}^{(H)} = \emptyset$ when $|h_j - k_j| > N_0$ for some $j=1,\ldots,n$; see [7] and [11].

Definition 2.5 (Partition of unity). For a fixed $H>1$, $\phi^{(H)}$ is the set of all the sequences $\{\varphi_{h,\lambda}\} \subset C_0^\infty$ such that, for every $h \in \mathbb{Z}_+^n$, $\lambda \in \mathbb{E}$, supp $\varphi_{h,\lambda} \subset P_{h,\lambda}^{(H)}$, $\sum_{h\in\mathbb{Z}_+^n,\lambda\in\mathbb{E}} \varphi_{h,\lambda}(\xi) = 1$ and, for any $\alpha \in \mathbb{Z}_+^n$ there exists a positive constant C_α such that

$$|\partial_\xi^\alpha \varphi_{h,\lambda}(\xi)| \leq C_\alpha 2^{-h\cdot\alpha}, \qquad \xi \in \mathbb{R}^n, \qquad h \in \mathbb{Z}_+^n, \qquad \lambda \in \mathbb{E}.$$

Remark 2.6. In the remaining part of the present paper we will always consider $h = (h_1,\ldots,h_n) \in \mathbb{Z}_+^n$ and $\lambda = (\lambda_1,\ldots,\lambda_n) \in \mathbb{E} = \{-1,1\}^n$.

Definition 2.7. For any integer $N \geq 0$, we define $XM_E(N)$ as the class of all the expansions

$$\sigma(x,\xi) := \sum_{h,\lambda} d_{h,\lambda}(x) \psi_{h,\lambda}(\xi), \tag{2.3}$$

whose terms $d_{h,\lambda} \in L^\infty(\mathbb{R}^n) \cap X$ and $\psi_{h,\lambda} \in C_0^\infty(\mathbb{R}^n)$ satisfy for some $M>0$, $H>1$ and $C>0$

$$\|d_{h,\lambda}\|_\infty < M; \qquad \|d_{h,\lambda}\| < M; \qquad \operatorname{supp} \psi_{h,\lambda} \subset P_{h,\lambda}^{(H)};$$

$$|\partial^\alpha \psi_{h,\lambda}(\xi)| < C2^{-h\cdot\alpha}, \quad \text{for any } \xi \in \mathbb{R}^n \quad \text{and} \quad |\alpha| \leq N.$$

We say that $\sigma(x,\xi)$ is an *elementary symbol.*

Remark 2.8. The expansion in (2.3) is trivially convergent since for any fixed $\xi \in \mathbb{R}^n$ all but a finite number of terms vanish. Moreover $XM_E(N) \subset XM^0_\Lambda(N)$, for any weight function $\Lambda(\xi)$, if $N \geq 0$.

Proposition 2.9. *Provided that $N \geq n+1$, any symbol $a(x,\xi) \in XM^0_\Lambda(N)$ may be written as an expansion of elementary symbols $a_m(x,\xi) \in XM_E(N-n-1)$, $m \in \mathbb{Z}^n$, in the following way:*

$$a(x,\xi) = \sum_{m\in\mathbb{Z}^n} \frac{1}{(1+|m|)^{n+1}} a_m(x,\xi),$$

where $|m| = |m_1| + \cdots + |m_n|$ and the expansion is absolutely convergent in $L^\infty(\mathbb{R}^n_x \times \mathbb{R}^n_\xi)$.

By means of a simple application of the Leibnitz rule and (2.2) we obtain the following

Lemma 2.10. *For $\{\varphi_{h,\lambda}\} \in \Phi^{(H)}$ and $a(x,\xi) \in XM^0_\Lambda(N)$, $N \geq 0$, let us set*

$$a_{h,\lambda}(x,\xi) := \varphi_{h,\lambda}(\xi) a(x,\xi). \tag{2.4}$$

Then for any $x, \xi \in \mathbb{R}^n$:

$$a_{h,\lambda}(x,\xi) = 0 \qquad \textit{for} \quad \xi \notin P^{(H)}_{h,\lambda};$$

$$|\partial^\alpha_\xi a_{h,\lambda}(x,\xi)| \leq C2^{-\alpha\cdot h}, \quad \|\partial^\alpha_\xi a_{h,\lambda}(\cdot,\xi)\| \leq C2^{-\alpha\cdot h}, \qquad |\alpha| \leq N.$$

Lemma 2.11. *For any $N \in \mathbb{N}$, $\phi \in C^N_0(\mathbb{R}^n)$ and $m \in \mathbb{Z}^n$ we can find a linear partial differential operator $\mathcal{M}(D)$ of order N such that:*

$$(1+|m|)^N \int \phi(x) e^{-im\cdot x}\, dx = \int e^{-im\cdot x} \mathcal{M}(D)\phi(x)\, dx.$$

Proof. Let us assume, without loss of generality, that the negative components of $m \in \mathbb{Z}^n$ are $m_{j_1}, \dots, m_{j_i}$, $i \leq n$. Then setting $\epsilon_{m,\alpha} := (-1)^{\alpha_{j_1}+\cdots+\alpha_{j_i}}$ for $\alpha := (\alpha_1,\dots,\alpha_n) \in \mathbb{Z}^n_+$ and using the multi-index notation we have:

$$\begin{aligned}(1+|m|)^N &= \textstyle\sum_{k=0}^N \binom{N}{k} \sum_{\alpha_1+\cdots+\alpha_n=k} \frac{k!}{\alpha_1!\dots\alpha_n!} |m_1|^{\alpha_1}\dots|m_n|^{\alpha_n} \\ &= \textstyle\sum_{|\alpha|\leq N} \binom{N}{|\alpha|} \frac{|\alpha|!}{\alpha!} \epsilon_{m,\alpha} m^\alpha.\end{aligned}$$

Integrating by parts we have for $(1+|m|)^N \int \phi(x) e^{-im\cdot x}\, dx$ the following expression :

$$\sum_{|\alpha|\leq N} \binom{N}{|\alpha|} \frac{|\alpha|!}{\alpha!} \epsilon_{m,\alpha} \int e^{-im\cdot x} D^\alpha_x \phi(x) dx.$$

Then by setting $\mathcal{M}(D) = \sum_{|\alpha|\leq N} \binom{N}{|\alpha|} \frac{|\alpha|!}{\alpha!} \epsilon_{m,\alpha} D^\alpha_x$ we end the proof. □

Proof of Proposition 2.9. By means of a partition of unity $\{\varphi_{h,\lambda}\} \in \Phi^{(H)}$ and (2.4) we have:

$$a(x,\xi) = \sum_{h,\lambda} a_{h,\lambda}(x,\xi), \qquad x,\xi \in \mathbb{R}^n. \tag{2.5}$$

Let us set now

$$b_{h,\lambda}(x,\xi) := a_{h,\lambda}(x, 2^{h_1+\theta_{h_1}}\xi_1 + \lambda_1 c_{h_1}, ..., 2^{h_n+\theta_{h_n}}\xi_n + \lambda_n c_{h_n}),$$

where $\theta_0 = 1$, $\theta_t = 0$ when $t \neq 0$; $c_0 = 1$, $c_t = \frac{2}{3}2^t$ when $t \neq 0$. For some $K = K(H)$ and any $x, \xi \in \mathbb{R}^n$ it follows from Lemma 2.10 that:

$$\begin{aligned} &b_{h,\lambda}(x,\xi) = 0, \text{ when } \xi \notin Q_1 := \left[-\tfrac{K}{2}, \tfrac{K}{2}\right]^n; \\ &|\partial_\xi^\alpha b_{h,\lambda}(x,\xi)| \leq C_N, \qquad \|\partial_\xi^\alpha b_{h,\lambda}(.,\xi)\| \leq C_N, \qquad |\alpha| \leq N. \end{aligned} \tag{2.6}$$

Arguing now as in Coifman-Meyer [[3], Ch. II, §9], we can choose the constant $K > 1$ so that $Q_1 \subset [-\pi,\pi]^n$ and set for any $h \in \mathbb{Z}_+^n$ and $\lambda \in \mathbb{E}$

$$B_{h,\lambda}(x,\xi) := \sum_{m \in \mathbb{Z}^n} b_{h,\lambda}(x, \xi - 2m\pi), \qquad x,\xi \in \mathbb{R}^n,$$

where for fixed $x, \xi \in \mathbb{R}^n$ all the terms but one in the expansion vanish. Then $B_{h,\lambda}(x,\xi)$ is the 2π-periodic function in the ξ variable obtained by extending $b_{h,\lambda}(x,\xi)$ on each n-cube of $\mathbb{R}^n_\xi$ with sides of length 2π.
Moreover if $\phi(\xi) \in C_0^\infty(\mathbb{R}^n)$ vanishes outside $[-\pi,\pi]^n$ and it is identically equal to 1 on Q_1, we have

$$b_{h,\lambda}(x,\xi) = \phi(\xi) B_{h,\lambda}(x,\xi), \qquad x,\xi \in \mathbb{R}^n. \tag{2.7}$$

We can write $B_{h,\lambda}(x,\xi)$ in terms of its Fourier expansion:

$$B_{h,\lambda}(x,\xi) = \sum_{m \in \mathbb{Z}^n} e^{im\cdot\xi} \int_{[-\pi,\pi]^n} b_{h,\lambda}(x,\eta) e^{-im\cdot\eta}\, d\eta,$$

with convergence in $L^2([-\pi,\pi]^n)$ with respect to ξ.
Since for fixed $x \in \mathbb{R}^n$, $b_{h,\lambda}(x,\cdot)$ satisfies the conditions of Lemma 2.11, we can write

$$B_{h,\lambda}(x,\xi) = \sum_{m \in \mathbb{Z}^n} \frac{1}{(1+|m|)^N} e^{im\cdot\xi} d^m_{h,\lambda}(x), \tag{2.8}$$

where for some $\mathcal{M}(D)$ of order N:

$$d^m_{h,\lambda}(x) = \int e^{-im\cdot\eta} \mathcal{M}(D_\eta) b_{h,\lambda}(x,\eta)\, d\eta.$$

In view of (2.6), for some positive constant C depending only on N and the dimension n, it yields for every $m \in \mathbb{Z}^n$:

$$\|d^m_{h,\lambda}\|_\infty \leq C \quad ; \quad \|d^m_{h,\lambda}\| \leq C. \tag{2.9}$$

For any $m \in \mathbb{Z}^n$ let us set $\phi_m(\xi) := (1+|m|)^{-(N-n-1)} e^{im\cdot\xi}\phi(\xi)$, then it follows from (2.7) and (2.8):

$$b_{h,\lambda}(x,\xi) = \sum_{m\in\mathbb{Z}^n} \frac{1}{(1+|m|)^{n+1}} \phi_m(\xi) d^m_{h,\lambda}(x), \qquad x,\xi \in \mathbb{R}^n,$$

with absolute convergence in $L^\infty\left(\mathbb{R}^{2n}\right)$.
By the change of variables $\zeta_j = 2^{h_j+\theta_{h_j}}\xi_j + \lambda_j c_{h_j}$, $j = 1, ..., n$, where θ_{h_j} and c_{h_j} are defined above, we get:

$$a_{h,\lambda}(x,\zeta) = \sum_{m\in\mathbb{Z}^n} \frac{1}{(1+|m|)^{n+1}} d^m_{h,\lambda}(x)\psi_{m,h,\lambda}(\zeta), \quad x,\zeta \in \mathbb{R}^n, \tag{2.10}$$

where $\psi_{m,h,\lambda}(\zeta) = \phi_m(2^{-h_1-\theta_{h_1}}(\zeta_1 - \lambda_1 c_{h_1}), ..., 2^{-h_1-\theta_{h_1}}(\zeta_1 - \lambda_1 c_{h_1}))$ and the expansion is absolutely convergent in $L^\infty(\mathbb{R}^{2n})$.
We can then find $H' > 1$ and a positive constant C_N only depending on N and the dimension n such that for any $m \in \mathbb{Z}^n$:

$$\begin{aligned} &\operatorname{supp}\psi_{m,h,\lambda} \subset P^{(H')}_{h,\lambda}; \\ &|\partial^\alpha_\xi \psi_{m,h,\lambda}(\xi)| \le C_N 2^{-h\cdot\alpha}, \quad |\alpha| \le N-n-1,\ \xi \in \mathbb{R}^n. \end{aligned} \tag{2.11}$$

Re placing now in (2.5) the expansion (2.10) we obtain for any $x, \xi \in \mathbb{R}^n$:

$$a(x,\xi) = \sum_{h\in\mathbb{Z}^n_+,\lambda\in\mathbb{E}} \sum_{m\in\mathbb{Z}^n} \frac{1}{(1+|m|)^{n+1}} d^m_{h,\lambda}(x)\psi_{m,h,\lambda}(\xi), \tag{2.12}$$

which in view of (2.9) and (2.11) realizes to be absolutely convergent in $L^\infty(\mathbb{R}^{2n})$
Thus, by changing the order of the two sums in (2.12), we can write

$$a(x,\xi) = \sum_{m\in\mathbb{Z}^n} \frac{1}{(1+|m|)^{n+1}} a_m(x,\xi), \qquad x,\xi \in \mathbb{R}^n.$$

Thanks once again to (2.9) and (2.11) we can conclude that the expansion is absolutely convergent in $L^\infty(\mathbb{R}^{2n})$; moreover for every $m \in \mathbb{Z}^n$:

$$a_m(x,\xi) = \sum_{h,\lambda} d^m_{h,\lambda}(x)\psi_{m,h,\lambda}(\xi) \in XM_E(N-n-1). \tag{2.13}$$

□

In the last part of this section we are interested in finding some symbols which satisfy (2.2) for an effective finite number of derivatives.

Proposition 2.12. *Let us consider a measurable function $\Psi : \mathbb{R}^n_x \times \mathbb{R}_t \mapsto \mathbb{C}$, which admits N derivatives with respect to t, it belongs to a Banach space X with respect to x and it satisfies for some positive constant C_k:*

$$|\partial^k_t \Psi(x,t)| \le C_k(1+|t|)^{-k}, \quad k = 0, \dots, N \quad (x,t) \in \mathbb{R}^{n+1}; \tag{2.14}$$

$$\|\partial^k_t \Psi(\cdot,t)\| \le C_k(1+|t|)^{-k}, \quad k = 0, \dots, N \quad t \in \mathbb{R}. \tag{2.15}$$

For any weight function $\Lambda(\xi)$ *it follows:*

$$a(x,\xi) = \Psi\left(x, \prod_{j=1}^{n}(1+\xi_j^2)^{\frac{1}{2}}\right) \in XM_\Lambda^0(N).$$

Proof. For any fixed $x \in \mathbb{R}^n$ let us set $\psi(t) = \Psi(x,t)$. Provided that $|\alpha| \le N$, by derivation of the composite functions we have for $\partial_\xi^\alpha a(x,\xi)$ the following expression:

$$\sum_{k=1}^{|\alpha|} \psi^{(k)}\left(\prod_{j=1}^{n}(1+\xi_j^2)^{\frac{1}{2}}\right) \sum_{\substack{\alpha^1+\dots+\alpha^k=\alpha \\ \alpha^J>0, J=1,\dots,k}} C_{\alpha,k,\alpha^J} \prod_{J=1}^{k}\left[\partial_\xi^{\alpha^J} \prod (1+\xi_j^2)^{\frac{1}{2}}\right]. \qquad (2.16)$$

From (2.14) it follows

$$|\psi^{(k)}\left(\prod_{j=1}^{n}(1+\xi_j^2)^{\frac{1}{2}}\right)| \le C_k \prod_{j=1}^{n}(1+\xi_j^2)^{-\frac{k}{2}}, \qquad (2.17)$$

and for any $J = 1,\dots,k$:

$$|\partial_\xi^{\alpha^J} \prod_{j=1}^{n}(1+\xi_j^2)^{\frac{1}{2}}| = \prod_{j=1}^{n} |\frac{d^{\alpha_j^J}}{d\xi_j^{\alpha_j^J}}(1+\xi_j^2)^{\frac{1}{2}}| \le C_{\alpha^J} \prod_{j=1}^{n}(1+\xi_j^2)^{\frac{1}{2}-\frac{\alpha_j^J}{2}}. \qquad (2.18)$$

Since $\alpha^1 + \dots + \alpha^k = \alpha$, and $|\alpha| \le N$, from (2.16)-(2.18) it directly follows that $a(x,\xi) \in XM_\Lambda^0(N)$. □

Example. Let us consider $z(x) \in L^\infty(\mathbb{R}^n) \cap Y$, where Y is a Banach algebra for the pointwise multiplication, with respect to the norm $\|\cdot\| = \|\cdot\|_Y$. Let moreover $\phi : [0,\infty) \mapsto \mathbb{C}$ be a N-times differentiable function which satisfies for a positive constant C, depending only on ϕ and N:

$$\left|\phi^{(k)}(t)\right| < C(1+t)^{-k}, \quad k = 0,1,\dots,N. \qquad (2.19)$$

Set now

$$\Xi(x,t) = \sum_{r=1}^{\infty} \frac{1}{r!} z^r(x)\phi\left(\frac{t}{r!}\right). \qquad (2.20)$$

Since for any $k = 0,1,\dots,N$ and $(x,t) \in \mathbb{R}^{n+1}$

$$\left|\frac{d^k}{dt^k}\left[z(x)^r\phi\left(\frac{t}{r!}\right)\right]\right| < \frac{1}{(r!)^k}\left|\phi^{(k)}\left(\frac{t}{r!}\right)\right| |z(x)|^r < C_k(1+t)^{-k}|z(x)|^r, \qquad (2.21)$$

and in the same way

$$\left\|\frac{d^k}{dt^k}\left[z(\cdot)^r\phi\left(\frac{t}{r!}\right)\right]\right\| < \frac{1}{(r!)^k}\left|\phi^{(k)}\left(\frac{t}{r!}\right)\right| \|z(\cdot)\|^r < C_k(1+t)^{-k}\|z(\cdot)\|^r, \qquad (2.22)$$

the expansion in (2.20) is absolutely and uniformly convergent both in the norms $\|\cdot\|_{L^\infty(\mathbb{R}^{n+1})}$ and $\|\cdot\|$.

We can then conclude that $\Xi(t,x)$ is N-times differentiable with bounded derivatives with respect to t, uniformly with respect to x and, thanks to (2.21), (2.22), it satisfies the conditions in (2.14), (2.15).

Remark 2.13. For some $0<\rho<1$ let us consider the smooth function on $\mathbb{R}_+$: $\phi(t)=(t+1)^{-1}I_\rho(t)$, where $I_\rho(t)=\int_0^t[\cos(1+\tau)^{1-\rho}+2]\,d\tau$. Since

$$\frac{d}{dt}\phi(t)=-(1+t)^{-2}I_\rho(t)+(1+t)^{-1}\left[\cos(1+t)^{1-\rho}+2\right] \tag{2.23}$$

and

$$\frac{d^2}{dt^2}\phi(t)=\frac{2}{(1+t)^3}I_\rho(t)-\frac{2}{(1+t)^2}\left[\cos(1+t)^{1-\rho}+2\right]-\frac{1-\rho}{(1+t)^{1+\rho}}\sin(1+t)^{1-\rho}, \tag{2.24}$$

it easily follows that $\phi(t)$ satisfies the assumption in (2.19) for $k=0,1$.
On the other hand, if we choose now a sequence $\{t_k\}_{k=1}^\infty$ such that $(1+t_k)^{1-\rho}=\frac{3\pi}{2}+2k\pi$, it is not difficult to verify that

$$\lim_{k\to+\infty}t_k^2\frac{d^2}{dt^2}\phi(t_k)=+\infty.$$

Then $\phi(t)$ does not satisfy (2.19) for $k=2$.
At the end we can explicitly consider a symbol which does not satisfy (2.2) for derivatives of order greater or equal than 2. For the sake of simplicity let us argue now in $\mathbb{R}^2_\xi$ and set:

$$\Phi(\xi_1,\xi_2)=\phi\left((1+\xi_1^2)^{\frac{1}{2}}(1+\xi_2^2)^{\frac{1}{2}}\right).$$

By restriction on the ξ_1 axis, we have:

$$\partial^2_{\xi_1}\Phi(t,0)=\frac{d^2}{dt^2}\phi(\langle t\rangle)\left(\frac{d}{dt}\langle t\rangle\right)^2+\frac{d}{dt}\phi(\langle t\rangle)\frac{d^2}{dt^2}\langle t\rangle,$$

where $\langle t\rangle=(1+t^2)^{\frac{1}{2}}$. Then from (2.23), (2.24) we obtain

$$(1+t^2)\partial^2_{\xi_1}\Phi(t,0)\geq\varphi_1(t)+\varphi_2(t),$$

where

$$\varphi_1(t)=\frac{1-2\langle t\rangle}{\langle t\rangle^3(1+\langle t\rangle)^2}-\frac{6t^2}{(1+\langle t\rangle)^2}+\frac{2t^2\langle t\rangle}{(1+\langle t\rangle)^3}$$

is bounded, while the term

$$\varphi_2(t)=-\frac{(1-\rho)t^2}{(1+\langle t\rangle)^{1+\rho}}\sin\left((1+\langle t\rangle)^{1-\rho}\right)$$

is not bounded since $0<\rho<1$.
We can then conclude that $(1+\xi_1^2)\partial^2_{\xi_1}\Phi(\xi_1,\xi_2)$ is not bounded and then the symbol $\Psi(x,\xi)=z(x_1,x_2)\phi(\xi_1,\xi_2)$, with $z(x_1,x_2)\in L^\infty(\mathbb{R}^2)\cap X$, does not satisfy (2.2) for derivatives of order $N\geq 2$.

3. Characterization of Weighted Sobolev Spaces

From this moment we will write $\mathcal{F}_{x\to\xi}u(\xi) = \hat{u}(\xi)$ for the usual Fourier transform of a distribution $u \in \mathcal{S}'(\mathbb{R}^n)$ and $\mathcal{F}^{-1}_{\xi\to x}$ for its inverse transform.
Moreover for any given function $m(\xi)$ on $\mathbb{R}^n$ and $u \in \mathcal{S}'(\mathbb{R}^n)$ we set $m(D)u(x) = \mathcal{F}^{-1}_{\xi\to x}(m(\xi)\hat{u}(\xi))$, provided that the involved expressions make sense.
Let us consider now $\{\varphi_{h,\lambda}\} \in \phi^{(H)}$, $H > 1$, $1 < p < \infty$, $s \in \mathbb{R}$; we can then define, for any weight function $\Lambda(\xi)$, the following norms:

$$\|u\|_{B^{s,\Lambda}_{p,q}} := \|\{\Lambda(c^{(H)}_{h,\lambda})^s u_{h,\lambda}\}\|_{\ell^q(L^p)} := \left(\sum_{h,\lambda} \|\Lambda(c^{(H)}_{h,\lambda})^s u_{h,\lambda}\|^q_p\right)^{\frac{1}{q}}, \tag{3.1}$$

for $1 \leq q < \infty$, (modification for $q = \infty$);

$$\|u\|_{F^{s,\Lambda}_{p,q}} := \|\{\Lambda(c^{(H)}_{h,\lambda})^s u_{h,\lambda}\}\|_{L^p(\ell^q)} := \left\|\left(\sum_{h,\lambda} |\Lambda(c^{(H)}_{h,\lambda})^s u_{h,\lambda}|^q\right)^{\frac{1}{q}}\right\|_p, \tag{3.2}$$

for $1 < q < \infty$.
Here and later on, for any $u \in \mathcal{S}'(\mathbb{R}^n)$, $u_{h,\lambda} = \varphi_{h,\lambda}(D)u$ and $c^{(H)}_{h,\lambda}$ is the center of the n-interval $P^{(H)}_{h,\lambda}$.
We denote by $B^{s,\Lambda}_{p,q}$ and $F^{s,\Lambda}_{p,q}$ the Banach spaces given by the Schwartz distributions whose norms in (3.1) and (3.2) are respectively bounded.
It may be shown that for different choices of the systems $\{\varphi_{h,\lambda}(\xi)\} \in \Phi^{(H)}$ the norms in (3.1), and (3.2) are equivalent.
If we introduce now for any weight function Λ, $1 < p < \infty$ and $s \in \mathbb{R}$, the *weighted Sobolev space*:

$$H^{s,p}_\Lambda := \{u \in \mathcal{S}'(\mathbb{R}^n) \quad ; \quad \Lambda^s(D)u \in L^p\},$$

equipped with the natural norm $\|u\|_{H^{s,p}_\Lambda} := \|\Lambda^s(D)u\|_{L^p}$, it follows that

$$H^{s,p}_\Lambda = F^{s,\Lambda}_{p,2}, \quad 1 < p < \infty \tag{3.3}$$

and moreover for some positive constants c_1, c_2:

$$c_1\|u\|_{F^{s,\Lambda}_{p,2}} \leq \|u\|_{H^{s,p}_\Lambda} \leq c_2\|u\|_{F^{s,\Lambda}_{p,2}}. \tag{3.4}$$

For the detailed proofs and any comments the reader can see Triebel [11].

We conclude this section with some embeddings and technical properties which realize to be useful in the following.

Proposition 3.1. *Let $\Lambda(\xi)$ be a weight function, $1 < p < \infty$ and $s \in \mathbb{R}$. Then the following inclusions hold with continuous embedding:*

$$B^{s,\Lambda}_{p,\min(p,q)} \subset F^{s,\Lambda}_{p,q} \subset B^{s,\Lambda}_{p,\max(p,q)}, \quad \textit{if } 1 < q < \infty; \tag{3.5}$$

$$\mathcal{S}(\mathbb{R}^n) \subset B^{s,\Lambda}_{p,q_1} \subset B^{s,\Lambda}_{p,q_2} \subset \mathcal{S}'(\mathbb{R}^n), \quad \textit{if } 1 \le q_1 < q_2 \le \infty; \tag{3.6}$$

$$\mathcal{S}(\mathbb{R}^n) \subset F^{s,\Lambda}_{p,q_1} \subset F^{s,\Lambda}_{p,q_2} \subset \mathcal{S}'(\mathbb{R}^n), \quad \textit{if } 1 < q_1 < q_2 < \infty. \tag{3.7}$$

Moreover for any $\epsilon > 0$, we have:

$$B^{s+\epsilon,\Lambda}_{p,q_1} \subset B^{s,\Lambda}_{p,q_2}, \quad \textit{if } 1 \le q_1, q_2 \le \infty; \tag{3.8}$$

$$F^{s+\epsilon,\Lambda}_{p,q_1} \subset F^{s,\Lambda}_{p,q_2}, \quad \textit{if } 1 < q_1, q_2 < \infty. \tag{3.9}$$

Proof. The inclusions (3.5)-(3.7) are proved in Triebel [[13], Theorem 6.1/2]. In order to prove (3.8), (3.9) let us remark that for any sequence $\{b_{h,\lambda}\}$ of positive numbers , $1 \le q_1, q_2 < \infty$ and $H > 1$, the following estimate holds:

$$\begin{aligned} &\left(\sum_{h,\lambda} \Lambda(c^{(H)}_{h,\lambda})^{sq_2} b^{q_2}_{h,\lambda}\right)^{\frac{1}{q_2}} \le A_{q_2,\epsilon} \sup_{h,\lambda} \Lambda(c^{(H)}_{h,\lambda})^{s+\epsilon} b_{h,\lambda} \\ &\le A_{q_2,\epsilon} C_{q_1} \left(\sum_{h,\lambda} \Lambda(c^{(H)}_{h,\lambda})^{(s+\epsilon)q_1} b^{q_1}_{h,\lambda}\right)^{\frac{1}{q_1}} \end{aligned} \tag{3.10}$$

where $A_{q_2,\epsilon} := \left(\sum_{h,\lambda} \Lambda(c^{(H)}_{h,\lambda})^{-\epsilon q_2}\right)^{\frac{1}{q_2}}$ is finite. In order to show the convergence of such a series let us notice that the center $c^{(H)}_{h,\lambda}$ of the cube $P^{(H)}_{h,\lambda}$ has coordinates $c^{(H)}_{h,\lambda} = (\lambda_1 c_{1,H} 2^{h_1}; \dots; \lambda_n c_{n,H} 2^{h_n})$ and the numbers $c_{j,H}$, $1 \le j \le n$ are equal to either $H + \frac{1}{2H}$ or $H - \frac{1}{2H}$.
It follows from the assumption 1. in Definition 2.1 that there exists a positive constant C_H such that $\Lambda(c^{(H)}_{h,\lambda}) \ge C_H\left(1 + \sum_{l=1}^n 2^{h_l}\right)^{\mu_0}$. Thus we obtain

$$\begin{aligned} \sum_{h,\lambda} \Lambda(c^{(H)}_{h,\lambda})^{-\epsilon q_2} &\le C'_{H,\epsilon,q_2} \sum_\lambda \sum_h \frac{1}{\left(1+\sum_{l=1}^n 2^{h_l}\right)^{\mu_0 \epsilon q_2}} \\ &\le C'_{H,\epsilon,q_2} 2^n \sum_{h_1=0}^\infty \frac{1}{(2^{h_1})^{\mu_0\epsilon q_2}} \cdots \sum_{h_n=0}^\infty \frac{1}{(2^{h_n})^{\mu_0\epsilon q_2}}, \end{aligned}$$

where $\sum_{h_j=0}^\infty \frac{1}{(2^{h_j})^{\mu_0\epsilon q_2}} < \infty$. We immediately get inclusion (3.8) by setting $b_{h,\lambda} = \|u_{h,\lambda}\|_p$, while inclusion (3.9) follows by setting $b_{h,\lambda} = |u_{h,\lambda}(x)|$ and taking the L^p norm of the first and last sides of (3.10). □

Remark 3.2. The characterization of Sobolev spaces given by (3.3) and (3.4), jointly with the continuous embeddings in (3.5), gives

$$B^{s,\Lambda}_{p,\min(p,2)} \subset H^s_{p,\Lambda} \subset B^{s,\Lambda}_{p,\max(p,2)}, \quad \text{with continuous embedding.}$$

The following Lemma is a slight modification of Nikol'skij inequality given in Triebel [13].

Lemma 3.3. *Let us consider* $\alpha \in \mathbb{Z}_+^n$, $1 \le p \le q \le \infty$ *and* $H>1$. *Then there exists a constant* $C_\alpha > 0$, *only depending on* α, p, q, n *and* H *such that*

$$\|D^\alpha f\|_q \le C_\alpha 2^{h\cdot\alpha + \left(\frac{1}{p}-\frac{1}{q}\right)|h|}\|f\|_p, \tag{3.11}$$

for every function $f \in L^p(\mathbb{R}^n)$ *such that supp* $\hat{f} \subset P_{h,\lambda}^{(H)}$.

Proof. For the sake of simplicity we can assume that $h_j > 0$ for all j. Let us define the function $g_h(x) := (2^{-h_1}x_1, \dots, 2^{-h_n}x_n)$. From well-known properties of the Fourier transform, it follows that $\hat{g_h}(\xi) = 2^{|h|}\hat{f}(2^{h_1}\xi_1, \dots, 2^{h_n}\xi_n)$, $\xi \in \mathbb{R}^n$; thus supp $\hat{g_h} \subset P^{(H)} := [-2H, 2H]^n$
From the Nikol'skij inequalities we know there exists a constant C_α, only depending on α, p, q, n and the compact $P^{(H)}$ such that:

$$\|D^\alpha g_h\|_q \le C_\alpha \|g_h\|_p. \tag{3.12}$$

But $D^\alpha g_h(x) = 2^{-h\cdot\alpha}(D^\alpha f)(2^{-h_1}x_1, \dots, 2^{-h_n}x_n)$, whence $\|g_h\|_p = 2^{|h|\frac{1}{p}}\|f\|_p$ and $\| D^\alpha g_h\|_q = 2^{-h\cdot\alpha}2^{|h|\frac{1}{q}}\|D^\alpha f\|_q$. Thus the inequality (3.11) follows from (3.12) by re placing the previous expressions for $\|D^\alpha g_h\|_q$ and $\|g_h\|_p$. □

Proposition 3.4. *For* $1 < p < \infty$ *and* $s \in \mathbb{R}$ *we can find a constant* $M = M(p, n, s, H)$, *such that:*

$$\|u_{h,\lambda}\|_\infty < M\|u\|_{H_\Lambda^{s,p}}\Lambda(C_{h,\lambda}^{(H)})^{-s}2^{\frac{|h|}{p}}, \qquad u \in \mathcal{S}'(\mathbb{R}^n). \tag{3.13}$$

Proof. From Proposition 3.1 it follows that $H_\Lambda^{s,p} \subset B_{p,\max(p,2)}^{s,\Lambda} \subset B_{p,\infty}^{s,\Lambda}$ with continuous embedding; this just means that, given a system $\{\varphi_{h,\lambda}\} \in \Phi^{(H)}$, any $u \in H_\Lambda^{s,p}$ fulfills the following estimates:

$$\|u_{h,\lambda}\|_p \le M'\|u\|_{H_\Lambda^{s,p}}\Lambda(c_{h,\lambda}^{(H)})^{-s}, \quad h \in \mathbb{Z}_+^n, \lambda \in \mathbb{E},$$

with some positive M' depending on p, H, s, the dimension n and independent of u. To get estimates (3.13), it suffices now to apply Lemma 3.3 with $q = \infty$, $\alpha = 0$ and $f(x) = u_{h,\lambda}(x)$.
These estimates are clearly trivial when the distribution u does not belong to $H_\Lambda^{s,p}$. □

4. Continuity in Weighted Sobolev Spaces

For any weight function $\Lambda(\xi)$ and $a(x,\xi) \in XM_\Lambda^m(N)$, $m \in \mathbb{R}$ we can define as usual the pseudodifferential operator:

$$a(x,D)u(x) := (2\pi)^{-n}\int e^{ix\cdot\xi}a(x,\xi)\hat{u}(\xi)d\xi, \qquad u \in \mathcal{S}(\mathbb{R}^n).$$

This section is devoted to prove the following continuity result for pseudo-differential operators whose symbol belongs to the class $H_\Lambda^{r,p}M_\Lambda^m(N)$:

Theorem 4.1. *For any weight function $\Lambda(\xi)$, let $a(x,\xi)$ be a symbol in $H^{r,p}_\Lambda M^m_\Lambda(N)$ with $r > \frac{n}{(1-\delta)\mu_0 p}$, $N \geq 2n+1$, $1<p<\infty$ and $m \in \mathbb{R}$. Then:*

$$a(x,D) : H^{s+m,p}_\Lambda \longmapsto H^{s,p}_\Lambda, \quad \textit{continuously for any } 0 \leq s \leq r.$$

In order to prove Theorem 4.1 the following remarks realize to be useful.

1) For any $m, s \in \mathbb{R}$, $1<p<\infty$, $\Lambda^m(D)$ is bounded from $H^{s+m,p}_\Lambda$ into $H^{s,p}_\Lambda$ and $a(x,D)\Lambda^{-m}(D)$ has symbol $a(x,\xi)\Lambda^{-m}(\xi) \in H^{r,p}_\Lambda M^0_\Lambda(N)$. Assuming that Theorem 4.1 holds in the case $m=0$, we obtain that the operator $a(x,D) = (a(x,D)\Lambda^{-m}(D))\Lambda^m(D)$ is bounded from $H^{s+m,p}_\Lambda$ into $H^{s,p}_\Lambda$, when $0 \leq s \leq r$.

2) Proposition 2.9 and the Dominated Convergence Theorem assure that for any $a(x,\xi) \in H^{r,p}_\Lambda M^0_\Lambda(N)$, we can write for $u \in \mathcal{S}'(\mathbb{R}^n)$:

$$a(x,D)u(x) = \sum_{m\in\mathbb{Z}^n} \frac{1}{(1+|m|)^{n+1}} a_m(x,D)u(x), \quad |m| = \sum_{j=1}^n |m_j|, \tag{4.1}$$

where $a_m(x,\xi) \in H^{r,p}_\Lambda M_E(N-n-1)$ and now we have $N-n-1 \geq n$. Let us then prove at a first time Theorem 4.1 for $\sigma(x,\xi) \in H^{r,p}_\Lambda M^0_E(N)$, with $N \geq n$.

3) For any $\sigma(x,\xi) \in H^{s,p}_\Lambda M^0_E(N)$, $N \in \mathbb{N}$, we can write:

$$\sigma(x,D)u(x) = \sum_{h,\lambda} d_{h,\lambda}(x) u_{h,\lambda}(x), \qquad u \in \mathcal{S}(\mathbb{R}^n), \tag{4.2}$$

with absolute convergence in $L^\infty(\mathbb{R}^n)$. In fact since $\sum_{h,\lambda} |d_{h,\lambda}(x)||\psi_{h,\lambda}(\xi)| \leq MCN_1$ for any $x, \xi \in \mathbb{R}^n$ ($N_1 = N_1(H)$ is the number of terms in the sum which do not vanish), (4.2) holds with pointwise absolute convergence. Moreover for any $T>0$ and some suitable positive constant C we also have:

$$\|u_{h,\lambda}\|_\infty \leq C \frac{C_T}{\left(1 + \frac{1}{H^2}\sum_{j=1}^n \chi_{h_j} 2^{2h_j}\right)^T} \leq CC_T a_{h_1} \dots a_{h_n}, \tag{4.3}$$

where $C_T := \int((1+|\xi|^2)^T |\hat{u}(\xi)|\, d\xi$ is finite, $\chi_h = 0$ if $h=0$, $\chi_h = 1$, if $h>0$ and $a_h = 1$ for $h=0$, $a_h = \frac{1}{\frac{1}{H^2} 2^{\frac{2hM}{n}}}$ for $h>0$.

Thus

$$\sum_{h,\lambda} \|d_{h,\lambda}\|_\infty \|u_{h,\lambda}\|_\infty \leq M2^n C_T \sum_{h_1=0}^\infty a_{h_1} \cdots \sum_{h_n=0}^\infty a_{h_n} < \infty$$

shows that (4.2) is absolutely convergent in $L^\infty(\mathbb{R}^n)$.
Similarly we can write for $v \in \mathcal{S}'(\mathbb{R}^n)$ and $\{\varphi_{h,\lambda}\} \in \phi^{(H)}$:

$$v = \sum_{h,\lambda} \varphi_{h,\lambda}(D)v = \sum_{h,\lambda} v_{h,\lambda}, \text{ with convergence in } \mathcal{S}'(\mathbb{R}^n). \tag{4.4}$$

4) For $\{\psi_{k,\epsilon}(\xi)\} \in \Phi^{(K)}$, $K>1$, let us set $d^{k,\epsilon}_{h,\lambda}(x) := \psi_{k,\epsilon}(D) d_{h,\lambda}(x)$ and consider

$$a(x,D)u(x) = \sum_{h,\lambda}\sum_{k,\epsilon} d^{k,\epsilon}_{h,\lambda}(x) u_{h,\lambda}(x). \tag{4.5}$$

It follows from Proposition 3.4 that $\|d^{k,\epsilon}_{h,\lambda}\|_\infty < M\Lambda(c^{(K)}_{k,\epsilon})^{-\left(r-\frac{n}{\mu_0 p}\right)}$, for some $M > 0$. Then using (4.3) and provided that $r > \frac{n}{\mu_0 p}$ we can end that the expansion in (4.5) is absolutely convergent in $L^\infty(\mathbb{R}^n_x)$.

5) Thanks to the absolute convergence we can change each other the terms of the expansion in (4.5), agreeing to a useful order. Let us introduce at first some notations. Namely for a fixed $N_0 \in \mathbb{N}$ and any $j \in \mathbb{Z}_+$ we set:

$$E^{(N_0)}_{1,j} := \begin{cases} \emptyset\,, & j \le N_0; \\ \mathbb{Z}_+ \cap [0, j - N_0[\,, & j > N_0; \end{cases} \tag{4.6}$$

$$E^{(N_0)}_{2,j} := \mathbb{Z}_+ \cap [j - N_0, j + N_0[; \tag{4.7}$$

$$E^{(N_0)}_{3,j} := \mathbb{Z}_+ \cap [j + N_0, \infty[. \tag{4.8}$$

For $A := \{1, 2, \dots, n\}$ and $B := \{1, 2, 3\}$, let us consider B^A, that is the set of all the functions $\omega : A \mapsto B$. For any $h \in \mathbb{Z}^n_+$ and $\omega \in B^A$ we set $E^{(N_0)}_{\omega,h} := \prod_{i=1}^n E^{(N_0)}_{\omega(i),h_i}$. Agreeing with the previous notation we can write:

$$\sigma(x, D)u(x) = \sum_{\omega\in B^A} \sum_{h,\lambda} \sum_{k\in E^{(N_0)}_{\omega,h},\ \epsilon\in\mathbb{E}} d^{k,\epsilon}_{h,\lambda}(x) u_{h,\lambda}(x).$$

6) For every $h, k \in \mathbb{Z}^n_+$ and $\lambda, \epsilon \in \mathbb{E}$: $\operatorname{supp}(\widehat{d^{k,\epsilon}_{h,\lambda} u_{h,\lambda}}) \subset P^{(K)}_{k,\epsilon} + P^{(H)}_{h,\lambda}$, for some $K, H > 1$.
Given $r, s \in \mathbb{Z}_+$ and $\kappa, \delta \in \{-1, 1\}$, we can say that the n-intervals $P^{(H)}_{h,\lambda}$ and $P^{(K)}_{k,\epsilon}$ are obtained as superposition of n intervals of the type $L^{(H)}_{r,\kappa}$ and $L^{(K)}_{s,\delta}$, where

$$L^{(H)}_{r,\kappa} := \left\{ t \in \mathbb{R}\,,\ \frac{1}{H} 2^r \eta_r \le \kappa t \le H 2^{r+1} \right\},$$

with $\eta_r = 1$ if $r = 0$ and $\eta_r = 1$ if $r > 0$ (analogous definition for $L^{(K)}_{s,\delta}$). Therefore we are reduced to argue on the sums $L^{(H)}_{r,\kappa} + L^{(K)}_{s,\delta}$.
We need now the following technical lemma.

Lemma 4.2. ([7], Lemma 7.1) *Let us consider $r, s \in \mathbb{Z}_+$, $\kappa, \delta \in \{-1, 1\}$, H, K greater than 1. For any N_0 positive integer, $N_0 > \log_2(2HK)$, we can always find two positive constants T, M such that $T > H + K$, $\frac{1}{T} < \min\{\frac{1}{K} - \frac{2H}{2^{N_0}}, \frac{1}{H} - \frac{2K}{2^{N_0}}\}$ and $M > 2^{N_0+1}K + 2H$, which fulfill the following statements:*

(a) *if $s \in E^{(N_0)}_{1,r}$ and $r > N_0$ then*

$$L^{(H)}_{r,\kappa} + L^{(K)}_{s,\delta} \subset \left\{ \theta \in \mathbb{R} : \frac{2^r \eta_r}{T} \le \kappa\theta \le T 2^{r+1} \right\} =: L^{(T)}_{r,\kappa}; \tag{4.9}$$

(b) *if* $s \in E_{2,r}^{(N_0)}$ *then*

$$L_{r,\kappa}^{(H)} + L_{s,\delta}^{(K)} \subset \{\theta \in \mathbb{R} : \ |\theta| \le M2^r\} =: [-M2^r, M2^r]; \tag{4.10}$$

(c) *if* $s \in E_{3,r}^{(N_0)}$ *then*

$$L_{r,\kappa}^{(H)} + L_{s,\delta}^{(K)} \subset \left\{\theta \in \mathbb{R} : \frac{2^s \eta_s}{T} \le \delta\theta \le T2^{s+1}\right\} =: L_{s,\delta}^{(T)}. \tag{4.11}$$

It then follows that $\widehat{d_{h,\lambda}^{k,\epsilon} u_{h,\lambda}}$ is supported in the product of n real interval of the type (4.9)-(4.11). This suggests to split B^A in the following way:

$$C_1 := \{\omega \in B^A : \omega(A) = \{1\}\}; \qquad C_2 := \{\omega \in B^A : \omega(A) = \{2\}\};$$

$$C_3 := \{\omega \in B^A : \omega(A) = \{3\}\}; \qquad C_4 := \{\omega \in B^A : \omega(A) = \{1,2\}\};$$

$$C_5 := \{\omega \in B^A : \omega(A) = \{1,3\}\}; \qquad C_6 := \{\omega \in B^A : \omega(A) = \{2,3\}\};$$

$$C_7 := \{\omega \in B^A : \ \omega(A) = \{1,2,3\}\}.$$

The sets C_1, C_2 and C_3 reduce to a single constant function ω, while C_4-C_7 contain several functions, for any dimension $n \ge 2$.
For any $\sigma(x, D) \in H_\Lambda^{r,p} M_E(N)$ we can write

$$\sigma(x, D)u(x) = \sum_{j=1}^{7} T_j u(x), \quad u \in \mathcal{S}'(\mathbb{R}^n),$$

where for $j = 1, \dots, 7$:

$$T_j u(x) := \sum_{\omega \in C_j} \sum_{h,\lambda} \sum_{k \in E_{\omega,h}^{(N_0)},\ \epsilon \in \mathbb{E}} d_{h,\lambda}^{k,\epsilon}(x) u_{h,\lambda}(x), \qquad u \in \mathcal{S}(\mathbb{R}^n)$$

In the following we will work under the conditions obtained step by step in the remarks 1)-6). Let us then assume $\sigma(x,\xi) \in H_\Lambda^{r,p} M_E^0(N)$, with $1 < p < \infty$, $\Lambda(\xi)$ weight function, $N \ge n$ and $r > \frac{n}{\mu_0 p}$.

Proposition 4.3.

$$T_1 : H_\Lambda^{s,p} \mapsto H_\Lambda^{s,p}, \qquad \textit{continuously for any } s \in \mathbb{R};$$

$$T_2 : H_\Lambda^{s,p} \mapsto H_\Lambda^{s+r-\frac{n}{\mu_0 p}-\theta,p}, \quad \textit{continuously for any } s > -r + n/\mu_0 p \textit{ and } 0 < \theta < s + r - n/\mu_0 p; \tag{4.12}$$

$$T_3 : H^{s-r+\theta+\frac{n}{\mu_0 p},p} \mapsto H^{s,p}, \quad \textit{continuously for any } s \le r \textit{ and } \theta > 0. \tag{4.13}$$

Proposition 4.3 is proved in details in [[7], Propositions 7.1,7.2,7.3]. Its proof essentially relies on a vector valued version of the Marcinckiewicz-Lizorkin Lemma about Fourier multipliers quoted in the Introduction and the results of § 3.

Remark 4.4. Let us notice that, under the assumption $r > \frac{n}{\mu_0 p}$, we may always find a suitably small $\theta > 0$ such that $H_\Lambda^{s+r-\frac{n}{\mu_0 p}-\theta,p} \subset H_\Lambda^{s,p} \subset H_\Lambda^{s-r+\theta+\frac{n}{\mu_0 p},p}$ continuously, then from (4.12) and (4.13) it follows:

$$T_2 : H_\Lambda^{s,p} \mapsto H_\Lambda^{s,p}, \quad \text{continuously for every } s > -r + \frac{n}{\mu_0, p};$$

$$T_3 : H_\Lambda^{s,p} \longmapsto H_\Lambda^{s,p}, \quad \text{continuously for every } s \le r.$$

Let us remark moreover that any operators T_j, $j = 4, ..., 7$, may be expressed as a finite sum of operators with the following form

$$Ru(x) = \sum_{h,\lambda} \sum_{\substack{k \in E_h^{(N_0),n_1,n_2,\pi}, \\ \lambda \in \mathbb{E}}} d_{h,\lambda}^{k,\epsilon}(x) u_{h,\lambda}(x), \qquad u \in \mathcal{S}(\mathbb{R}^n). \tag{4.14}$$

Here n_1, n_2 are integers such that $0 \le n_1 \le n_2 \le n$ and two at least of these inequalities must be strict; π is any permutation of the set $\{1, 2, ..., n\}$, $E_{1,h_{\pi(j)}}^{(N_0)}$, $E_{2,h_{\pi(j)}}^{(N_0)}$, $E_{3,h_{\pi(j)}}^{(N_0)}$ are defined by (4.6) - (4.8) and

$$E_h^{(N_0),n_1,n_2,\pi} := \prod_{j=1}^{n_1} E_{1,h_{\pi(j)}}^{(N_0)} \times \prod_{j=n_1+1}^{n_2} E_{2,h_{\pi(j)}}^{(N_0)} \times \prod_{j=n_2+1}^{n} E_{3,h_{\pi(j)}}^{(N_0)}.$$

We are at the moment reduced to the following Proposition, where it is a crucial point that the weight function $\Lambda(\xi)$ satisfies the δ condition in (2.1), for some $0 \le \delta < 1$.

Proposition 4.5. ([7], Proposition 7.4) *Any operator R defined by* (4.14) *is bounded from $H_\Lambda^{s,p}$ to itself for any $0 \le s \le r$, provided that $r > \frac{n}{(1-\delta)\mu_0 p}$.*

Proof of Theorem 4.1. Thanks to Propositions 4.3, 4.5 we immediately get the statement for $\sigma(x,\xi) \in H_\Lambda^{r,p} M_E^0(N)$, provided that $N \ge n$. More precisely we can show that for any $0 \le s \le r$ and $1 < p < \infty$:

$$\|\sigma(x,D)u\|_{H_\Lambda^{s,p}} \le C \sup_{h,\lambda} \|d_{h,\lambda}\|_{H_\Lambda^{r,p}} \|u\|_{H_\Lambda^{s,p}}, \qquad u \in \mathcal{S}(\mathbb{R}^n),$$

where the constant $C > 0$ depends only on r, s, p and n.
Let us take now an arbitrary symbol $a(x,\xi)$ in $H_\Lambda^{r,p} M_\Lambda^0(N)$ with $N \ge 2n + 1$; thanks to (4.1) and the previous results for elementary symbols we obtain that, for every $0 \le s \le r$, $1 < p < \infty$, there exists a positive constant C such that, for any $u \in \mathcal{S}(\mathbb{R}^n)$,

$$\|a(x,D)u\|_{H_\Lambda^{s,p}} \le C\|u\|_{H_\Lambda^{s,p}} \sum_{m \in \mathbb{Z}^n} \frac{1}{(1+|m|)^{n+1}} \sup_{h,\lambda} \|d_{h,\lambda}^m\|_{H_\Lambda^{r,p}}. \tag{4.15}$$

It is crucial at the moment to notice that the constant C in (4.15) depends only r, s, p and the dimension n, but it is independent of $m \in \mathbb{Z}^n$.
Since the sequences $\{d_{h,\lambda}^m\}_{h,\lambda}$ are bounded in $H_\Lambda^{r,p}$ uniformly in $m \in \mathbb{Z}^n$, the expansion $\sum_{m \in \mathbb{Z}^n} \frac{1}{(1+|m|)^{n+1}}$ is finite and $\mathcal{S}(\mathbb{R}^n)$ is dense in $H_\Lambda^{s,p}$, it follows that $a(x,D)$

is $H^{s,p}_\Lambda$ bounded.
At the end if the symbol $a(x,\xi) \in H^{r,p}_\Lambda M^m_\Lambda(N)$ has an arbitrary order m, we easily reduce to the case of order zero as already noticed in this section. □

References

[1] M. Beals and M.C. Reeds, Microlocal regularity theorems for nonsmooth pseudodifferential operators and applications to nonlinear problems, *Trans. Amer. Math. Soc.* **285** (1984), 159–184.

[2] J.M. Bony, Calcul simbolique et propagation des singularités pour les équations aux dérivées partielles non lineaires, *Ann. Sc. Ec. Norm. Sup.* **14** (1981), 161–205

[3] R. Coifman and Y. Meyer, Au Delà des Opérateurs Pseudo-Differentiels", *Astérisque* **57**, Soc. Math. France, 1978.

[4] C. Fefferman, L^p bounds for pseudo-differential operators, *Israel J. Math.* **14** (1973), 413–417.

[5] G. Garello, Pseudo-differential operators with symbols in weighted Sobolev spaces and regularity for nonlinear partial differential equations, *Math. Nachr.* **239-240** (2001), 62–79.

[6] G. Garello and A. Morando, L^p-bounded pseudo-differential operators and regularity for multi-quasi-elliptic equations, *Integral Equations Operator Theory*, to appear.

[7] G. Garello and A. Morando, L^p boundedness for pseudo-differential operators with nonsmooth symbols and applications, *Boll. Un. Mat. It.*, to appear.

[8] J. Marschall, Pseudo-differential operators with non-regular symbols of the class $S^m_{\rho,\delta}$, *Comm. Partial Differential Equations* **12** (1987), 921–965.

[9] J. Marschall, Pseudo-differential operators with coefficients in Sobolev spaces, *Trans. Amer. Math. Soc.* **307** (1988), 335–361.

[10] M.E. Taylor, *Pseudo-Differential Operators*, Princeton Univ. Press, 1981.

[11] H. Triebel, General function spaces, III: spaces $B^{g(x)}_{p,q}$ and $F^{g(x)}_{p,q}$, $1 < p < \infty$: basic properties, *Anal. Math.* **3** (1977), 221–249.

[12] H. Triebel, General Function Spaces, IV: spaces $B^{g(x)}_{p,q}$ and $F^{g(x)}_{p,q}$, $1 < p < \infty$: special properties, *Anal. Math.* **3** (1977), 299–315.

[13] H. Triebel, *Theory of Function Spaces*, Birkhäuser Verlag, Basel, 1983.

Gianluca Garello
Dipartimento di Matematica, Università di Torino, Via Carlo Alberto 10, I-10123, Torino
Italy
e-mail: `gianluca.garello@unito.it`

Alessandro Morando
Dipartimento di Matematica, Università di Brescia, Via Valotti 9, I-25133, Brescia
Italy
e-mail: `morando@ing.unibs.it`

Operator Theory:
Advances and Applications, Vol. 155, 107–116

Symmetry-Breaking for Wigner Transforms and L^p-Boundedness of Weyl Transforms

M.W. Wong

Abstract. Using the symmetric formula for the Wigner transform, we give L^p-boundedness results for Weyl transforms with symbols in $L^1(\mathbb{R}^n \times \mathbb{R}^n)$. Breaking the symmetry of the Wigner transform, we obtain new results for Weyl transforms with symbols in $L^p(\mathbb{R}^n \times \mathbb{R}^n)$, $1 \le p \le 2$. Applications to localization operators on the Weyl-Heisenberg group are given for the case when the admissible wavelet is the ground state given by the normalized Gaussian function on $\mathbb{R}^n$.

Mathematics Subject Classification (2000). Primary 47G30.

Keywords. Weyl transforms, Wigner transforms, pseudo-differential operators.

1. Weyl Transforms

Let $f \in L^p(\mathbb{R}^n)$ and $g \in L^{p'}(\mathbb{R}^n)$, where $1 \le p \le \infty$ and p' is the conjugate index of p. Then the Wigner transform $W(f,g)$ of f and g is defined by

$$W(f,g)(x,\xi) = (2\pi)^{-n/2} \int_{\mathbb{R}^n} e^{-i\xi\cdot y} f\left(x+\frac{y}{2}\right) \overline{g\left(x-\frac{y}{2}\right)}\, dy, \quad x,\xi \in \mathbb{R}^n. \tag{1.1}$$

It is obvious from the symmetric formula (1.1) that

$$W(g,f) = \overline{W(f,g)}.$$

It also follows from the symmetric formula (1.1) and Hölder's inequality that $W(f,g) \in L^\infty(\mathbb{R}^n \times \mathbb{R}^n)$ and

$$\|W(f,g)\|_{L^\infty(\mathbb{R}^n\times\mathbb{R}^n)} \le (2\pi)^{-n/2} 2^n \|f\|_{L^p(\mathbb{R}^n)} \|g\|_{L^{p'}(\mathbb{R}^n)}. \tag{1.2}$$

Using the symmetric formula (1.1), it can be proved easily that if φ and ψ are functions in the Schwartz space $\mathcal{S}(\mathbb{R}^n)$ on $\mathbb{R}^n$, then the Wigner transform of φ and ψ is in the Schwartz space $\mathcal{S}(\mathbb{R}^n \times \mathbb{R}^n)$ on $\mathbb{R}^n \times \mathbb{R}^n$. Now, let σ be a tempered

This research has been partially supported by the Natural Sciences and Engineering Research Council of Canada under Grant OPG0008562.

distribution on $\mathbb{R}^n \times \mathbb{R}^n$, and let $\varphi \in \mathcal{S}(\mathbb{R}^n)$. Then we define $W_\sigma \varphi$ to be the tempered distribution on $\mathbb{R}^n$ by

$$(W_\sigma \varphi, \psi) = (2\pi)^{-n/2} \int_{\mathbb{R}^n} \int_{\mathbb{R}^n} \sigma(x,\xi) W(\varphi,\psi)(x,\xi)\, dx\, d\xi, \quad \psi \in \mathcal{S}(\mathbb{R}^n),$$

where (F, G) is defined by

$$(F,G) = \int_{\mathbb{R}^n} F(x)\overline{G(x)}\, dx$$

for all measurable functions F and G on $\mathbb{R}^n$, provided that the integral exists. We call W_σ the Weyl transform associated with the symbol σ. It should be noted that if σ is a symbol in $\mathcal{S}(\mathbb{R}^n \times \mathbb{R}^n)$, then $W_\sigma \varphi$ is a function in $\mathcal{S}(\mathbb{R}^n)$ for all φ in $\mathcal{S}(\mathbb{R}^n)$.

The most basic problem in the study of Weyl transforms is to seek good conditions on the symbols σ to guarantee that the corresponding Weyl transforms are bounded linear operators from $L^p(\mathbb{R}^n)$ into $L^p(\mathbb{R}^n)$, $1 \leq p \leq \infty$.

The following result is an abridged version of Theorem 14.3 in the book [14] by Wong.

Theorem 1.1. *Let $\sigma \in L^p(\mathbb{R}^n \times \mathbb{R}^n)$, $1 \leq p \leq 2$. Then $W_{\hat{\sigma}}$ is a bounded linear operator from $L^2(\mathbb{R}^n)$ into $L^2(\mathbb{R}^n)$ and*

$$\|W_{\hat{\sigma}}\|_{B(L^2(\mathbb{R}^n))} \leq (2\pi)^{-1/p} \|\sigma\|_{L^p(\mathbb{R}^n \times \mathbb{R}^n)},$$

where $\hat{\sigma}$ is the Fourier transform of σ.

In Theorem 1.1 and elsewhere in this paper, the notation $\| \; \|_{B(L^p(\mathbb{R}^n))}$ is used to denote the norm in the Banach algebra of all bounded linear operators from $L^p(\mathbb{R}^n)$ into $L^p(\mathbb{R}^n)$, $1 \leq p \leq \infty$; the Fourier transform $\hat{F}$ of a function F in $L^1(\mathbb{R}^N)$, also denoted by $\mathcal{F}F$, is defined by

$$\hat{F}(\zeta) = (2\pi)^{-N/2} \int_{\mathbb{R}^N} e^{-iz\cdot\zeta} F(z)\, dz, \quad \zeta \in \mathbb{R}^N.$$

In Section 2, we recall a more or less well-known sufficient condition to ensure that the Weyl transform is a bounded linear operator from $L^p(\mathbb{R}^n)$ into $L^p(\mathbb{R}^n)$ for $1 < p < \infty$. The condition can be traced back to the study of classical pseudo-differential operators pioneered by Hörmander in [10] and can be found, for instance, in the books [11] by Hörmander, [13] by Stein and [15] by Wong. In Section 3, we give results on the L^p-boundedness of Weyl transforms with symbols in $L^1(\mathbb{R}^n \times \mathbb{R}^n)$ using the formula (1.1) for the Wigner transform. These are analogues of the corresponding results for Daubechies operators in the paper [5] by Boggiatto and Wong. Breaking the symmetry of the Wigner transform in Section 4, we give in Section 5 new results on the L^p-boundedness of Weyl transforms with symbols in $L^p(\mathbb{R}^n \times \mathbb{R}^n)$, $1 \leq p \leq 2$. We give in Section 6 applications to Daubechies operators, which are localization operators on the Weyl-Heisenberg group, to supplement the corresponding L^p-boundedness results in the paper [5] by Boggiatto and Wong. For the sake of brevity and elegance, only the ground state, *i.e.*, the normalized Gaussian function on $\mathbb{R}^n$, is used as the admissible wavelet for the localization operators.

2. Pseudo-Differential Operators

We recall here a well-known set of conditions originating from the study of classical pseudo-differential operators. To this end, let $m \in \mathbb{R}$. Then S^m is the set of all C^∞ functions on $\mathbb{R}^n \times \mathbb{R}^n$ such that for all multi-indices α and β, there exists a positive constant $C_{\alpha,\beta}$ for which

$$|(D_x^\alpha D_\xi^\beta \sigma)(x,\xi)| \le C_{\alpha,\beta}(1+|\xi|)^{m-|\beta|}, \quad x,\xi \in \mathbb{R}^n.$$

A function in S^m is known as a classical symbol. Let $\sigma \in S^m$. Then we define the pseudo-differential operator T_σ on $\mathcal{S}(\mathbb{R}^n)$ by

$$(T_\sigma \varphi)(x) = (2\pi)^{-n/2} \int_{\mathbb{R}^n} e^{ix\cdot\xi} \sigma(x,\xi)\hat{\varphi}(\xi)\, d\xi, \quad x \in \mathbb{R}^n,$$

for all φ in $\mathcal{S}(\mathbb{R}^n)$. It is then easy to prove that T_σ maps $\mathcal{S}(\mathbb{R}^n)$ into $\mathcal{S}(\mathbb{R}^n)$ continuously. It is another easy matter to extend the definition of T_σ from the relatively small space $\mathcal{S}(\mathbb{R}^n)$ to the rather large space $\mathcal{S}'(\mathbb{R}^n)$ of all tempered distributions on $\mathbb{R}^n$. To wit, let $u \in \mathcal{S}'(\mathbb{R}^n)$. Then for all $\varphi \in \mathcal{S}(\mathbb{R}^n)$, we define

$$(T_\sigma u)(\varphi) = u(\overline{T_\sigma^* \overline{\varphi}}),$$

where T_σ^* is the formal adjoint of T_σ. Then it can be shown that $T_\sigma : \mathcal{S}'(\mathbb{R}^n) \to \mathcal{S}'(\mathbb{R}^n)$ is continuous. In fact, we have the following much deeper result, which is Theorem 11.9 in the book [15] by Wong.

Theorem 2.1. *Let $\sigma \in S^m$, $m \in \mathbb{R}$. Then $T_\sigma : H^{s,p} \to H^{s-m,p}$ is a bounded linear operator for $-\infty < s < \infty$ and $1 < p < \infty$.*

In Theorem 2.1, $H^{s,p}$ is the L^p-Sobolev space of order s defined by

$$H^{s,p} = \{u \in \mathcal{S}'(\mathbb{R}^n) : J_{-s}u \in L^p(\mathbb{R}^n)\}$$

for $-\infty < s < \infty$ and $1 < p < \infty$, where J_{-s} is the pseudo-differential operator with symbol σ_{-s} given by

$$\sigma_{-s}(\xi) = (1+|\xi|^2)^{s/2}, \quad \xi \in \mathbb{R}^n.$$

We note that for $1 < p < \infty$, $H^{0,p} = L^p(\mathbb{R}^n)$, and for $s \le t$, the Sobolev embedding theorem gives

$$H^{t,p} \subseteq H^{s,p}$$

and

$$\|u\|_{s,p} \le \|u\|_{t,p}, \quad u \in H^{t,p}.$$

A detailed proof of Theorem 2.1 can be found in Section 5 of Chapter 6 of the book [13] by Stein or in Chapters 10 and 11 of the book [15] by Wong. Theorem 2.1 gives the following result on the L^p-boundedness of Weyl transforms.

Theorem 2.2. *Let $\sigma \in S^m$, $m \in \mathbb{R}$. Then $W_\sigma : H^{s,p} \to H^{s-m,p}$ is a bounded linear operator for $-\infty < s < \infty$ and $1 < p < \infty$.*

Theorem 2.2 follows immediately from Theorem 2.1 and a well-known result, which states that there exists a symbol τ in S^m such that $W_\sigma = T_\tau$. A good reference is Corollary 2.43 on page 102 of the book [7] by Folland.

3. Symbols in $L^1(\mathbb{R}^n \times \mathbb{R}^n)$

We can now give the main results on the L^p-boundedness of Weyl transforms with symbols in $L^1(\mathbb{R}^n \times \mathbb{R}^n)$ based on the symmetric formula (1.1) for the Wigner transform.

Theorem 3.1. *Let $\sigma \in L^1(\mathbb{R}^n \times \mathbb{R}^n)$. Then for $1 \le p < \infty$, $W_\sigma : L^p(\mathbb{R}^n) \to L^p(\mathbb{R}^n)$ is a bounded linear operator and*

$$\|W_\sigma\|_{B(L^p(\mathbb{R}^n))} \le \pi^{-n}\|\sigma\|_{L^1(\mathbb{R}^n\times\mathbb{R}^n)}.$$

Proof. For all φ in $\mathcal{S}(\mathbb{R}^n)$ and g in $L^{p'}(\mathbb{R}^n)$, using (1.2), we get

$$\begin{aligned} |(W_\sigma\varphi, g)| &= (2\pi)^{-n/2}\left|\int_{\mathbb{R}^n}\int_{\mathbb{R}^n}\sigma(x,\xi)W(\varphi,g)(x,\xi)\,dx\,d\xi\right| \\ &\le (2\pi)^{-n/2}\|\sigma\|_{L^1(\mathbb{R}^n\times\mathbb{R}^n)}\|W(\varphi,g)\|_{L^\infty(\mathbb{R}^n\times\mathbb{R}^n)} \\ &\le \pi^{-n}\|\sigma\|_{L^1(\mathbb{R}^n\times\mathbb{R}^n)}\|\varphi\|_{L^p(\mathbb{R}^n)}\|g\|_{L^{p'}(\mathbb{R}^n)}. \end{aligned}$$

Using the fact that the dual of $L^p(\mathbb{R}^n)$ is $L^{p'}(\mathbb{R}^n)$ for $1 \le p < \infty$, we get

$$\|W_\sigma\varphi\|_{L^p(\mathbb{R}^n)} \le \pi^{-n}\|\sigma\|_{L^1(\mathbb{R}^n\times\mathbb{R}^n)}\|\varphi\|_{L^p(\mathbb{R}^n)}.$$

□

The usual method of defining the Weyl transform on $L^p(\mathbb{R}^n)$, $1 \le p < \infty$, is to define it initially on the Schwartz space $\mathcal{S}(\mathbb{R}^n)$ and then extend it to the whole space $L^p(\mathbb{R}^n)$ by a density argument. This method fails for $L^\infty(\mathbb{R}^n)$ simply because $\mathcal{S}(\mathbb{R}^n)$ is not dense in $L^\infty(\mathbb{R}^n)$. But using Theorem 3.1 and a duality argument, we can define the Weyl transform W_σ on $L^\infty(\mathbb{R}^n)$ at least in the case when σ is a symbol in $L^1(\mathbb{R}^n \times \mathbb{R}^n)$. To wit, using Theorem 3.1, we know that for $1 \le p < \infty$, the adjoint $W_\sigma^t : L^{p'}(\mathbb{R}^n) \to L^{p'}(\mathbb{R}^n)$ of $W_\sigma : L^p(\mathbb{R}^n) \to L^p(\mathbb{R}^n)$ is a bounded linear operator and

$$\|W_\sigma^t\|_{B(L^{p'}(\mathbb{R}^n))} = \|W_\sigma\|_{B(L^p(\mathbb{R}^n))}.$$

Now, for all f in $L^{p'}(\mathbb{R}^n)$ and g in $L^p(\mathbb{R}^n)$, we have

$$\begin{aligned} (W_\sigma^t f, g) &= (f, W_\sigma g) = \overline{(W_\sigma g, f)} \\ &= (2\pi)^{-n/2}\int_{\mathbb{R}^n}\int_{\mathbb{R}^n}\overline{\sigma(x,\xi)\,W(g,f)(x,\xi)}\,dx\,d\xi \\ &= (2\pi)^{-n/2}\int_{\mathbb{R}^n}\int_{\mathbb{R}^n}\overline{\sigma(x,\xi)}W(f,g)(x,\xi)\,dx\,d\xi \\ &= (W_{\overline{\sigma}} f, g). \end{aligned}$$

So, $W_\sigma^t = W_{\overline{\sigma}}$. Therefore for every symbol σ in $L^1(\mathbb{R}^n \times \mathbb{R}^n)$, it is then natural to define the Weyl transform $W_\sigma : L^\infty(\mathbb{R}^n) \to L^\infty(\mathbb{R}^n)$ to be the adjoint of $W_{\overline{\sigma}} : L^1(\mathbb{R}^n) \to L^1(\mathbb{R}^n)$ and we have proved the following theorem.

Theorem 3.2. *Let $\sigma \in L^1(\mathbb{R}^n \times \mathbb{R}^n)$. Then $W_\sigma : L^\infty(\mathbb{R}^n) \to L^\infty(\mathbb{R}^n)$ is a bounded linear operator and*

$$\|W_\sigma\|_{B(L^\infty(\mathbb{R}^n))} \leq \pi^{-n} \|\sigma\|_{L^1(\mathbb{R}^n \times \mathbb{R}^n)}.$$

4. Symmetry-Breaking

While the formula for the Wigner transform given by (1.1) is beautiful for its symmetry, it is the breaking of this symmetry that gives further results on the L^p-boundedness of Weyl transforms. Breaking the symmetry, we get the following formula for the Wigner transform. It is just Lemma 4.3.1 in the book [9] by Gröchenig and follows immediately from (1.1) and a change of variable in the integral.

Lemma 4.1. *For all φ and ψ in $\mathcal{S}(\mathbb{R}^n)$,*

$$W(\varphi, \psi)(x, \xi) = 2^n e^{2ix\cdot\xi} (\mathcal{F}(\varphi \overline{T_{-2x}\tilde{\psi}}))(2\xi), \quad x, \xi \in \mathbb{R}^n,$$

where

$$(T_{-2x}\tilde{\psi})(q) = \psi(2x - q), \quad q \in \mathbb{R}^n.$$

Using the symmetry-broken formula for the Wigner transform provided by the preceding lemma, we can now give an estimate for the norm $\|W(f,g)\|_{L^{p'}(\mathbb{R}^n \times \mathbb{R}^n)}$ in terms of $\|f\|_{L^p(\mathbb{R}^n)}$ and $\|g\|_{L^{p'}(\mathbb{R}^n)}$ for all f in $L^p(\mathbb{R}^n)$ and g in $L^{p'}(\mathbb{R}^n)$ whenever $1 \leq p \leq 2$. The range $1 \leq p \leq 2$ suggests that the Hausdoff-Young inequality for the Fourier transform should be at our disposal. To recall, the Hausdorff-Young inequality states that for $1 \leq p \leq 2$, there exists a positive constant C_p such that

$$\|\mathcal{F}\varphi\|_{L^{p'}(\mathbb{R}^n)} \leq C_p \|\varphi\|_{L^p(\mathbb{R}^n)}, \quad \varphi \in S(\mathbb{R}^n).$$

Thus, for $1 \leq p \leq 2$, the Fourier transform $\mathcal{F} : L^p(\mathbb{R}^n) \to L^{p'}(\mathbb{R}^n)$ is a bounded linear operator and its norm is denoted again by C_p.

Theorem 4.2. *For $1 \leq p \leq 2$,*

$$\|W(\varphi, \psi)\|_{L^{p'}(\mathbb{R}^n \times \mathbb{R}^n)} \leq 2^{-n} 4^{n/p} C_p \|\varphi\|_{L^p(\mathbb{R}^n)} \|\psi\|_{L^{p'}(\mathbb{R}^n)}, \quad \varphi, \psi \in \mathcal{S}(\mathbb{R}^n).$$

Proof. Let φ and ψ be functions in $\mathcal{S}(\mathbb{R}^n)$. Then, by Lemma 4.1,

$$\begin{aligned} &\|W(\varphi, \psi)\|_{L^{p'}(\mathbb{R}^n \times \mathbb{R}^n)} \\ = &\left\{ \int_{\mathbb{R}^n} \int_{\mathbb{R}^n} |W(\varphi, \psi)(x, \xi)|^{p'} dx\, d\xi \right\}^{1/p'} \\ = &\ 2^n \left\{ \int_{\mathbb{R}^n} \int_{\mathbb{R}^n} |(\mathcal{F}(\varphi \overline{T_{-2x}\tilde{\psi}}))(2\xi)|^{p'} d\xi\, dx \right\}^{1/p'}. \end{aligned}$$

Hence, using the Hausdorff-Young inequality,

$$\begin{aligned}
&\|W(\varphi,\psi)\|_{L^{p'}(\mathbb{R}^n\times\mathbb{R}^n)}\\
=\;& 2^{n/p}\left\{\int_{\mathbb{R}^n}\int_{\mathbb{R}^n}|(\mathcal{F}(\varphi\overline{T_{-2x}\tilde{\psi}}))(\eta)|^{p'}d\eta\,dx\right\}^{1/p'}\\
\le\;& 2^{n/p}C_p\left\{\int_{\mathbb{R}^n}\left(\int_{\mathbb{R}^n}|\varphi(y)(T_{-2x}\tilde{\psi})(y)|^p dy\right)^{p'/p}dx\right\}^{1/p'}\\
=\;& 2^{n/p}C_p\left\{\left\{\int_{\mathbb{R}^n}\left(\int_{\mathbb{R}^n}|\varphi(y)(T_{-2x}\tilde{\psi})(y)|^p dy\right)^{p'/p}dx\right\}^{p/p'}\right\}^{1/p}.
\end{aligned}$$

So, by Minkowski's inequality in integral form, we get

$$\begin{aligned}
&\|W(\varphi,\psi)\|_{L^{p'}(\mathbb{R}^n\times\mathbb{R}^n)}\\
\le\;& 2^{n/p}C_p\left\{\int_{\mathbb{R}^n}\left(\int_{\mathbb{R}^n}|\varphi(y)|^{p'}|(T_{-2x}\tilde{\psi})(y)|^{p'}dx\right)^{p/p'}dy\right\}^{1/p}\\
=\;& 2^{n/p}C_p\left\{\int_{\mathbb{R}^n}|\varphi(y)|^p\left\{\int_{\mathbb{R}^n}|(T_{-2x}\tilde{\psi})(y)|^{p'}dx\right\}^{p/p'}dy\right\}^{1/p}\\
=\;& 2^{n/p}C_p\left\{\int_{\mathbb{R}^n}|\varphi(y)|^p\left\{\int_{\mathbb{R}^n}|\psi(2x-y)|^{p'}dx\right\}^{p/p'}dy\right\}^{1/p}\\
=\;& 2^{-n}4^{n/p}C_p\left\{\int_{\mathbb{R}^n}|\varphi(y)|^p\left(\int_{\mathbb{R}^n}|\psi(z)|^{p'}dz\right)^{p/p'}dy\right\}^{1/p}\\
=\;& 2^{-n}4^{n/p}C_p\left\{\int_{\mathbb{R}^n}|\varphi(y)|^p\|\psi\|^p_{L^{p'}(\mathbb{R}^n)}dy\right\}^{1/p}\\
=\;& 2^{-n}4^{n/p}C_p\|\varphi\|_{L^p(\mathbb{R}^n)}\|\psi\|_{L^{p'}(\mathbb{R}^n)}.
\end{aligned}$$

□

Remark 4.3. The classical estimate for the norm C_p is obtained by interpolating the boundedness of the Fourier transforms $\mathcal{F}: L^1(\mathbb{R}^n)\to L^\infty(\mathbb{R}^n)$ and $\mathcal{F}: L^2(\mathbb{R}^n)\to L^2(\mathbb{R}^n)$, and

$$C_p\le(2\pi)^{n(1/2-1/p)}.$$

In the paper [1] by Beckner, it is shown that

$$C_p=\left\{\frac{\left(\frac{p}{2\pi}\right)^{1/p}}{\left(\frac{p'}{2\pi}\right)^{1/p'}}\right\}^{n/2}.$$

5. Symbols in $L^p(\mathbb{R}^n \times \mathbb{R}^n)$, $1 \leq p \leq 2$

Theorem 5.1. *Let $\sigma \in L^p(\mathbb{R}^n \times \mathbb{R}^n)$, $1 \leq p \leq 2$. Then for all $\rho \in [p, p']$, $W_\sigma : L^\rho(\mathbb{R}^n) \to L^\rho(\mathbb{R}^n)$ is a bounded linear operator and*

$$\|W_\sigma\|_{B(L^\rho(\mathbb{R}^n))} \leq (2\pi)^{-n/2} 2^{-n} 4^{n/p} C_p \|\sigma\|_{L^p(\mathbb{R}^n \times \mathbb{R}^n)}, \quad \rho \in [p, p'].$$

Proof. Let $\sigma \in L^p(\mathbb{R}^n \times \mathbb{R}^n)$, $1 \leq p \leq 2$. Then, by Theorem 4.2 and Hölder's inequality, we get for all φ in $\mathcal{S}(\mathbb{R}^n)$ and g in $L^{p'}(\mathbb{R}^n)$,

$$\begin{aligned}
|(W_\sigma \varphi, g)| &= (2\pi)^{-n/2} \left| \int_{\mathbb{R}^n} \int_{\mathbb{R}^n} \sigma(x, \xi) W(\varphi, g)(x, \xi)\, dx\, d\xi \right| \\
&\leq (2\pi)^{-n/2} \|\sigma\|_{L^p(\mathbb{R}^n \times \mathbb{R}^n)} \|W(\varphi, g)\|_{L^{p'}(\mathbb{R}^n \times \mathbb{R}^n)} \\
&\leq (2\pi)^{-n/2} 2^{-n} 4^{n/p} C_p \|\sigma\|_{L^p(\mathbb{R}^n \times \mathbb{R}^n)} \|\varphi\|_{L^p(\mathbb{R}^n)} \|g\|_{L^{p'}(\mathbb{R}^n)}.
\end{aligned}$$

Using the fact that the dual of $L^p(\mathbb{R}^n)$ is equal to $L^{p'}(\mathbb{R}^n)$ for $1 \leq p \leq 2$, we get

$$\|W_\sigma \varphi\|_{L^p(\mathbb{R}^n)} \leq (2\pi)^{-n/2} 2^{-n} 4^{n/p} C_p \|\sigma\|_{L^p(\mathbb{R}^n \times \mathbb{R}^n)} \|\varphi\|_{L^p(\mathbb{R}^n)}, \quad \varphi \in \mathcal{S}(\mathbb{R}^n).$$

Now, the adjoint of $W_{\overline{\sigma}} : L^p(\mathbb{R}^n) \to L^p(\mathbb{R}^n)$ is the bounded linear operator $W_\sigma : L^{p'}(\mathbb{R}^n) \to L^{p'}(\mathbb{R}^n)$ and

$$\|W_\sigma\|_{B(L^{p'}(\mathbb{R}^n))} \leq (2\pi)^{-n/2} 2^{-n} 4^{n/p} C_p \|\sigma\|_{L^p(\mathbb{R}^n \times \mathbb{R}^n)}.$$

By an interpolation of the results for $L^p(\mathbb{R}^n)$ and $L^{p'}(\mathbb{R}^n)$, we complete the proof of the theorem. □

6. Daubechies Operators

Let us begin with a recall of the definition of the Weyl-Heisenberg group. Let $\mathbb{R}^n \times \mathbb{R}^n = \{(q, p) : q, p \in \mathbb{R}^n\}$ and let $\mathbb{Z}$ be the set of all integers. Let $(WH)^n = \mathbb{R}^n \times \mathbb{R}^n \times \mathbb{R}/2\pi\mathbb{Z}$. Then we define the binary operation $\cdot$ on $(WH)^n$ by

$$(q_1, p_1, t_1) \cdot (q_2, p_2, t_2) = (q_1 + q_2, p_1 + p_2, t_1 + t_2 + q_1 \cdot p_2)$$

for all (q_1, p_1, t_1) and (q_2, p_2, t_2) in $(WH)^n$, where t_1, t_2 and $t_1 + t_2 + q_1 \cdot p_2$ are cosets in the quotient group $\mathbb{R}/2\pi\mathbb{Z}$ in which the group law is addition modulo 2π. With respect to the binary operation $\cdot$, $(WH)^n$ is a non-abelian group in which $(0, 0, 0)$ is the identity element and the inverse element of (q, p, t) is $(-q, -p, -t + q \cdot p)$ for all (q, p, t) in $(WH)^n$. If we identify $\mathbb{R}/2\pi\mathbb{Z}$ with the interval $[0, 2\pi]$, then $(WH)^n$ can be identified with $\mathbb{R}^n \times \mathbb{R}^n \times [0, 2\pi]$. It is a locally compact and Hausdorff group on which the left and right Haar measure is the Lebesgue measure $dq\, dp\, dt$. It is a unimodular group known as the Weyl-Heisenberg group.

Let $U(L^2(\mathbb{R}^n))$ be the group of all unitary operators on $L^2(\mathbb{R}^n)$ and let $\pi : (WH)^n \to U(L^2(\mathbb{R}^n))$ be the mapping defined by

$$(\pi(q, p, t) f)(x) = e^{i(p \cdot x - q \cdot p + t)} f(x - q), \quad x \in \mathbb{R}^n,$$

for all (q,p,t) in $(WH)^n$ and f in $L^2(\mathbb{R}^n)$. Then π is a square-integrable representation of $(WH)^n$ on $L^2(\mathbb{R}^n)$ and every function φ in $L^2(\mathbb{R}^n)$ with $\|\varphi\|_{L^2(\mathbb{R}^n)} = 1$ satisfies the admissibility condition that

$$\int_0^{2\pi} \int_{\mathbb{R}^n} \int_{\mathbb{R}^n} |(\varphi, \pi(q,p,t)\varphi)|^2 dq\, dp\, dt < \infty.$$

Thus, every function φ in $L^2(\mathbb{R}^n)$ with $\|\varphi\|_{L^2(\mathbb{R}^n)} = 1$ is an admissible wavelet for the square-integrable representation π of $(WH)^n$ on $L^2(\mathbb{R}^n)$. The book [16] by Wong contains the basic facts on the Weyl-Heisenberg group $(WH)^n$ and the representation π of $(WH)^n$ on $L^2(\mathbb{R}^n)$.

Let φ be a function in $L^2(\mathbb{R}^n)$ such that $\|\varphi\|_{L^2(\mathbb{R}^n)} = 1$. Then the wavelet constant c_φ of φ is defined by

$$c_\varphi = \int_0^{2\pi} \int_{\mathbb{R}^n} \int_{\mathbb{R}^n} |(\varphi, \pi(q,p,t)\varphi)|^2 dq\, dp\, dt.$$

In fact, $c_\varphi = (2\pi)^{n+1}$.

Let $F \in L^1(\mathbb{R}^n \times \mathbb{R}^n)$. Then the bounded linear operator $D_{F,\varphi} : L^2(\mathbb{R}^n) \to L^2(\mathbb{R}^n)$ defined by

$$(D_{F,\varphi} f, g) = (2\pi)^{-n} \int_{\mathbb{R}^n} \int_{\mathbb{R}^n} F(q,p)(f, \varphi_{q,p})(\varphi_{q,p}, g)\, dq\, dp$$

for all f and g in $L^2(\mathbb{R}^n)$, where

$$\varphi_{q,p}(x) = e^{ip\cdot x}\varphi(x-q), \quad x \in \mathbb{R}^n,$$

for all q and p in $\mathbb{R}^n$, is used by Daubechies [6] as a filter in signal analysis. In fact, it can be proved that

$$\begin{aligned} &(D_{F,\varphi} f, g) \\ = \; & \frac{1}{c_\varphi} \int_0^{2\pi} \int_{\mathbb{R}^n} \int_{\mathbb{R}^n} F(q,p)(f, \pi(q,p,t)\varphi)(\pi(q,p,t)\varphi, g)\, dq\, dp\, dt \end{aligned}$$

for all f and g in $L^2(\mathbb{R}^n)$. Thus, the linear operator $D_{F,\varphi} : L^2(\mathbb{R}^n) \to L^2(\mathbb{R}^n)$, which is called the Daubechies operator in the book [16], is in fact the localization operator from $L^2(\mathbb{R}^n)$ into $L^2(\mathbb{R}^n)$ associated to the symbol F and the admissible wavelet φ for the representation π of $(WH)^n$ on $L^2(\mathbb{R}^n)$. It is also the same as the windowed Fourier transform used by Gabor [8] in time-frequency analysis in which the window is the admissible wavelet φ.

Daubechies operators are also the same as pseudo-differential operators with anti-Wick symbols studied by Boggiatto, Buzano and Rodino [2], Boggiatto and Cordero [3], Boggiatto and Cordero [4], and Shubin [12].

Using the theory explained in the book [16] by Wong, Daubechies operators $D_{F,\varphi} : L^2(\mathbb{R}^n) \to L^2(\mathbb{R}^n)$ can be shown, among other things, to be bounded linear operators for all F in $L^p(\mathbb{R}^n \times \mathbb{R}^n)$, $1 \le p \le \infty$. Results on boundedness of Daubechies operators on $L^p(\mathbb{R}^n)$, $p \neq 2$, are relatively recent and can be found in the paper [5] by Boggiatto and Wong.

Notwithstanding the fact that the results in this section are valid for arbitrary admissible wavelets for the square-integrable representation π of the Weyl-Heisenberg group $(WH)^n$ on $L^2(\mathbb{R}^n)$, our focus is only on the admissible wavelet given by the normalized Gaussian function φ, where

$$\varphi(x) = \pi^{-n/4} e^{-|x|^2/2}, \quad x \in \mathbb{R}^n.$$

The following result, which is actually Theorem 17.1 in the book [14], tells us that Daubechies operators are Weyl transforms. This connection gives us new results on the L^p-boundedness of Daubechies operators.

Theorem 6.1. *Let $F \in L^p(\mathbb{R}^n \times \mathbb{R}^n)$, $1 \le p \le \infty$. Then*

$$D_{F,\varphi} = W_{F*\Lambda},$$

where

$$\Lambda(x,\xi) = \pi^{-n} e^{-(|x|^2+|\xi|^2)}, \quad x,\xi \in \mathbb{R}^n,$$

and $F * \Lambda$ is the convolution of F with Λ given by

$$(F * \Lambda)(x,\xi) = \int_{\mathbb{R}^n} \int_{\mathbb{R}^n} F(x-y, \xi-\eta)\, \Lambda(y,\eta)\, dy\, d\eta, \quad x,\xi \in \mathbb{R}^n.$$

The following result is a consequence of Theorems 3.1 and 6.1, and is the analogue of Theorems 2.4 and 2.6 in [5].

Theorem 6.2. *Let $F \in L^1(\mathbb{R}^n \times \mathbb{R}^n)$. Then for $1 \le p \le \infty$, $D_{F,\varphi} : L^p(\mathbb{R}^n) \to L^p(\mathbb{R}^n)$ is a bounded linear operator and*

$$\|D_{F,\varphi}\|_{B(L^p(\mathbb{R}^n))} \le \pi^{-n} \|F\|_{L^1(\mathbb{R}^n \times \mathbb{R}^n)}.$$

Proof. By Theorem 3.1, Theorem 6.1 and Young's inequality, we get

$$\begin{aligned} \|D_{F,\varphi}\|_{B(L^p(\mathbb{R}^n))} &\le \pi^{-n} \|F * \Lambda\|_{L^1(\mathbb{R}^n \times \mathbb{R}^n)} \\ &\le \pi^{-n} \|\Lambda\|_{L^1(\mathbb{R}^n \times \mathbb{R}^n)} \|F\|_{L^1(\mathbb{R}^n \times \mathbb{R}^n)}. \end{aligned}$$

Since $\|\Lambda\|_{L^1(\mathbb{R}^n \times \mathbb{R}^n)} = 1$, the proof is complete. □

The following theorem is a consequence of Theorems 5.1 and 6.1, and is the counterpart of Theorem 3.2 in [5].

Theorem 6.3. *Let $F \in L^p(\mathbb{R}^n \times \mathbb{R}^n)$, $1 \le p \le 2$. Then for all ρ in $[p, p']$, $D_{F,\varphi} : L^\rho(\mathbb{R}^n) \to L^\rho(\mathbb{R}^n)$ is a bounded linear operator and*

$$\|D_{F,\varphi}\|_{B(L^\rho(\mathbb{R}^n))} \le (2\pi)^{-n/2} 2^{-n} 4^{n/p} C_p \|F\|_{L^p(\mathbb{R}^n \times \mathbb{R}^n)}.$$

Proof. By Theorem 5.1, Theorem 6.1, Young's inequality, and the fact that $\|\Lambda\|_{L^1(\mathbb{R}^n \times \mathbb{R}^n)} = 1$,

$$\begin{aligned} \|D_{F,\varphi}\|_{B(L^\rho(\mathbb{R}^n))} &\le (2\pi)^{-n/2} 2^{-n} 4^{n/p} C_p \|F * \Lambda\|_{L^p(\mathbb{R}^n \times \mathbb{R}^n)} \\ &\le (2\pi)^{-n/2} 2^{-n} 4^{n/p} C_p \|F\|_{L^p(\mathbb{R}^n \times \mathbb{R}^n)}, \end{aligned}$$

as asserted. □

Acknowledgment. The author is grateful to the referee for pointing out a useful reference and several comments on the first version of the paper.

References

[1] W. Beckner, Inequalities in Fourier analysis, *Ann. Math.* **102** (1975), 159–182.

[2] P. Boggiatto, E. Buzano and L. Rodino, *Global Hypoellipticity and Spectral Theory*, Academie-Verlag, 1996.

[3] P. Boggiatto and E. Cordero, Anti-Wick quantization with symbols in L^p spaces, *Proc. Amer. Math. Soc.* **130** (2002), 2679–2685.

[4] P. Boggiatto and E. Cordero, Anti-Wick quantization of tempered distributions, in *Progress in Analysis*, Editors: H.G.W. Begehr, R.P. Gilbert and M.W. Wong, World Scientific, 655–662.

[5] P. Boggiatto and M.W. Wong, Two-wavelet localization operators on $L^p(\mathbb{R}^n)$ for the Weyl-Heisenberg group, *Integral Equations Operator Theory* **49** (2004), 1–10.

[6] I. Daubechies, Time-frequency localization operators: a geometric phase space approach, *IEEE Trans. Inform. Theory* **34** (1988), 605–612.

[7] G.B. Folland, *Harmonic Analysis in Phase Space*, Princeton University Press, 1989.

[8] D. Gabor, Theory of communications, *J. Inst. Elec. Eng.* (London) **93** (1946), 429–457.

[9] K. Gröchenig, *Foundations of Time-Frequency Analysis*, Birkhäuser, 2001.

[10] L. Hörmander, Pseudo-differential operators and hypoelliptic equations, in *Singular Integrals*, Proc. Symp. Pure Math. Vol. X, American Mathematical Society, 1967, 138–183.

[11] L. Hörmander, *The Analysis of Linear Partial Differential Operators III*, Springer-Verlag, 1985.

[12] M.A. Shubin, *Pseudodifferential Operators and Spectral Theory*, Springer-Verlag, 1987.

[13] E.M. Stein, *Harmonic Analysis: Real-Variable Methods, Orthogonality, and Oscillatory Integrals*, Princeton University Press, 1993.

[14] M.W. Wong, *Weyl Transforms*, Springer-Verlag, 1998.

[15] M.W. Wong, *An Introduction to Pseudo-Differential Operators*, Second Edition, World Scientific, 1999.

[16] M.W. Wong, *Wavelet Transforms and Localization Operators*, Birkhäuser, 2002.

M.W. Wong
Department of Mathematics and Statistics
York University
4700 Keele Street
Toronto, Ontario M3J 1P3
Canada
e-mail: mwwong@pascal.math.yorku.ca

Operator Theory:
Advances and Applications, Vol. 155, 117–130

Pseudo-Differential Operators and Schatten-von Neumann Classes

Ernesto Buzano and Fabio Nicola

Abstract. For a Hörmander's symbol class $S(m, g)$, it is proved that the weight m is in $L^p(\mathbb{R}^{2n})$, with $1 \leqslant p < \infty$, if and only if all pseudo-differential operators with Weyl symbol in $S(m, g)$ are in the Schatten-von Neumann class $S_p(L^2)$.

Mathematics Subject Classification (2000). Primary 47B10; Secondary 35S05.

Keywords. Schatten-von Neumann classes, hypoellipticity, pseudo-differential operators, Weyl-Hörmander calculus.

1. Introduction and Statement of Main Result

In this paper we study the decay of the singular values of a class of compact pseudo-differential operators on $L^2(\mathbb{R}^n)$. We deal with linear operators of the form

$$\mathcal{W}_a u(x) := \int e^{i(x-y)\cdot\xi} a\left(\tfrac{1}{2}(x+y), \xi\right) u(y)\, dy\, đ\xi, \tag{1.1}$$

where as usual $đ\xi = (2\pi)^{-n} d\xi$. The function $a(x, \xi)$, with $(x, \xi) \in \mathbb{R}^{2n}$, is called the *Weyl symbol of* $\mathcal{W}_a$, and the mapping

$$a \mapsto \mathcal{W}_a \tag{1.2}$$

is the so-called *Weyl quantization.* At first one considers symbols, say, in the Schwartz space $\mathcal{S}(\mathbb{R}^{2n})$; then one sees that the mapping (1.2) extends to the temperate distributions $\mathcal{S}'(\mathbb{R}^{2n})$ and takes values in the space of all linear continuous mappings $\mathcal{S}(\mathbb{R}^n) \to \mathcal{S}'(\mathbb{R}^n)$. Moreover, when the symbols are smooth functions satisfying convenient growth estimates at infinity, a complete symbolic calculus is available for the corresponding operators, see e.g. Hörmander's book [9], Chapter XVIII.

We are interested in conditions on the Weyl symbol a which ensure that $\mathcal{W}_a$ is a compact operator in $L^2(\mathbb{R}^n)$ with a prescribed asymptotics for its singular values, i.e. it belongs to a given Schatten-von Neumann class.

Recall that the singular values $s_j(A)$, $j = 1, 2, \dots$ of a compact operator A on a Hilbert space H are the eigenvalues of $|A| := (A^*A)^{1/2}$.

A compact operator A on H is in the *Schatten-von Neumann class* $S_p(H)$, with $1 \leqslant p < \infty$, if the sequence of its *singular values* is in ℓ^p.

$S_p(H)$ is a Banach algebra with respect to the norm

$$\|A\|_{S_p} := \Big\{\sum_{j=1}^{\infty} s_j(A)^p\Big\}^{1/p}.$$

The elements of $S_2(X)$ are the *Hilbert-Schmidt operators,* while $S_1(X)$ is the algebra of *trace class operators.*

One also set $S_\infty(H) = \mathcal{K}(H)$, the ideal of compact operators on H in the algebra of bounded operators $\mathcal{B}(H)$, with the usual norm. We refer to Simon [14], Schatten [12], for the basic theory of these Banach algebras.

As it is well-known (see for example Folland [3]), we have

$$a \in L^1(\mathbb{R}^{2n}) \implies \mathcal{W}_a \text{ is compact,}$$

and

$$a \in L^2(\mathbb{R}^{2n}) \iff \mathcal{W}_a \text{ is Hilbert-Schmidt.}$$

More in general, by complex interpolation, we have a well defined continuous mapping (see [10])

$$L^p(\mathbb{R}^{2n}) \ni a \mapsto \mathcal{W}_a \in S_{p'}(L^2), \qquad 1 \leqslant p \leqslant 2, \quad \frac{1}{p} + \frac{1}{p'} = 1,$$

which shows, in particular, that $\mathcal{W}_a$ is compact if $a \in L^p(\mathbb{R}^{2n})$, for $1 \leqslant p \leqslant 2$.

Without additional conditions one cannot expect that symbols in $L^p(\mathbb{R}^{2n})$, with $p > 2$, define bounded operators on $L^2(\mathbb{R}^n)$, let alone in $S_p(L^2)$. In fact, already in the case $p = \infty$, it is easily seen that for the bounded symbol $a(x,\xi) = e^{ix\xi}$ it turns out that $\mathcal{W}_a(L^2) \not\subset L^2$. More in general, Simon showed in [15] that there are no estimates of the form $\|\mathcal{W}_a\|_{\mathcal{B}(L^2)} \leqslant C\|a\|_{L^p}$ when $p > 2$.

If we instead require L^∞-bounds for a and its derivatives, then $\mathcal{W}_a$ becomes bounded on $L^2(\mathbb{R}^n)$, as established by the celebrated Calderón-Vaillancourt Theorem. Similarly, if the symbol with a certain number of its derivatives are in $L^1(\mathbb{R}^{2n}) \cap L^\infty(\mathbb{R}^{2n})$, the corresponding operator is shown to be trace class, see Robert [13], Theorem (II-53), and Toft [17], where more general properties of Schatten-von Neumann operators in the point of view of the Weyl calculus are given.

A variety of other sufficient conditions can be found in the papers by Hörmander [7, 8], Daubechies [2], Heil, Ramanathan and Topiwala [6], Gröchenig and Heil [5], where techniques from Gabor Analysis and modulation spaces are employed, as well. In this connection, Gröchenig's book [4] is considered a reference work concerning analysing pseudo-differential operators in terms of time-frequency analysis. Related problems in [5] and [6] have also been considered by Sjöstrand [16], using methods which are more close to "traditional" approaches in pseudo-differential calculus.

In this paper we find *necessary and sufficient conditions* so that $\mathcal{W}_a \in S_p(L^2)$, when a is a symbol in Hörmander classes $S(m,g)$, we describe below: refer to [9], Chapter XVIII and [8] for further details.

We employ the following notation. Given two functions $f, g : X \to \mathbb{R}$,

$$f(x) \prec g(x), \qquad \forall x \in X,$$

(or simply $f \prec g$) means that there exists a constant $C > 0$ such that $f(x) \leqslant Cg(x)$ for all $x \in X$.

An *admissible metric* is a measurable function $g : (x,\xi) \mapsto g_{x,\xi}$ of $\mathbb{R}^{2n}$ into the set of positive definite quadratic forms on $\mathbb{R}^{2n}$, which is slowly varying, σ-temperate, and satisfies the uncertainty principle.

A *g-weight* is a positive measurable function $m : \mathbb{R}^{2n} \to \mathbb{R}_+$, which is g-continuous and (σ, g)-temperate.

We say that a smooth function $a : \mathbb{R}^{2n} \to \mathbb{C}$ is a *symbol* if for some g-weight m

$$\sup_{(x,\xi)} \frac{|a|_k^g(x,\xi)}{m(x,\xi)} < \infty, \qquad \text{for all } k \in \mathbb{N},$$

where $|a|_0^g(x,\xi) := |a(x,\xi)|$ and

$$|a|_k^g(x,\xi) := \sup_{(t_j,\tau_j)} \frac{\left|a^{(k)}\big((x,\xi);(t_1,\tau_1),\ldots,(t_k,\tau_k)\big)\right|}{g_{x,\xi}(t_1,\tau_1)^{1/2}\cdots g_{x,\xi}(t_k,\tau_k)^{1/2}}, \qquad \text{for } k \geqslant 1.$$

We denote by $S(m,g)$ the class of all symbols of weight m and metric g. $S(m,g)$ is a Frechét space with respect to the seminorms:

$$\|a\|_{k;S(m,g)} := \sup_{(x,\xi)} \frac{\|a\|_k^g(x,\xi)}{m(x,\xi)}, \qquad (k \in \mathbb{N}),$$

where

$$\|a\|_k^g(x,\xi) := \sup_{j \leqslant k} |a|_j^g(x,\xi).$$

Whenever $a \in S(m,g)$, the operator $\mathcal{W}_a$ is continuous on the Schwartz class $\mathcal{S}(\mathbb{R}^n)$, and has a continuous extension to the tempered distributions $\mathcal{S}'(\mathbb{R}^n)$. Moreover $\mathcal{W}_a$ is closable in $L^2(\mathbb{R}^n)$: we denote by $\overline{\mathcal{W}_a}$ its closure.

Let

$$h(x,\xi) := \left(\sup_{(t,\tau)} \frac{g_{x,\xi}(t,\tau)}{g^\sigma_{x,\xi}(t,\tau)}\right)^{1/2},$$

where g^σ is the *dual quadratic form:*

$$g^\sigma_{x,\xi}(t,\tau) := \sup_{g_{x,\xi}(y,\eta)=1} \sigma\big((t,\tau);(y,\eta)\big)^2,$$

with respect to the *standard symplectic form* $\sigma = \sum_{i=1}^n d\xi_i \wedge dx_i$ in $\mathbb{R}^{2n}$.

We say that the metric g satisfies the *strong uncertainty principle* if there exists a positive constant δ such that

$$h(x,\xi) \prec (1 + |x| + |\xi|)^{-\delta}, \qquad \forall\, (x,\xi).$$

Our result is the following

Theorem 1.1. *Consider an admissible metric g satisfying the strong uncertainty principle. For any g-weight m and all* $1 \leqslant p < \infty$*, we have*

$$m \in L^p(\mathbb{R}^{2n}) \iff \Psi(m,g) \subset S_p(L^2), \tag{1.3}$$

where

$$\Psi(m,g) = \{\mathcal{W}_a \ : \ a \in S(m,g)\}.$$

We prove this theorem in section 3.

When $p = 1$ the implication " $\Longrightarrow$ " in (1.3) has been already proved by Hörmander ([7], Theorem 3.9) under more general assumptions.

After submission of the manuscript for publication in the ISAAC Proceedings, Toft [18] informed us that he obtained the equivalence (1.3) for every p by interpolation, without assuming the strong uncertainty principle. However, we think that our approach has still some interest because it is simple and natural.

2. Hypoelliptic Symbols and Complex Powers

We present a definition of hypoelliptic symbol which generalizes the one given by Tulovskiĭ and Shubin in [19]:

Definition 2.1. A symbol $a \in S(m,g)$ is called *g-hypoelliptic* if there exists a positive constant R such that

(i) for all $k \in \mathbb{N}$ we have [1]

$$|a|_k^g(x,\xi) \prec |a(x,\xi)|, \qquad \text{for } |(x,\xi)| \geqslant R. \tag{2.1}$$

(ii) there exists a g-weight m_0 such that

$$|a(x,\xi)| \succ m_0(x,\xi), \qquad \text{for } |(x,\xi)| \geqslant R, \tag{2.2}$$

When $m_0 = m$ we say that the symbol a is *g-elliptic*. We denote by $HS(m, m_0; g)$ the class of g-hypoelliptic symbols belonging to $S(m,g)$ and satisfying (2.2).

It is easily seen that the sums and (pointwise) products of hypoelliptic symbols are in turn hypoelliptic. We emphasize that in Definition 2.1 we do *not* require a to be slowly varying or temperate.

In the sequel, we need the following composition theorem which is proved in [9], (Theorem 18.5.4):

Theorem 2.2. *Given two symbols* $a \in S(m_1,g)$ *and* $b \in S(m_2,g)$*, we have that* $\mathcal{W}_a\mathcal{W}_b$ *is a pseudo-differential operator with symbol*

$$a\#b \in S(m_1 m_2, g)$$

[1] $|(x,\xi)|^2 = |x|^2 + |\xi|^2 = \sum_{i=1}^n x_i^2 + \sum_{i=1}^n \xi_i^2$.

such that

$$R_N(a,b) := a\#b - \sum_{j=0}^{N} \frac{\{a,b\}_j}{(2i)^j\, j!} \in S(m_1 m_2 h^{N+1}, g), \tag{2.3}$$

for all $N \in \mathbb{N}$, where $\{a,b\}_0 = ab$, and

$$\{a,b\}_j = \left[\left(\sum_{i=0}^{n} \left(\frac{\partial}{\partial \xi_i}\frac{\partial}{\partial y_i} - \frac{\partial}{\partial x_i}\frac{\partial}{\partial \eta_i} \right) \right)^j a(x,\xi) b(y,\eta) \right]_{\substack{y=x\\ \eta=\xi}},$$

for $j > 0$. More precisely, for each N, $k \in \mathbb{N}$ there exists an integer $l_{N,k}$ such that

$$\|R_N(a,b)\|_{k;S(m_1m_2h^{N+1},g)} \prec \|a\|_{l_{N,k};S(m_1,g)} \|b\|_{l_{N,k};S(m_2,g)}$$

for all $a \in S(m_1, g)$ and all $b \in S(m_2, g)$. □

Furthermore we have the following technical result, see [1] for the proof.

Lemma 2.3. *Given two smooth functions*

$$a,\, b : \mathbb{R}^{2n} \to \mathbb{C},$$

we have

$$\left| \{a,b\}_j \right|_k^g \leqslant (2n)^j \sum_{l=1}^{k} \binom{k}{l} |a|_{j+l}^g \, |b|_{j+k-l}^g \, h^j,$$

for all k, $j \in \mathbb{Z}_+$. □

An important property of hypoelliptic symbols is that they possess an approximate parametrix, when a big power of h is small compared to m_0/m:

Proposition 2.4. *Consider a g-hypoelliptic symbol $a \in HS(m, m_0; g)$ and assume there exists $N_0 \in \mathbb{N}$ such that*

$$h^{N_0} \prec \frac{m_0}{m}. \tag{2.4}$$

Then for all $N \in \mathbb{N}$ there exists a symbol $q = q_N \in S(m_0^{-1}, g)$ such that

$$q\#a - 1 \in S(h^{N+1}, g). \tag{2.5}$$

Moreover for all $k \in \mathbb{N}$ we have

$$|q|_k^g (x,\xi) \prec \frac{1}{|a(x,\xi)|}, \qquad \text{for } |(x,\xi)| \geqslant 2R, \tag{2.6}$$

where R is a positive constant such that the estimates (2.1) and (2.2) are satisfied.

Proof. The proof is a slight generalization of an argument due to Hörmander ([7], Lemma 3.1).

Let $\chi : \mathbb{R}^{2n} \to \mathbb{R}$ be a smooth non negative function such that

$$\chi(x,\xi) = \begin{cases} 0, & \text{if } |(x,\xi)| \leqslant R, \\ 1, & \text{if } |(x,\xi)| \geqslant 2R \end{cases} \tag{2.7}$$

and define[2]

$$r := 1 - \left(a^{-1}\chi\right) \# a. \tag{2.8}$$

From the estimate (18.4.4) of [9], for all $k \in \mathbb{Z}_+$ we have

$$\left|a^{-1}\right|_k^g (x,\xi) \prec \frac{1}{|a(x,\xi)|}\left(\sum_{j=1}^{k}\left(\frac{|a|_j^g(x,\xi)}{|a(x,\xi)|}\right)^{1/j}\right)^k, \qquad \text{for } |(x,\xi)| \geqslant R,$$

which, by hypoellipticity, implies for all $k \in \mathbb{N}$:

$$\left|a^{-1}\right|_k^g (x,\xi) \prec \frac{1}{|a(x,\xi)|}, \qquad \text{for } |(x,\xi)| \geqslant R.$$

Because

$$r = -\sum_{j=1}^{N_0} \frac{\left\{a^{-1}\chi, a\right\}_j}{(2i)^j j!} - R_{N_0}\left(a^{-1}\chi, a\right),$$

and h is a g-weight, from Lemma 2.3, Theorem 2.2, and (2.4) we obtain

$$r \in S\left(h + \frac{m h^{N_0+1}}{m_0}, g\right) \subset S(h,g). \tag{2.9}$$

Now let

$$r^{\#0} := 1,$$

and

$$r^{\#j} := \underbrace{r\#r\#\cdots\#r}_{j-\text{times}}, \qquad (j \geqslant 1). \tag{2.10}$$

Then from Theorem 2.2 and (2.9) we have

$$r^{\#j} \in S(h^j, g), \qquad \forall j \in \mathbb{N}.$$

Therefore, from Lemma 2.3 and Theorem 2.2 for all k, $j \in \mathbb{N}$ we obtain

$$\begin{aligned}
&\left|r^{\#j}\#(a^{-1}\chi)\right|_k^g (x,\xi) \prec \\
&\quad \prec \sum_{i=0}^{N_0} \left|\left\{r^{\#j}, a^{-1}\chi\right\}_i\right|_k^g (x,\xi) + \left|R_{N_0}\left(r^{\#j}, a^{-1}\chi\right)\right|_k^g (x,\xi) \\
&\quad \prec \frac{h(x,\xi)^j}{a(x,\xi)} + \frac{h(x,\xi)^{j+N_0+1}}{m_0(x,\xi)} \\
&\quad \prec \frac{h(x,\xi)^j}{a(x,\xi)} + \frac{h(x,\xi)^j}{m(x,\xi)} \\
&\quad \prec \frac{h(x,\xi)^j}{a(x,\xi)}, \qquad \text{for } |(x,\xi)| \geqslant 2R.
\end{aligned} \tag{2.11}$$

[2] We mean that $a(x,\xi)^{-1}\chi(x,\xi) = 0$ when $a(x,\xi) = 0$ and $\chi(x,\xi) = 0$.

For all $N \in \mathbb{N}$ define

$$q := \sum_{j=0}^{N} r^{\#j} \# \left(a^{-1}\chi\right).$$

From the estimate (2.11) we have (2.6) and $q \in S(m_0^{-1}, g)$. Finally, from (2.8) we have

$$\begin{aligned} q\#a &= \sum_{j=0}^{N} r^{\#j} \# \left(a^{-1}\chi\right) \#a \\ &= \sum_{j=0}^{N} r^{\#j} \# (1-r) = 1 - r^{\#(N+1)}, \end{aligned}$$

which implies $q\#a - 1 \in S(h^{N+1}, g)$. □

When h is small enough, we can show that hypoelliptic operators have essentially only one closed extension:

Theorem 2.5. *Consider a g-hypoelliptic symbol $a \in HS(m, m_0; g)$ and assume there exists $N_0 \in \mathbb{N}$ such that*

$$h^{N_0} \prec \frac{\inf\{1, m_0\}}{m}. \tag{2.12}$$

Then the minimal and the maximal extension of $\mathcal{W}_a$ to $L^2(\mathbb{R}^n)$ coincide.

Proof. The proof is a slight generalization of an argument due to Hörmander.

Denote by $\widetilde{\mathcal{W}}_a$ the extension of $\mathcal{W}_a$ to $\mathcal{S}'(\mathbb{R}^n)$.

We have to show that for any $u \in \mathcal{S}'(\mathbb{R}^n)$ such that $\widetilde{\mathcal{W}}_a u \in L^2(\mathbb{R}^n)$, there exists a sequence u_j in $\mathcal{S}(\mathbb{R}^n)$ such that

$$u_j \to u \quad \text{and} \quad \widetilde{\mathcal{W}}_a u_j \to \widetilde{\mathcal{W}}_a u \qquad \text{in } L^2(\mathbb{R}^n).$$

Thanks to Theorem 2.4 there exists $q \in S(m_0^{-1}, g)$ such that

$$r := 1 - q\#a \in S(h^{N_0+1}, g).$$

Using the partition of unity associated to the metric g in Lemma 18.4.4 of [9], one constructs a sequence of symbols $\chi_j \in \mathcal{S}(\mathbb{R}^{2n})$ which is bounded in $S(1, g)$ and converges to 1 in the $\mathcal{C}^\infty$ topology.

Then we set

$$u_j = \widetilde{\mathcal{W}}_{\chi_j} u.$$

We have that $u_j \in \mathcal{S}(\mathbb{R}^n)$, because $\chi_j \in \mathcal{S}(\mathbb{R}^{2n})$.

From the definition of r we have

$$\begin{aligned} \widetilde{\mathcal{W}}_a u_j &= \widetilde{\mathcal{W}}_a \widetilde{\mathcal{W}}_{\chi_j} \widetilde{\mathcal{W}}_q \widetilde{\mathcal{W}}_a u + \widetilde{\mathcal{W}}_a \widetilde{\mathcal{W}}_{\chi_j} \widetilde{\mathcal{W}}_r u \\ &= \widetilde{\mathcal{W}}_{a\#\chi_j\#q} \widetilde{\mathcal{W}}_a u + \widetilde{\mathcal{W}}_{a\#\chi_j\#r} u. \end{aligned}$$

Thanks to Theorem 2.2 and Lemma 2.3, for all $k \in \mathbb{N}$ there exists l_k, $l'_k \in \mathbb{N}$, $l_k \leqslant l'_k$, such that we have the estimates:

$$\begin{aligned}
|a\#\chi_j\#q|_k^g &\leqslant \sum_{l=0}^{N_0} \frac{1}{2^l\, l!} \left|\{a\#\chi_j, q\}_l\right|_k^g + |R_{N_0}(a\#\chi_j, q)|_k^g \\
&\prec \|a\#\chi_j\|_{l_k}^g \|q\|_{l_k}^g + \frac{mh^{N_0+1}}{m_0} \\
&\prec \left(\|a\|_{l'_k}^g \|\chi_j\|_{l'_k}^g + mh^{N_0+1}\right) \|q\|_{l_k}^g + \frac{mh^{N_0+1}}{m_0} \\
&\prec \left(|a| + mh^{N_0+1}\right) |a|^{-1} + \frac{mh^{N_0+1}}{m_0} \\
&\prec 1 + \frac{mh^{N_0+1}}{m_0} \prec 1,
\end{aligned}$$

for all $|(x,\xi)| \geqslant R$ and $j \in \mathbb{Z}_+$.

It follows that

$$\sup_{j\in\mathbb{Z}_+} \|a\#\chi_j\#q\|_{k;S(1,g)} < \infty.$$

Moreover from (2.12) and Theorem 2.2 we have also

$$\begin{aligned}
\|a\#\chi_j\#r\|_{k;S(1,g)} &\leqslant \|a\#\chi_j\#r\|_{k;S(mh^{N_0+1},g)} \\
&\prec \|a\|_{l_k;S(m,g)} \|\chi\|_{l_k;S(1,g)} \|r\|_{l_k;S(h^{N_0+1},g)},
\end{aligned}$$

for all $j \in \mathbb{Z}_+$. This means that the sequences $a\#\chi_j\#q$ and $a\#\chi_j\#r$ belong to a bounded set in $S(1,g)$.

Finally, from Theorem 18.5.4 of [9] we have that $a\#\chi_j\#q$ and $a\#\chi_j\#r$ converge pointwise to $a\#q$ and $a\#r$.

The last part of the proof is dependent of the following lemma, due to Hörmander ([7], Lemma 3.3):

Lemma 2.6. *If ϕ_j is a bounded sequence in $S(1,g)$, converging pointwise to $\phi \in S(1,g)$, then $\widetilde{W}_{\phi_j} u \to \widetilde{W}_\phi u$ in $L^2(\mathbb{R}^n)$ for all $u \in L^2(\mathbb{R}^n)$.* □

End of the proof of Theorem 2.5. From Lemma 2.6 we obtain that

$$\begin{aligned}
u_j &= \widetilde{W}_{\chi_j} u \to \widetilde{W}_1 u = u, \\
\widetilde{W}_a u_j &= \widetilde{W}_{a\#\chi_j\#q} \widetilde{W}_a u + \widetilde{W}_{a\#\chi_j\#r} u \to \widetilde{W}_{a\#q} \widetilde{W}_a u + \widetilde{W}_{a\#r} u \\
&= \widetilde{W}_{a\#q\#a+a\#r} u = \widetilde{W}_{a\#(q\#a+r)} u = \widetilde{W}_a u,
\end{aligned}$$

with convergence in $L^2(\mathbb{R}^n)$. □

Corollary 2.7. *Under the same hypotheses of Theorem 2.5 we have that $\overline{W}_{\bar{a}}$ is the adjoint of $\overline{W}_a$. In particular $\overline{W}_a$ is self-adjoint iff a is real-valued.*

Now we recall from [1] and [11] some results on complex powers of a non-negative operator.

Definition 2.8 (Komatsu). A closed operator A on a Banach space X is called *non-negative* if

(i) $(-\infty, 0)$ is contained in the resolvent set of A;

(ii) $\sup_{\lambda\in\mathbb{R}_+} \lambda \left\|(A+\lambda I)^{-1}\right\|_{\mathcal{B}(X)} < \infty.$

For example, when X is a Hilbert space, A is non-negative if

$$(Au, u)_X \geqslant 0, \qquad \text{for all } u \in \mathcal{D}(A).$$

Set $\mathbb{C}_+ := \{z \in \mathbb{C} \;:\; \operatorname{Re} z > 0\}$ and

$$\gamma_k(z) := \frac{\Gamma(k)}{\Gamma(z)\Gamma(k-z)} = \frac{(k-1)!\sin \pi z}{(k-1-z)\cdots(1-z)\pi}, \tag{2.13}$$

for $k \in \mathbb{Z}_+$ and $z \in \mathbb{C}\setminus\mathbb{Z}$.

The proof of Proposition 2.9 and Theorem 2.11 below are given in [11].

Proposition 2.9. *Consider a non-negative operator A on a Banach space X. Given $z \in \mathbb{C}_+$, and $u \in \mathcal{D}(A^{[\operatorname{Re} z]+1})$ we have that the integral*[3]

$$I^z_{A,k}u := \gamma_k(z)\int_0^\infty \lambda^{z-1}\left[A\,(A+\lambda I)^{-1}\right]^k u\, d\lambda,$$

is convergent for all integers $k > \operatorname{Re} z$.

Moreover these integrals are independent of k:

$$I^z_{A,k+1}u = I^z_{A,k}u, \qquad \forall\, k > \operatorname{Re} z.$$

□

Definition 2.10 (Balakhrishnan). Given a non-negative operator A on a Banach space X and a complex number $z \in \mathbb{C}_+$, define a new operator J^z_A on X as

$$\begin{cases} \mathcal{D}\left(J^z_A\right) := \mathcal{D}\left(A^{[\operatorname{Re} z]+1}\right), \\ J^z_A u := I^z_{A,k}u, \qquad \text{for any } k > \operatorname{Re} z. \end{cases}$$

Theorem 2.11. *Assume that A is a non-negative, densely defined operator on a Banach space X, then*

$$A^z := \overline{J^z_A}, \qquad z \in \mathbb{C}_+$$

is the unique *family of operators which enjoys the following set of properties:*

(i) $\mathcal{D}(J^z_A) \subset \mathcal{D}(A^z)$;

(ii) A^z *is closed;*

(iii) $A^k = \underbrace{AA\cdots A}_{k-times}$;

(iv) $A^zA^w = A^{z+w}$;

(v) *(Spectral Mapping Theorem) the spectrum of A^z is given by*

$$\sigma(A^z) = \{\lambda^z \;:\; \lambda \in \sigma(A)\}.$$

---

[3] The complex power λ^z is the principal branch $\lambda^z = \exp\big(z\,(\log|\lambda| + i\arg\lambda)\big)$, with $-\pi < \arg\lambda \leqslant \pi$.

(vi) *For all* $u \in \mathcal{D}(A^n)$, *with* $n \in \mathbb{Z}_+$, *the mapping*

$$z \mapsto A^z u$$

is analytic in the strip $\{z \in \mathbb{C} \,:\, 0 < \operatorname{Re} z < n\}$. □

Assume now that A is the closure in $L^2(\mathbb{R}^n)$ of a pseudo-differential operator $\mathcal{W}_a$ and is non-negative. Next theorem shows that under suitable hypotheses $\overline{\mathcal{W}_a}^z$ is pseudo-differential.

Theorem 2.12. *Consider an admissible metric* g *satisfying the strong uncertainty principle and a* g*-hypoelliptic symbol* $a \in HS(m, m_0; g)$ *such that*

$$\operatorname{Re} a(x,\xi) \geqslant -R\,|\operatorname{Im} a(x,\xi)|\,,$$

for $|(x,\xi)| \geqslant R$, *where* R *is a positive constant such that estimates* (2.1) *and* (2.2) *are satisfied. Assume that the closure* $\overline{\mathcal{W}_a}$ *of* $\mathcal{W}_a$ *in* $L^2(\mathbb{R}^n)$ *is* non-negative.

Let $a_0 = (|a|\,\chi + 1 - \chi)\, e^{i(\arg a)\chi}$, *where* $\arg a = 2\arctan \frac{\operatorname{Im} a}{\operatorname{Re} a + |a|}$ *and* $\chi \in C^\infty(\mathbb{R}^{2n})$ *is a real-valued function satisfying* $0 \leqslant \chi(x,\xi) \leqslant 1$, $(x,\xi) \in \mathbb{R}^{2n}$ *and* (2.7).

Then for all $z \in \mathbb{C}_+$ *there exists a* g*-hypoelliptic symbol*

$$a^{\#z} \in HS\left(m^{\operatorname{Re} z}, m_0^{\operatorname{Re} z}; g\right)$$

such that

(i) *for all* $k \in \mathbb{N}$ *and* $z \in \mathbb{C}_+$ *we have*

$$\left|a^{\#z} - a_0^z\right|_k^g (x,\xi) \prec |a_0(x,\xi)|^{\operatorname{Re} z}\, h(x,\xi), \qquad \forall\,(x,\xi); \tag{2.14}$$

(ii) *for all* $z \in \mathbb{C}_+$ *we have* $\overline{\mathcal{W}_a}^z = \overline{\mathcal{W}}_{a^{\#z}}$.

Proof. We give some indications on the proof of this theorem. Details will appear in [1].

First we show that for all $\lambda > 0$ the operator $\mathcal{W}_a + \lambda : \mathcal{S}(\mathbb{R}^n) \to \mathcal{S}(\mathbb{R}^n)$ has a continuous inverse which is a g-hypoelliptic pseudo-differential operator with Weyl symbol $\tilde{a}_\lambda$ such that for all k, $l \in \mathbb{N}$ we have the estimate

$$\left|(a\#\tilde{a}_\lambda)^{\#k} - \left(\frac{a_0}{a_0 + \lambda}\right)^k\right|_l^g \prec \left|\frac{a_0 h}{a_0 + \lambda}\right|^k, \qquad \forall\,(x,\xi),\ \forall\,\lambda > 0, \tag{2.15}$$

where $\tilde{a}_\lambda^{\#k} := \underbrace{\tilde{a}_\lambda \# \cdots \# \tilde{a}_\lambda}_{k\text{-times}}$.

Then we define

$$a^{\#z}(x,\xi) := \gamma_k(z) \int_0^\infty \lambda^{z-1}\, (a\#\tilde{a}_\lambda)^{\#k}\,(x,\xi)\, d\lambda$$

where k is an integer greater than $\operatorname{Re} z > 0$ and we show that the left-hand side is independent of k and the definition of $a^{\#z}$ is consistent with the definition of $a^{\#k} = a\# \cdots \#a$ by composition.

Integration of (2.15) yields (2.14). Because $h(x,\xi) \to 0$ as $|x| + |\xi| \to \infty$, (2.14) implies in particular that $a^{\#z}$ is a g-hypoelliptic symbol.

In order to complete the proof, it remains to show that

$$\overline{\mathcal{W}}_{a^{\#z}} = \left(\overline{\mathcal{W}}_a\right)^z, \qquad \forall\, z \in \mathbb{C}_+. \tag{2.16}$$

It suffices to prove this identity on $\mathcal{S}$. But then (2.16) becomes

$$\mathcal{W}_{a^{\#z}} u(x) = \gamma_k(z) \int_0^\infty \lambda^{z-1} \mathcal{W}_{(a\#\tilde{a}_\lambda)^{\#k}} u(x)\, d\lambda, \tag{2.17}$$

for all $k > \operatorname{Re} z > 0$, all $u \in \mathcal{S}$ and all $x \in \mathbb{R}^n$.

Now $a^{\#z}$ is the (pointwise) limit of a sequence of Riemann sums

$$\gamma_k(z) \sum_{j=1}^m \lambda_j^{z-1} \left(a\#\tilde{a}_{\lambda_j}\right)^{\#k} (x,\xi) \Delta\gamma_j,$$

which are bounded in $S(h^{-N}, g)$ for a large N. This implies that the Riemann sums

$$\gamma_k(z) \sum_{j=1}^m \lambda_j^{z-1} \mathcal{W}_{\left(a\#\tilde{a}_{\lambda_j}\right)^{\#k}} u(x) \Delta\gamma_j, \tag{2.18}$$

converge to $\mathcal{W}_{a^{\#z}} u(x)$. Because we know that (2.18) converge also to the right-hand side of (2.17), the proof is complete. □

3. Proof of Theorem 1.1

First we prove a version of Theorem 1.1 for hypoelliptic operators:

Proposition 3.1. *Consider an admissible metric g satisfying the strong uncertainty principle. Then for all g-hypoelliptic symbols a we have*

$$a \in L^p(\mathbb{R}^{2n}) \iff \overline{\mathcal{W}}_a \in S_p(L^2).$$

Proof. From Corollary 2.7 we have that the operator $\mathcal{W}_{\bar{a}} \mathcal{W}_a$ is essentially self-adjoint and its closure is non-negative. Moreover, the symbol $\bar{a}$ is hypoelliptic and

$$\bar{a}\#a = |a|^2 + b$$

with

$$|b|_k^g (x,\xi) \leqslant C_k \, |a(x,\xi)|^2 \, h(x,\xi), \tag{3.1}$$

for all $k \in \mathbb{N}$, $|(x,\xi)| \geqslant R$. Indeed, if a is, say, in $HS(m, m_0; g)$, then by Theorem 2.2 and Lemma 2.3 for all $k \in \mathbb{N}$, $N \in \mathbb{N}$ there exists $l_{k,N}$ such that

$$\begin{aligned} |b|_k^g(x,\xi) &\leqslant \sum_{l=1}^N \frac{1}{2^l l!} |\{\bar{a}, a\}|_k^g(x,\xi) + |R_N(\bar{a}, a)|_k^g(x,\xi) \prec \\ &\prec \|\bar{a}\|_{l_{k,N}}^g(x,\xi) \|a\|_{l_{k,N}}^g(x,\xi) h(x,\xi) + m(x,\xi)^2 h(x,\xi)^{N+1} \prec \\ &\prec |a(x,\xi)|^2 h(x,\xi) + m(x,\xi)^2 h(x,\xi)^{N+1}, \end{aligned} \tag{3.2}$$

for $|(x,\xi)| \geqslant R$. Since the weights m and m_0 are temperate, the functions $m^{\pm 1}$, $m_0^{\pm 1}$ have a polynomial growth; hence we see that if N is large enough, by the strong uncertainty principle, $m(x,\xi)^2h(x,\xi)^N \prec m_0(x,\xi)^2 \prec |a(x,\xi)|^2$ for $|(x,\xi)| \geqslant R$, and therefore (3.2) implies (3.1). Because $h(x,\xi) \to 0$ as $|(x,\xi)| \to \infty$, we have that $\bar{a}\#a$ is g-hypoelliptic. It follows therefore from Theorem 2.12 that the powers of $\overline{\mathcal{W}}_{\bar{a}}\overline{\mathcal{W}}_a$ are pseudo-differential operators. In particular

$$\left|\overline{\mathcal{W}}_a\right|^{p/2} = \left|\overline{\mathcal{W}}_{\bar{a}}\overline{\mathcal{W}}_a\right|^{p/4} = \overline{\mathcal{W}}_{(\bar{a}\#a)^{\#(p/4)}}.$$

On the other hand, by (i) of Theorem 2.12,

$$(\bar{a}\#a)^{\#(p/4)} = |a|^{p/2} + b_p,$$

with

$$|b_p(x,\xi)| \leqslant \tilde{C}_p\, |a(x,\xi)|^{p/2}\, h(x,\xi), \quad |(x,\xi)| \geqslant R.$$

Because $h(x,\xi) \to 0$ as $|(x,\xi)| \to \infty$, we obtain that

$$(\bar{a}\#a)^{\#(p/4)} \in L^2(\mathbb{R}^{2n}) \iff a \in L^p(\mathbb{R}^{2n}).$$

Now it is well-known that a pseudo-differential operator is Hilbert-Schmidt if and only if its symbol is in $L^2(\mathbb{R}^{2n})$. Thus we have shown that $\left|\overline{\mathcal{W}}_a\right|^{p/2}$ is Hilbert-Schmidt if and only if $a \in L^p(\mathbb{R}^{2n})$.

On the other hand, by Spectral Mapping Theorem, we have that the singular values of $\left|\overline{\mathcal{W}}_a\right|^{p/2}$ are those of $\overline{\mathcal{W}}_a$ to the power $p/2$:

$$s_j\left(\left|\overline{\mathcal{W}}_a\right|^{p/2}\right) = s_j\left(\overline{\mathcal{W}}_a\right)^{p/2}.$$

Thus

$$\left|\overline{\mathcal{W}}_a\right|^{p/2} \in S_2(L^2) \iff \overline{\mathcal{W}}_a \in S_p(L^2).$$

□

Now we may prove Theorem 1.1.

Assume that $\Psi(m,g) \subset S_p(L^2)$. It is known (see for example [9], pg. 143) that there exists a symbol $\tilde{m} \in S(m,g)$ such that

$$C^{-1}m \leqslant \tilde{m} \leqslant Cm$$

for a suitable constant $C > 1$. In particular $\tilde{m}$ is g-elliptic and $\mathcal{W}_{\tilde{m}}$ belongs to $S_p(L^2)$. Then by Proposition 3.1 we have $\tilde{m} \in L^p(\mathbb{R}^{2n})$, which implies $m \in L^p(\mathbb{R}^{2n})$.

Assume now that

$$m \in L^p(\mathbb{R}^{2n}), \quad \text{i.e. } \tilde{m} \in L^p(\mathbb{R}^{2n})$$

and let $a \in S(m,g)$. We have to show that $\mathcal{W}_a \in S_p(L^2)$.

By linearity we may assume that a is real and non-negative. Moreover, by hypothesis we know that

$$\begin{aligned} |\tilde{m}+a|_k^g &\leqslant |\tilde{m}+a|_{k;S(m,g)}\, m \\ &\leqslant |\tilde{m}+a|_{k;S(m,g)}\, (\tilde{m}+a) \end{aligned}$$

for all $k \in \mathbb{N}$.

Thus $\tilde{m} + a$ is g-elliptic and therefore, by Proposition 3.1, $\mathcal{W}_{\tilde{m}+a} \in S_p(L^2)$, for $\tilde{m} + a \in L^p(\mathbb{R}^{2n})$. Since $\mathcal{W}_{\tilde{m}} \in S_p$, it follows that

$$\mathcal{W}_a = \mathcal{W}_{a+\tilde{m}} - \mathcal{W}_{\tilde{m}} \in S_p.$$

This conclude the proof of Theorem 1.1. □

Notice that in the proof of Theorem 1.1 we did not apply Proposition 3.1 in its full generality, but only for symbols which are *g-elliptic*. On the other hand, Proposition 3.1 seems interesting in its own right, because the *hypoelliptic* symbol $a \in L^p(\mathbb{R}^{2n})$ is not assumed to belong to a class $S(m, g)$ associated with a weight m in $L^p(\mathbb{R}^{2n})$.

Acknowledgments. It is a pleasure for us to express our gratitude to Joachim Toft and Luigi Rodino for helpful discussions.

References

[1] E. Buzano and F. Nicola, Complex powers of hypoelliptic operators, in preparation.

[2] I. Daubechies, On the distributions corresponding to bounded operators in the Weyl quantization, *Comm. Math. Phys.* **75** (1980), 229–238.

[3] G.B. Folland, *Harmonic Analysis in Phase Space*, Princeton University Press, 1989.

[4] K. H. Gröchenig, *Foundations of Time-Frequency Analysis*, Birkhäuser, Boston, 2001.

[5] K.H. Gröchenig and C. Heil, Modulation spaces and pseudo-differential operators, *Integral Equations Operator Theory* **34** (1999), 439–457.

[6] J. Ramanathan, C. Heil and P. Topiwala, Singular values of compact pseudo-differential operators, *J. Funct. Anal.* **150** (1997), 426–452.

[7] L. Hörmander, On the asymptotic distribution of the eigenvalues of pseudo-differential operators in $\mathbb{R}^n$, *Ark. Mat.* **17** (1979), 297–313.

[8] ———, The Weyl calculus of pseudo-differential operators, *Comm. Pure Appl. Math.* **32** (1979), 359–443.

[9] ———, *The Analysis of Linear Partial Differential Operators III*, Springer-Verlag, Berlin, 1985.

[10] R. Howe, Quantum mechanics and partial differential equations, *J. Funct. Anal.* **38** (1980), 188–254.

[11] C. Martínez Carracedo and M. Sanz Alix, *The Theory of Fractional Powers of Operators*, North-Holland, Amsterdam, 2001.

[12] R. Schatten, *Norm Ideals of Completely Continuous Operators*, Springer-Verlag, Berlin-Göttingen-Heidelberg, 1960.

[13] D. Robert, *Autour de l'Approximation Semi-Classique*, Birkhäuser, Boston, MA, 1987.

[14] B. Simon, *Trace Ideals and Their Applications I*, Cambridge University Press, 1979.

[15] ———, The Weyl transform and L^p functions on phace space, *Proc. Amer. Math. Soc.* **116** (1992), 1045–1047.

[16] J. Sjöstrand, Wiener type algebras of pseudo-differential operators, *Séminaire Equations aux Dérivées Partielles, Ecole Polytechnique*, 1994/1995, Exposé n° IV.

[17] J. Toft, Continuity properties for non-commutative convolution algebras with applications in pseudo-differential calculus, *Bull. Sci. Math.* **126** (2002), 115–142.

[18] ———, Personal communication.

[19] V.N. Tulovskiĭ and M.A. Shubin, On the asymptotic distribution of eigenvalues of pseudo-differential operators in $\mathbb{R}^n$, *Math. USSR Sbornik* **21** (1973), 565–583.

Ernesto Buzano and Fabio Nicola
Dipartimento di Matematica
Università di Torino
Via Carlo Alberto 10
10123 Torino
Italy
e-mail: `buzano@dm.unito.it`
`nicola@dm.unito.it`.

Operator Theory:
Advances and Applications, Vol. 155, 131–147

Localization Operators via Time-Frequency Analysis

E. Cordero and A. Tabacco

Abstract. A systematic overview of localization operators using a time-frequency approach is given. Sufficient and necessary regularity results for localization operators with symbols and windows living in various function spaces (such as L^p or modulation spaces) are discussed. Finally, an exact and an asymptotic product formulae are presented.

Mathematics Subject Classification (2000). Primary 35S05; Secondary 47G30.

Keywords. Localization operators, modulation spaces, time-frequency analysis, Weyl calculus, Wigner distributions, short-time Fourier transforms.

1. Introduction

Localization operators arise from pure and applied mathematics in connection with various areas of research. Depending on the field of application, these operators are known under the names of Wick, anti-Wick or Toeplitz operators, as well as wave packets, Gabor or short time Fourier transform multipliers. There is a host of articles on the subject written by mathematicians, physicists and engineers; we refer, for instance, to [2, 9, 10, 11, 12, 16, 24, 26, 30, 34].

In Section 2 we introduce localization operators by means of time-frequency analysis. Next, we recall the Weyl operators and their connection with localization operators. In Section 3, firstly, we introduce the spaces where windows and symbols live in, secondly, we present the regularity results known so far. Different time-frequency methods and techniques contribute to get such results. We sketch the ones used for sufficient regularity conditions. Our main interest focuses on the interplay between the roughness of the symbol and the time-frequency concentration of the window functions that build up the localization operator. Namely, depending on the function spaces where the symbol and windows lie, we derive regularity conditions for the corresponding operator. We shall start considering symbols in L^p-spaces, with $1 \leq p \leq \infty$, we go to potential Sobolev spaces and we end up with the so-called *modulation spaces*. For symbols in L^p-spaces, we recall results contained in [34] and we show how to get them by means of shorter proofs [4]. However, sometimes in the applications one needs rougher symbols, even tempered distributions rather than functions. Hence, the subsequent step is to consider potential Sobolev spaces W^p_s, with $s < 0$. To cover the whole space of tempered distributions $\mathcal{S}'$ we require other spaces. For instance, we need to have bounded measures as symbols to recapture Gabor multipliers [16]; therefore it is natural to choose modulation spaces as both symbol and windows spaces for getting localization operator regularity results. Moreover, the regularity results obtained show that the appropriate Banach spaces of tempered distributions for windows and symbols are precisely the modulation spaces. Furthermore, the optimality of the last spaces' choice is made clear by the necessary regularity conditions shown at the end of the section. In Section 4 we deal with the problem of the product of two localizations operators. We provide two different answers, namely, both an exact [13] and an asymptotic formula [5, 8], depending on the framework one wants to work with.

Throughout the paper, we shall use the notation $A \lesssim B$ to indicate $A \leq cB$ for a suitable constant $c > 0$, independent of the parameters A and B may depend on; whereas $A \asymp B$ means $A \leq cB$ and $B \leq kA$, for suitable $c, k > 0$.

2. Time-Frequency Analysis of Localization Operators

Their first introduction as *anti-Wick* operators is due to Berezin, in 1971. As a physicist, he introduced them by means of a quantization rule $a \mapsto A_a$, from a symbol a, defined on a phase space, to an operator A_a, acting on a suitable Hilbert space. The symbol a is called *anti-Wick* symbol, while the corresponding operator A_a is referred to as the *anti-Wick* operator associated to the symbol a (let us notice that other authors talk about *Wick quantization* rather than anti-Wick, see, e.g., [1, 26]).

It is worthwhile pointing out that the "canonical" *coherent states* occur implicitly in the definition of A_a. Coherent states are L^2-functions labelled by phase space points. Here the associated phase space is the time-frequency plane $\mathbb{R}^{2d}$. Therefore, in order to construct a family of coherent states, one starts by choosing a vector φ in $L^2(\mathbb{R}^d)$. The associated coherent states are then generated from φ by time-frequency plane translations. Namely, for every point $(x, \omega) \in \mathbb{R}^{2d}$, the coherent state $\varphi_{(x,\omega)}$ is defined by

$$\varphi_{(x,\omega)}(t) = e^{2\pi i t\omega}\varphi(t - x) = M_\omega T_x \varphi(t), \tag{1}$$

where M_ω is the frequency shift operator given by $M_\omega \varphi(t) = e^{2\pi i t\omega} f(t)$ and T_x stands for the time shift operator defined by $T_x f(t) = f(t - x)$. The classical choice for φ in the anti-Wick construction is given by the Gaussian function $\varphi(t) = 2^{d/4} e^{-\pi t^2}$. The resulting coherent states $\varphi_{(x,\omega)}$ are often called canonical coherent states in the physics literature or Gabor wave functions in the engineering literature. They are the building blocks of anti-Wick operators, in a sense that we shall explain later. These operators turn out to be useful not only in physics but also for the theory of PDE's. In this setting smooth and regular anti-Wick symbols have been used (e.g., belonging to Hörmander or Shubin classes). For a deeper understanding of this classical framework, we refer to [7], included in this volume. Here we pay attention to the need of describing and extracting features of a given function, the so-called "*signal*", according to engineering terminology.

The terminology *localization operators* appears for the first time in 1988, in a paper by Daubechies [10]. She introduced these operators to localize a signal both in time and frequency. Her primary motivations were applications in optics and signal analysis rather than PDE's. For instance, localization operators could be used to filter out noise from given (noisy) signals. Since then, they have been extensively investigated as an important mathematical tool in the applications mentioned above [11, 16, 27, 33, 34]. Therefore, let us come deeper into this subject. We shall make use of time-frequency representations, a method commonly used by engineers to represent and study features of a signal.

The localization operators can be defined by means of one of the most important and widely used time-frequency representation: the *short-time Fourier transform*, using once again the terminology of signal analysis. Why do we need a joint time-frequency information about a signal? A simple example gives the answer to our question. For a given square integrable signal f, its Fourier transform $\hat{f}$, the "spectrum" of f, is not sufficient to describe the physical phenomenon because it records which frequencies are present in the signal f, without saying anything about "when" those frequencies exist. The need to know how the spectral content is changing in time, that is, to understand what a time-varying spectrum is, has led to develop physical and mathematical ideas. Time-varying spectra are common in ordinary life. The pitch, which is the common word for frequencies of human speech, changes as we speak and produces the richness of sounds in our language. The short time Fourier transform is the most widely used method for studying a non-stationary signal, like human speech. The concept behind it is simple and powerful. Suppose we listen to a piece of music that lasts an hour and suppose that it begins with violins and it ends with drums. If we Fourier analyze the whole hour by $\hat{f}$, we see peaks at the frequencies corresponding to the violins and the drums. This will tell us that there were violins and drums but will not give us any indication of when the violins and drums were played. The most straightforward way to overcome this lack of information is to break up the time interval into small segments (e.g., minutes) and Fourier analyze each interval resolutions. This is done by considering a suitable window function φ, localized in a neighborhood of the origin (the initial time), and making it sliding along the time interval: the time-frequency information about the signal f in a neighborhood of the instant x is obtained by Fourier-analyzing the product of f with the window φ shifted at x (see Figure 1).

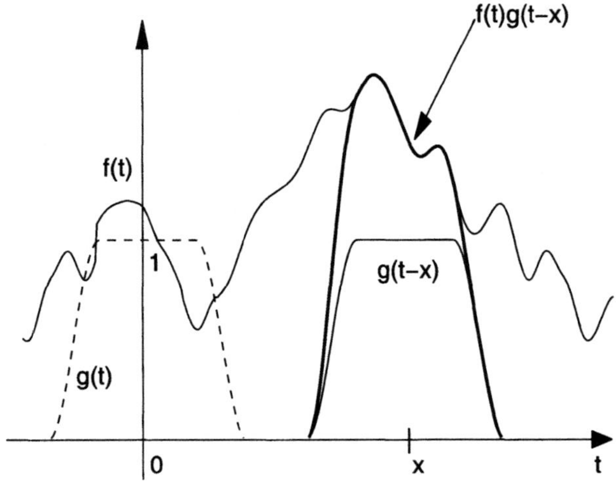

FIGURE 1. The short-time Fourier transform.

This is the reason why we talk about *short time* Fourier transform. The short-time Fourier transform $V_\varphi f$ of the signal $f \in L^2(\mathbb{R}^d)$, with respect to the window $\varphi \in L^2(\mathbb{R}^d)$, is given by

$$V_\varphi f(x,\omega) = \int_{\mathbb{R}^d} f(t)\,\overline{\varphi(t-x)}\,e^{-2\pi i\omega t}\,dt \;= \langle f, M_\omega T_x \varphi\rangle,$$

where the last equality expresses the L^2-inner product between the signal f and the coherent states of φ, but it can also regarded as a duality on a suitable pair of dual spaces $B' \times B$, extending the inner product on $L^2(\mathbb{R}^{2d})$.

Therefore, once we fix a non-zero analysis window $\varphi_1 \in L^2(\mathbb{R}^d)$, the short-time Fourier transform$V_{\varphi_1} f$ represents the time-frequency analysis of the signal f. If we want to come back to the original signal f, we consider a suitable synthesis window $\varphi_2 \in L^2(\mathbb{R}^d)$, such that the L^2-product of the windows φ_1 and φ_2 is not zero. The following reconstruction formula then holds:

$$f = \frac{1}{\langle \varphi_2, \varphi_1\rangle} \iint_{\mathbb{R}^{2d}} V_{\varphi_1} f(x,\omega) M_\omega T_x \varphi_2 \, dx d\omega, \quad \forall f \in L^2(\mathbb{R}^d).$$

The equality is to be understood in a L^2-sense, but it can have different meanings depending on the window spaces.

In applications, it is often desirable to consider only certain times and frequencies of a signal f, or to single out certain regions in the time-frequency distribution of f (one can consider, for instance, the problem of filtering out the noise of a signal). For this purpose, we fix a non-zero analysis window φ_1, we analyze f by means of $V_{\varphi_1} f$ and then we multiply the time-frequency representation $V_{\varphi_1} f$ by a suitable weight function a defined on the time-frequency plane. Finally, we reconstruct the signal f starting from $a(x,\omega)V_{\varphi_1} f(x,\omega)$, the modified time-frequency representation of f. What we get is a modified version of the original signal f, namely

$$A_a^{\varphi_1,\varphi_2} f = \iint_{\mathbb{R}^{2d}} a(x,\omega) V_{\varphi_1} f(x,\omega) M_\omega T_x \varphi_2 \, dx d\omega, \tag{2}$$

where the meaning of the previous integrals depends on the spaces where we have chosen the symbol a and the windows φ_1 and φ_2. Informally, we have just defined the time-frequency localization operator $A_a^{\varphi_1,\varphi_2}$, associated with the symbol a and with φ_1 and φ_2 as analysis and synthesis windows, respectively.

To prove regularity results for localization operators, it is often useful the weak definition of the previous integral, that is

$$\begin{aligned} \langle A_a^{\varphi_1,\varphi_2} f, g\rangle &= \iint_{\mathbb{R}^{2d}} a(x,\omega)\, V_{\varphi_1} f(x,\omega)\, \langle M_\omega T_x \varphi_2, g\rangle \, dx d\omega \qquad (3) \\ &= \langle a, \overline{V_{\varphi_1} f}\, V_{\varphi_2} g\rangle, \qquad \text{for } f, g \in \mathcal{S}(\mathbb{R}^d). \end{aligned}$$

The last equality shows that the action of a localization operator can be interpreted as the action of its symbol on the pointwise product of two short-time Fourier transforms. The localization operator can be seen as a generalization of the anti-Wick construction, that is recaptured for $\varphi_1(t) = \varphi_2(t) = 2^{d/4} e^{-t^2/2}$. The previous definition also extends Daubechies' one, in which the analysis and synthesis windows coincide. The time-frequency construction of a localization operator naturally suggests to replace the study of the linear anti-Wick quantization mapping by the investigation of the multilinear mapping

$$(a, \varphi_1, \varphi_2) \mapsto A_a^{\varphi_1,\varphi_2}. \tag{4}$$

The result we get is a host of sufficient regularity conditions. Among the methods we use to achieve the previous goal, there is the connection between anti-Wick operator and Weyl calculus. To introduce Weyl operators, we use again a time frequency approach. Indeed we give a weak definition by means of another important time-frequency representation, the cross-Wigner distribution, defined as

$$W(f,g)(x,\omega) = \int_{\mathbb{R}^d} e^{-2\pi i \omega t} f(x + \frac{t}{2}) \overline{g(x - \frac{t}{2})}\, dt,$$

where f, g are suitable functions, chosen such as the previous integral makes sense. Given a symbol σ on the phase space $\mathbb{R}^{2d}$, the Weyl operator L_σ is defined as

$$\langle L_\sigma f, g\rangle = \langle \sigma, W(g,f)\rangle, \qquad \forall f, g \in \mathcal{S}(\mathbb{R}^d). \tag{5}$$

The mapping

$$\sigma \mapsto L_\sigma$$

is another quantization rule, and it is called *Weyl quantization* or *Weyl transform* [13, 32].

A localization operator can always be written as a Weyl operator. Namely, if $W(\varphi_2, \varphi_1)$ is the cross-Wigner distribution of φ_2, φ_1, and L_σ is the Weyl transform with Weyl symbol σ, then we have (see [4, 18])

$$A_a^{\varphi_1,\varphi_2} = L_{a * W(\varphi_2,\varphi_1)}. \tag{6}$$

3. Regularity Results

In order to provide our result, we explicitly recall the definition of the function spaces we will work with. Secondly, we present sufficient regularity results and finally we show the (almost) optimality of the results obtained giving necessary conditions.

3.1. Function Spaces

If $1 \leq p < \infty$, then $\|f\|_p = \left(\int_{\mathbb{R}^d} |f(x)|^p\, dx\right)^{1/p}$ is the L^p-norm of f, and $L^p(\mathbb{R}^d)$, or just L^p, is the Banach space of all measurable functions f that have finite L^p-norm. For $p = \infty$, $L^\infty(\mathbb{R}^d)$ is the space of essentially bounded measurable functions with the norm $\|f\|_\infty = \operatorname{ess\,sup}_{x\in\mathbb{R}^d} |f(x)|$. In the next section we shall also make use of the weighted $\mathcal{F}L^p$-spaces. First, given a positive and continuous *weight function* m on $\mathbb{R}^d$, a function f belongs to the weighted L^p-spaces, i.e., $f \in L^p_m(\mathbb{R}^d)$ if and only if the pointwise product fm is in $\in L^p(\mathbb{R}^d)$. Next, we define the $\mathcal{F}L^p_m$-spaces of functions as

$$\mathcal{F}L^p_m(\mathbb{R}^d) = \{f \in \mathcal{S}'(\mathbb{R}^d) \,:\, \exists g \in L^p_m(\mathbb{R}^d),\ \hat{g} = f\}.$$

These spaces are Banach spaces equipped with the norm

$$\|f\|_{\mathcal{F}L^p_m} = \|g\|_{L^p_m}, \qquad \text{with } \hat{g} = f.$$

We now recall the definition of *potential Sobolev spaces*. Let $s \in \mathbb{R}$, the *Bessel kernel* is defined by

$$G_s = \mathcal{F}^{-1}\{(1 + |\cdot|^2)^{-s/2}\}. \tag{7}$$

Then the *potential Sobolev spaces* are given by

$$W^p_s = G_s * L^p(\mathbb{R}^d) = \{f \in \mathcal{S}',\, f = G_s * g,\, g \in L^p\}, \tag{8}$$

with norm $\|f\|_{W^p_s} = \|g\|_p$ (see [3, 29]).

The spaces that provide a measure of the phase space distribution of symbols and windows, are the so-called modulation spaces. For their basic properties we refer, for instance, to [20, Ch. 11-13] and the original literature quoted there.

Firstly, let us introduce the weight functions we shall deal with. Let v be a continuous, positive, even, submultiplicative weight function on $\mathbb{R}^{2d}$ (in short, a submultiplicative weight), i.e., $v(0) = 1$, $v(z) = v(-z)$, and $v(z_1+z_2) \leq v(z_1)v(z_2)$, for all $z, z_1, z_2 \in \mathbb{R}^{2d}$. A positive, even weight function m on $\mathbb{R}^{2d}$ is called *v-moderate* if $m(z_1 + z_2) \leq Cv(z_1)m(z_2)$ for all $z_1, z_2 \in \mathbb{R}^{2d}$. We shall mostly use the previous weights when working with modulation spaces, even though we define them for general weights.

Namely, let $g \in \mathcal{S}(\mathbb{R}^d)$ be a fixed non-zero window, m a non-negative, even weight function on $\mathbb{R}^{2d}$ of polynomial growth, and $1 \leq p, q \leq \infty$; the *modulation space* $M^{p,q}_m(\mathbb{R}^d)$ consists of all tempered distributions $f \in \mathcal{S}'(\mathbb{R}^d)$ such that $V_g f \in L^{p,q}_m(\mathbb{R}^{2d})$ (weighted mixed-norm spaces). The norm on $M^{p,q}_m$ is

$$\|f\|_{M^{p,q}_m} = \|V_g f\|_{L^{p,q}_m} = \left(\int_{\mathbb{R}^d} \left(\int_{\mathbb{R}^d} |V_g f(x,\omega)|^p m(x,\omega)^p\, dx\right)^{q/p} d\omega\right)^{1/q}.$$

If $p = q$, we write M^p_m instead of $M^{p,p}_m$, and if $m(z) \equiv 1$ on $\mathbb{R}^{2d}$, then we write $M^{p,q}$ and M^p for $M^{p,q}_m$ and $M^{p,p}_m$, respectively.

$M^{p,q}_m(\mathbb{R}^d)$ is a Banach space, whose definition is *independent of the choice of the window* $g \in \mathcal{S}(\mathbb{R}^d)$. Moreover, for any v-moderate weight m, and any non-zero $g \in \mathcal{S}(\mathbb{R}^d)$, then $\|V_g f\|_{L^{p,q}_m}$ are equivalent norms for $M^{p,q}_m(\mathbb{R}^d)$ (see [20, Thm. 11.3.7]). In particular, we will use the polynomial weights defined by

$$v_s(z) = v_s(x,\omega) = \langle z\rangle^s = (1+x^2+\omega^2)^{s/2}, \quad \tau_s(z) = \tau_s(x,\omega) = \langle \omega\rangle^s = (1+\omega^2)^{s/2},$$

where $z = (x,\omega) \in \mathbb{R}^{2d}$ and $s \in \mathbb{R}$.

Among the modulation spaces one can recognize well-known function spaces (see [4, 22, 28]):

(i) $L^2(\mathbb{R}^d) = M^2(\mathbb{R}^d)$.

(ii) Sobolev spaces: $H^s(\mathbb{R}^d) = \{f \,:\, \hat{f}(\omega)\langle\omega\rangle^s \in L^2(\mathbb{R}^d)\} = M^2_{\tau_s}(\mathbb{R}^d)$.

(iii) Shubin-Sobolev spaces:

$$Q_s(\mathbb{R}^d) = L^2_s(\mathbb{R}^d) \cap H^s(\mathbb{R}^d) = M^2_{v_s}(\mathbb{R}^d). \tag{9}$$

(iv) Feichtinger's algebra: $M^1(\mathbb{R}^d) = S_0(\mathbb{R}^d)$.

(v) Schwartz class: $\mathcal{S}(\mathbb{R}^d) = \bigcap_{s\geq 0} M^\infty_{v_s}(\mathbb{R}^d)$.

(vi) Space of tempered distributions: $\mathcal{S}'(\mathbb{R}^d) = \bigcup_{s\geq 0} M^\infty_{1/v_s}(\mathbb{R}^d)$.

For comparison we list the following embeddings between potential and modulation spaces [6].

Lemma 3.1. *For* $1 \leq p \leq \infty$ *and* $s \in \mathbb{R}$, *we have*

$$W^p_s(\mathbb{R}^d) \hookrightarrow M^{p,\infty}_{\tau_s}(\mathbb{R}^d).$$

3.2. Sufficient Conditions

To present our result, we need to recall the definition of Schatten-von Neumann classes. Consider a compact operator $A \in B(X)$, where X is a separable Hilbert space and $B(X)$ is the space of bounded linear operators on X. For $p \in [1,\infty)$, the *Schatten-von Neumann class* is defined as

$$S_p = \{A \in B(X), \text{ with } \|A\|_{S_p} := \sum_{k=1}^{\infty} s_k(A)^p < \infty\}, \tag{10}$$

where $s_k(A)$ are the *singular values* of A, i.e., the eigenvalues of the operator $\sqrt{A^*A}$. In particular, S_2 is the space of *Hilbert-Schmidt* operators, and S_1 is the space of *trace class* operators. For consistency, we define the class $S_\infty := B(X)$. Here we consider the Hilbert space $X = L^2(\mathbb{R}^d)$. We write $\mathcal{E}'(\mathbb{R}^{2d})$ for the subspace of tempered distributions of compact support on $\mathbb{R}^{2d}$.

The sufficient regularity conditions known so far can be easily summarized by means of the following table.

Results	Symbol $\mathbf{a}$	Windows φ_1	φ_2	Operator $\mathbf{A}_{\mathbf{a}}^{\varphi_1,\varphi_2}$
(i)	$L^p(\mathbb{R}^{2d})$	$L^2(\mathbb{R}^d)$	$L^2(\mathbb{R}^d)$	S_p
(ii)	$W^p_{-s}(\mathbb{R}^{2d})$, $s>0$	$M^1_{v_s}(\mathbb{R}^d)$	$M^1_{v_s}(\mathbb{R}^d)$	S_p
(iii)	$M^\infty_{1/\tau_s}(\mathbb{R}^{2d})$	$M^1_{v_s}(\mathbb{R}^d)$	$M^1_{v_s}(\mathbb{R}^d)$	$B(M^{p,q}(\mathbb{R}^d))$
(iv)	$H^{-s}(\mathbb{R}^{2d})$	$M^2_{v_s}(\mathbb{R}^d)$	$M^2_{v_s}(\mathbb{R}^d)$	S_2
(v)	$M^{p,\infty}_{1/\tau_s}(\mathbb{R}^{2d})$	$M^1_{v_s}(\mathbb{R}^d)$	$M^1_{v_s}(\mathbb{R}^d)$	S_p
(vi)	$\mathcal{E}'(\mathbb{R}^{2d})$	$\mathcal{S}(\mathbb{R}^d)$	$\mathcal{S}(\mathbb{R}^d)$	S_1

The result (i) is well known and first appeared in [34]; here we sketch a different and shorter proof of it. More details are contained in [4].

Theorem 3.2. *If $\varphi_1, \varphi_2 \in L^2(\mathbb{R}^d)$ and $a \in L^p(\mathbb{R}^{2d})$ for $1 \le p \le \infty$, then $A_a^{\varphi_1,\varphi_2} \in S_p$ and*

$$\|A_a^{\varphi_1,\varphi_2}\|_{S_p} \le \|a\|_p \,\|\varphi_1\|_2 \,\|\varphi_2\|_2 \,. \tag{11}$$

Proof. We prove the result for trace class $p=1$ and the boundedness for $p=\infty$ and then interpolate between these endpoints.

(a) $p=1$.
For $z=(x,\omega)\in\mathbb{R}^{2d}$, we consider the rank one operator

$$Q_z f = \langle f, M_\omega T_x \varphi_1\rangle M_\omega T_x \varphi_2.$$

Its trace class norm is

$$\|Q_z\|_{S_1} = \|\varphi_1\|_2 \,\|\varphi_2\|_2$$

and the mapping $z \mapsto Q_z$ is continuous from $\mathbb{R}^{2d}$ to S_1. Therefore the vector-valued integral defining $A_a^{\varphi_1,\varphi_2}$ as

$$A_a^{\varphi_1,\varphi_2} = \iint_{\mathbb{R}^{2d}} a(z) Q_z \, dz$$

is well-defined in S_1 (see, e.g., [25]). The generalized triangle inequality now yields

$$\|A_a^{\varphi_1,\varphi_2}\|_{S_1} \le \iint_{\mathbb{R}^{2d}} |a(z)| \,\|Q_z\|_{S_1} \, dz = \|a\|_1 \,\|\varphi_1\|_2 \,\|\varphi_2\|_2.$$

(b) $p=\infty$.
In this case we use the weak definition (3) of $A_a^{\varphi_1,\varphi_2}$ and estimate directly

$$\begin{aligned} |\langle A_a^{\varphi_1,\varphi_2} f, g\rangle| &= |\langle a, \overline{V_{\varphi_1} f}\, V_{\varphi_2} g\rangle| \le \|a\|_\infty \,\|V_{\varphi_1} f\, V_{\varphi_2} g\|_1 \\ &\le \|a\|_\infty \,\|V_{\varphi_1} f\|_2 \,\|V_{\varphi_2} g\|_2 = \|a\|_\infty \,\|\varphi_1\|_2 \,\|\varphi_2\|_2 \,\|f\|_2 \,\|g\|_2 \,. \end{aligned}$$

In the last equality we used the *isometry property* of the short-time Fourier transform [20, Chp.3], given by

$$\|V_\varphi f\|_{L^2(\mathbb{R}^{2d})} = \|f\|_{L^2(\mathbb{R}^d)} \|\varphi\|_{L^2(\mathbb{R}^d)} \qquad f,\varphi \in L^2(\mathbb{R}^d).$$

It follows that $A_a^{\varphi_1,\varphi_2}$ is bounded on $L^2(\mathbb{R}^d)$ and (11) holds for $p=\infty$.

The result follows now by complex interpolation. □

The results (ii)–(vi) were proved by a different method. Namely, thanks to relation (6), finding sufficient regularity conditions for localization operators can be translated into seeking conditions for Weyl operators. In this last framework, there are many papers and results on the topic. The most recent and general ones involve modulation spaces (see [6, 21]) and can be summarized as follows.

Theorem 3.3. (i) *If $\sigma \in M^{\infty,1}(\mathbb{R}^{2d})$, then L_σ is bounded on $M^{p,q}(\mathbb{R}^d)$, where $1 \le p, q \le \infty$, with a uniform estimate $\|L_\sigma\|_{op} \lesssim \|\sigma\|_{M^{\infty,1}}$ for the operator norm. In particular, L_σ is bounded on $L^2(\mathbb{R}^d)$.*

(ii) *If $1 \le p \le 2$ and $\sigma \in M^p(\mathbb{R}^{2d})$, then $L_\sigma \in S_p$ and $\|L_\sigma\|_{S_p} \lesssim \|\sigma\|_{M^p}$.*

(iii) *If $2 \le p \le \infty$ and $\sigma \in M^{p,p'}(\mathbb{R}^{2d})$, then $L_\sigma \in S_p$ and $\|L_\sigma\|_{S_p} \lesssim \|\sigma\|_{M^{p,p'}}$.*

Relation (6) provides the (Weyl) symbol of $A_a^{\varphi_1,\varphi_2}$, that is

$$\sigma = a * W(\varphi_2, \varphi_1)\,. \tag{12}$$

Therefore, if we want to use Theorem 3.3, we need to understand convolution relations between modulation spaces and properties of the Wigner distribution. This is made clear by the following two propositions, proved in [6]. Let us notice that the first result shows a convolution relation for modulation spaces in the style of Young's theorem.

Proposition 3.4. *Let v be a submultiplicative weight on $\mathbb{R}^{2d}$ and m be a v-moderate weight. We write $m_1(x) = m(x,0)$ and $m_2(\omega) = m(0,\omega)$ for the restrictions to $\mathbb{R}^d \times \{0\}$ and $\{0\} \times \mathbb{R}^d$, and likewise for v. Let $\nu(\omega) > 0$ be an arbitrary weight function on $\mathbb{R}^d$ and $1 \le p, q, r, s, t \le \infty$. If*

$$\frac{1}{p} + \frac{1}{q} - 1 = \frac{1}{r}, \qquad \text{and} \qquad \frac{1}{t} + \frac{1}{t'} = 1\,,$$

then

$$M^{p,st}_{m_1\otimes\nu}(\mathbb{R}^d) * M^{q,st'}_{v_1\otimes v_2\nu^{-1}}(\mathbb{R}^d) \hookrightarrow M^{r,s}_m(\mathbb{R}^d) \tag{13}$$

with norm inequality $\|f * h\|_{M^{r,s}_m} \lesssim \|f\|_{M^{p,st}_{m_1\otimes\nu}} \|h\|_{M^{q,st'}_{v_1\otimes v_2\nu^{-1}}}$.

Next, the modulation space norm of a cross-Wigner distribution is controlled by means of the modulation space norms of the corresponding windows.

Proposition 3.5. *If $1 \le p \le \infty$, $s \ge 0$, $\varphi_1 \in M^1_{v_s}(\mathbb{R}^d)$ and $\varphi_2 \in M^p_{v_s}(\mathbb{R}^d)$, then $W(\varphi_2, \varphi_1) \in M^{1,p}_{\tau_s}(\mathbb{R}^{2d})$, and*

$$\|W(\varphi_2, \varphi_1)\|_{M^{1,p}_{\tau_s}} \lesssim \|\varphi_1\|_{M^1_{v_s}} \|\varphi_2\|_{M^p_{v_s}}\,. \tag{14}$$

Based on the tools developed so far, it is now straightforward to prove the sufficient regularity conditions for localization operators contained in the following theorem (for details we refer to [6]).

Theorem 3.6. (i) *Let $s \ge 0$. If $1 \le p \le 2$, then the mapping $(a, \varphi_1, \varphi_2) \mapsto A_a^{\varphi_1,\varphi_2}$ is bounded from $M^{p,\infty}_{1/\tau_s}(\mathbb{R}^{2d}) \times M^1_{v_s}(\mathbb{R}^d) \times M^p_{v_s}(\mathbb{R}^d)$ into S_p; in other words,*

$$\|A_a^{\varphi_1,\varphi_2}\|_{S_p} \lesssim \|a\|_{M^{p,\infty}_{1/\tau_s}} \|\varphi_1\|_{M^1_{v_s}} \|\varphi_2\|_{M^p_{v_s}}\,.$$

(ii) *If* $2 \le p \le \infty$, *then the mapping* $(a, \varphi_1, \varphi_2) \mapsto A_a^{\varphi_1,\varphi_2}$ *is bounded from* $M^{p,\infty}_{1/\tau_s} \times M^1_{v_s} \times M^{p'}_{v_s}$ *into* S_p, *and*

$$\|A_a^{\varphi_1,\varphi_2}\|_{S_p} \lesssim \|a\|_{M^{p,\infty}_{1/\tau_s}} \|\varphi_1\|_{M^1_{v_s}} \|\varphi_2\|_{M^{p'}_{v_s}} .$$

(iii) *If* $a \in M^{\infty}_{1/\tau_s}(\mathbb{R}^{2d})$, $\varphi_1, \varphi_2 \in M^1_{v_s}(\mathbb{R}^d)$, *then* $A_a^{\varphi_1,\varphi_2}$ *is bounded on* $M^{p,q}(\mathbb{R}^d)$ *for all* $1 \le p, q \le \infty$, *and the operator norm satisfies the uniform estimate*

$$\|A_a^{\varphi_1,\varphi_2}\|_{op} \lesssim \|a\|_{M^{\infty}_{1/\tau_s}} \|\varphi_1\|_{M^1_{v_s}} \|\varphi_2\|_{M^1_{v_s}} .$$

Using the embedding result of Lemma 3.1 and the embedding $M^1_{v_s} \hookrightarrow M^p_{v_s}$, with $1 \le p \le \infty$ (see e.g., [20]), we derive the result *(ii)*:

Corollary 3.7. *Let* $a \in W^p_{-s}(\mathbb{R}^{2d})$ *for some* $s \ge 0$, $1 \le p \le \infty$, *and* $\varphi_1, \varphi_2 \in M^1_{v_s}(\mathbb{R}^d)$. *Then*

$$\|A_a^{\varphi_1,\varphi_2}\|_{S_p} \lesssim \|a\|_{W^p_{-s}} \|\varphi_1\|_{M^1_{v_s}} \|\varphi_2\|_{M^1_{v_s}} .$$

Let us end up this subsection saying few words about tempered distributions of compact support $\mathcal{E}'(\mathbb{R}^{2d})$. It is well-known that every $a \in \mathcal{E}'(\mathbb{R}^{2d})$ can be represented as

$$a = \sum_{|\alpha| \le n} \partial^\alpha f_\alpha \tag{15}$$

for compactly supported continuous functions f_α on $\mathbb{R}^{2d}$ [31, p. 263, Cor. 2]. For convenience and in contrast to the standard definition, we will call the integer n in (15) the order of $a \in \mathcal{E}'$. If we compute the short-time Fourier transform of a tempered distribution of order n, we get the estimate (see [6, Prop. 3.6])

$$|V_g a(x, \omega)| \le C_N \langle x \rangle^{-N} \langle \omega \rangle^n \quad \forall N \in \mathbb{N},$$

and, consequently, $a \in M^{1,\infty}_{1/\tau_n}(\mathbb{R}^d)$. If $\varphi_1, \varphi_2 \in M^1_{v_n}(\mathbb{R}^d)$, then Theorem 3.6*(i)* implies that $A_a^{\varphi_1,\varphi_2}$ is a trace class operator.

3.3. Necessary Conditions

In this section we show that the sufficient conditions obtained in Theorem 3.6 (for $p = \infty$ and $p = 2$) are essentially optimal. This investigation requires a different approach. The starting point is a local regularity property of the short-time Fourier transform, written in terms of Wiener amalgam spaces [14, 15, 17, 19]. Here we do not intend to come into this matter. Moreover, the proof of the necessary conditions is achieved by heavily working with short-time Fourier transform properties and it requires a broad background on the topic. Therefore, we limit ourselves to present the result, referring the reader to [6], for all the proofs and details.

Theorem 3.8. *Let* $a \in \mathcal{S}'(\mathbb{R}^{2d})$ *and* $s \ge 0$.

(i) *If there exists a constant* $C = C(a) > 0$, *depending only on* a, *such that*

$$\|A_a^{\varphi_1,\varphi_2}\|_{S_\infty} \le C \|\varphi_1\|_{M^1_{v_s}} \|\varphi_2\|_{M^1_{v_s}} \tag{16}$$

for all $\varphi_1, \varphi_2 \in \mathcal{S}(\mathbb{R}^d)$, *then* $a \in M^{\infty}_{1/\tau_s}$.

(ii) *If there exists a constant $C = C(a) > 0$, depending only on a, such that*

$$\|A_a^{\varphi_1,\varphi_2}\|_{S_2} \le C\,\|\varphi_1\|_{M^1}\|\varphi_2\|_{M^1} \tag{17}$$

for all $\varphi_1, \varphi_2 \in \mathcal{S}(\mathbb{R}^d)$, then $a \in M^{2,\infty}$.

Let us end up this section by giving an application of localization operators with rough symbols. Namely, symbols that are tempered distributions enable us to work with bounded measures of the form $a = \sum_{k\in\mathbb{Z}^{2d}} a_k\delta_k$, where $(a_k) \in \ell^\infty$ (the space of bounded sequences) and δ_k is the Dirac measure centered at the point k. If we substitute the previous measure to the function a in the definition (2) of a localization operator, we recapture the discrete *Gabor multiplier* [16], with a considerable improvement of the regularity results known so far.

4. Composition Formula

Given two localization operators, we aim at computing their product. It would be useful to express it in terms of localization operators. In the sequel we present two different product formulae. The former is an exact formula, i.e., the composition result is a localization operator as well. However, the formula holds only for Gaussian windows. The latter works for every pair of localization operators; in this case, the result is written as a sum of localization operators, plus a remainder term, in the Weyl form.

4.1. Exact Product

We reformulate the result of [13] according to the notation of [18, 20]. Furthermore, we provide regularity estimates in dependence of the function spaces the symbols of the factors belong to.

From then onward we consider the window functions $\varphi_1(t) = \varphi_2(t) = \varphi(t) = 2^{d/4}e^{-\pi t^2}$, $t \in \mathbb{R}^d$. In this case, the Wigner distribution of the Gaussian φ is a Gaussian as well: $W(\varphi,\varphi)(z) = 2^{d/2}e^{-2\pi z^2}$, $z \in \mathbb{R}^{2d}$. If we compute the Weyl symbol σ of the operator $A_a^{\varphi,\varphi}$, by means of relation (6) we get $\sigma(\zeta) = 2^{d/2}(a * e^{-2\pi z^2})(\zeta)$, $z, \zeta \in \mathbb{R}^{2d}$. In order to express the product in the localization form, firstly, we rewrite the factors in a Weyl form and then we use the product formula for Weyl operators [18, Chp. 3.2]. Namely, the composition of the Weyl operators L_σ and L_τ, with symbols σ and τ belonging to suitable function spaces, is a Weyl operator L_μ, with symbol $\mu = \sigma\natural\tau$, the twisted multiplication of the symbols σ and τ:

$$\sigma\natural\tau(\zeta) = 2^{2d}\iint_{\mathbb{R}^{2d}} \sigma(z)\tau(w)e^{4\pi i[\zeta-w,\zeta-z]}\,dz dw,$$

where the brackets $[\cdot,\cdot]$ express the *symplectic form* on $\mathbb{R}^{2d}$, i.e.,

$$[(z_1,z_2),(\zeta_1,\zeta_2)] = z_1\zeta_2 - z_2\zeta_1, \quad \text{with} \quad z = (z_1,z_2),\ \zeta = (\zeta_1,\zeta_2).$$

Secondly, we come back to localization operators by means of relation (6). Given the Weyl symbols $\sigma, \tau \in \mathcal{S}(\mathbb{R}^{2d})$, we are interested in the Fourier transform of the twisted multiplication $\sigma \sharp \tau$, that is

$$\mathcal{F}(\sigma \sharp \tau)(\zeta) = \hat{\sigma} \natural \hat{\tau}(\zeta),$$

where the twisted product $\natural$ is given by

$$\hat{\sigma} \natural \hat{\tau}(\zeta) = \iint_{\mathbb{R}^{2d}} \hat{\sigma}(z) \hat{\tau}(\zeta - z) e^{\pi i [z,\zeta]} \, dz. \tag{18}$$

For any $f, g \in \mathcal{S}(\mathbb{R}^{2d})$, we define the $\natural^b$ product by

$$f \natural^b g(\zeta) = \iint_{\mathbb{R}^{2d}} f(z) g(z - \zeta) e^{\pi (z\zeta + i[z,\zeta])} \, dz, \tag{19}$$

then the product of localization operators is given by the following result.

Theorem 4.1. *Let $a, b \in \mathcal{S}(\mathbb{R}^{2d})$. If there exists a symbol $c \in \mathcal{S}'(\mathbb{R}^{2d})$ so that*

$$\hat{c} = 2^{-d/2} \hat{a} \natural^b \hat{b}, \tag{20}$$

then we have

$$A_a^{\varphi,\varphi} A_b^{\varphi,\varphi} = A_c^{\varphi,\varphi}.$$

Proof. It is a straightforward consequence of relations (18) and (19). Indeed, we rewrite $A_a^{\varphi,\varphi} A_b^{\varphi,\varphi}$ in the Weyl form and we use relation (18) for the Weyl product: the result is the Weyl operator L_μ, where the Fourier transform of μ is given by

$$\begin{aligned} \hat{\mu}(\zeta) &= \mathcal{F}(a * W(\varphi,\varphi)) \natural \mathcal{F}(b * W(\varphi,\varphi)) \\ &= 2^{-d} \iint_{\mathbb{R}^{2d}} \hat{a}(z) \hat{b}(z - \zeta) e^{-(\pi/2) z^2} e^{-(\pi/2)(\zeta - z)^2} e^{\pi i [z,\zeta]} \, dz \\ &= 2^{-d} e^{-(\pi/2)\zeta^2} \iint_{\mathbb{R}^{2d}} \hat{a}(z) \hat{b}(z - \zeta) e^{\pi (z\zeta + i[z,\zeta])} \, dz. \end{aligned}$$

Hence, we have

$$\hat{\mu}(\zeta) = \hat{c}(\zeta) (2^{-d/2} e^{-(\pi/2)\zeta^2}) = \mathcal{F}(c * W(\varphi,\varphi))(\zeta),$$

where $\hat{c}$ is given by relation (20). □

Next, we want to obtain estimates related to the product $\natural^b$.

Lemma 4.2. *Let $f, g \in \mathcal{S}(\mathbb{R}^{2d})$ and $h = f \natural^b g$. Then, for every constant $\delta > 0$, we have*

$$|h(\zeta)| \lesssim \left(|f(z)| e^{(\pi/2)(2+\delta^2) z^2} * |g(z)| e^{(\pi/2\delta^2) z^2} \right)(\zeta), \qquad \forall \zeta \in \mathbb{R}^{2d}. \tag{21}$$

Proof. We have

$$\begin{aligned} e^{\pi z \zeta} &\leq e^{\pi |z| \, |\zeta|} \leq e^{\pi z^2} \, |\zeta - z| \, |z| = e^{\pi z^2} e^{\pi |\zeta - z|^2 / 2\delta^2} e^{\pi \delta^2 |z|^2 / 2} \\ &\leq e^{\pi (2+\delta^2)(z^2/2)} e^{\pi |\zeta - z|^2 / 2\delta^2}, \end{aligned}$$

then the result is obtained using relation (19). □

Theorem 4.3. *Let $\delta > 0$ be fixed. If $a \in \mathcal{F}L^p_{e^{(\pi/2)(2+\delta^2)z^2}}(\mathbb{R}^{2d})$, $b \in \mathcal{F}L^q_{e^{(\pi/2\delta^2)z^2}}(\mathbb{R}^{2d})$, with $1 \leq p, q \leq \infty$, then the product $A_a^{\varphi,\varphi} A_b^{\varphi,\varphi} = A_c^{\varphi,\varphi}$ gives a localization operator $A_c^{\varphi,\varphi}$ with symbol c satisfying*

$$c \in \mathcal{F}L^r(\mathbb{R}^{2d}), \qquad \textit{with} \quad \frac{1}{r} = \frac{1}{p} + \frac{1}{q} - 1. \tag{22}$$

Proof. It is a straightforward consequence of Young inequality applied to relation (21). □

Corollary 4.4. *Consider the symbols $a \in \mathcal{F}L^p_{e^{(\pi/2)(2+\delta^2)z^2}}(\mathbb{R}^{2d})$, $b \in \mathcal{F}L^q_{e^{(\pi/2\delta^2)z^2}}(\mathbb{R}^{2d})$ with*

$$\frac{1}{p} + \frac{1}{q} = \frac{3}{2},$$

then the product $A_a^{\varphi,\varphi} A_b^{\varphi,\varphi} = A_c^{\varphi,\varphi}$ gives an Hilbert-Schmidt localization operator.

Proof. Using Theorem 4.3, we get $\hat{c} \in \mathcal{F}L^2$. Therefore $c \in L^2$ and the corresponding localization operator $A_c^{\varphi,\varphi}$ is a Hilbert-Schmidt operator (see [6]). □

Let us observe that, using the same technique as in Corollary 4.4, one can also obtain Schatten-class property for the product of two localization operators.

4.2. Asymptotic Product

We end up this section by giving some hints about an asymptotic formula for the product. The motivation comes up from the previous subsection: the product of two localization operators with symbols belonging to some classes does not yield, in general, to a localization operator with symbol in the same class.

In the framework of PDE, the product of two localization operators has been written as a sum of localization operators plus a remainder term expressed in a Weyl, instead of a localization operator, form. The localization operators of this sum have symbols given by pointwise multiplications of the first symbol with derivatives of the second one, while the windows are the same as the factors, but only for Gaussian windows [1]. For the most general formula in the classical framework we refer to [7, 8]; the extension to modulation spaces is contained in [5] and is presented in the subsequent relation (23). In the PDE context, the operator symbols belong to smooth classes, such as the Shubin or the Hörmander classes [23, 28]. In the previous works it has been shown that the bigger the derivative order, the more regular the symbol is. Moreover, the Weyl symbol of the remainder term belongs to a class more regular than the localization symbol of the sum, so that the Weyl remainder could be omitted.

Our point of view is signal analysis rather than PDE. In this context, we deal with rougher symbols and windows, even tempered distributions rather than functions are allowed; the symbol classes do not enjoy the property *of increasing regularity with the derivative order*, typical of the classical case. Hence, the Weyl remainder symbol might be less regular than the localization symbols contained in the sum. It is clear that a different approach is required. The interest is focused on norm estimates for the operators. Since it is hopeless to find a more regular symbol for the remainder when rough symbols are considered, it is fundamental to understand when the localization operators and the Weyl remainder are bounded on suitable Banach spaces. Furthermore, for some applications (e.g., the characterization given by (24)), the Weyl term compactness is needed.

This approach allows to measure how much a localization operator affects the time-frequency concentration of a signal. As usual, the spaces that give time-frequency measures are the modulation spaces. How do we translate the classical product formula [1, 7, 8] in this context? How much can we extend it to hold for rough symbols and windows? How can pointwise symbol multiplication be made when it comes to tempered distributions?

The starting point is the product formula for localization operators contained in [8]. We translate the Shubin calculus into a corresponding one for symbols in modulation spaces. Let us sketch the main results obtained in [5].

Let $N \in \mathbb{N}$ be fixed and φ_i, $i = 1, \ldots, 4$, be non-zero windows in $\mathcal{S}(\mathbb{R}^d)$ (even rougher windows are allowed, with smoothness depending on the integer N). If we consider any rough symbol $a \in \bigcup_{s \geq 0} M^{\infty}_{1/v_s}(\mathbb{R}^{2d}) = \mathcal{S}'(\mathbb{R}^d)$ and we choose a symbol b smooth enough to make the pointwise multiplication $ab \in M^{\infty}_{1/\tau_s}(\mathbb{R}^{2d})$ (e.g., b lives in some suitable Wiener amalgam space), then the composition of the two operators $A_a^{\varphi_1,\varphi_2}$ and $A_b^{\varphi_3,\varphi_4}$ can be written as

$$A_a^{\varphi_1,\varphi_2} A_b^{\varphi_3,\varphi_4} = \sum_{|\alpha|=0}^{N-1} \frac{(-1)^{|\alpha|}}{\alpha!} A_{a\partial^\alpha b}^{\Phi_\alpha,\varphi_2} + L_\sigma. \tag{23}$$

Namely, the product is expressed in terms of a sum of localization operators having the first window Φ_α made of the factor windows $\varphi_1, \varphi_3, \varphi_4$ and depending on the derivative order (for the detailed expression we refer to [5, 8]); the remainder L_σ is a Weyl operator. Moreover, one has *Formula* (23) *is well-defined on every modulation space* $M^{p,q}$; *i.e., the operators on the right hand-side are all linear and bounded on every* $M^{p,q}$.

Also operator norm estimates, involving the time-frequency distribution of symbols and windows, can be easily provided. Here we end up recalling an important application of formula (23). Indeed, formula (23) enables us to get a nice characterization of the weighted modulation space $M^{p,q}_m(\mathbb{R}^d)$, where m is a v-moderate weight. Indeed, if we write the composition formula (23) in the case $N = 1$ and $b = 1/a$, where $a \in C^1(\mathbb{R}^{2d})$ is a positive weight function satisfying the following properties:

(i) $a(z) \asymp m(z)$;
(ii) for $j = 1, \ldots, 2d$, we have

$$\frac{(\partial_j a)(z) v(z)}{m^2(z)} \to 0 \qquad \text{for} \qquad |z| \to \infty,$$

then, the weighted modulations spaces $M^{p,q}_m$ can be defined as localization operator pre-images of the un-weighted $M^{p,q}$. In details, in [5, Thm.7.3], we get the following *characterization for weighted modulation spaces*:

$$M^{p,q}_m(\mathbb{R}^d) = (A^{\varphi,\varphi}_a)^{-1} M^{p,q}(\mathbb{R}^d), \tag{24}$$

for any non-zero window $\varphi \in \mathcal{S}(\mathbb{R}^d)$. Relation (24) represents a generalization of the classical result for Shubin-Sobolev spaces [28], recaptured in the framework of modulation spaces by means of (9).

Acknowledgments. The authors are grateful to the referee for useful comments on the first version of the paper. The first author is grateful to the organizers for their invitation to speak in the special session on pseudo-differential operators at the Fourth ISAAC Congress held at York University in Toronto.

References

[1] H. Ando and Y. Morimoto, Wick calculus and the Cauchy problem for some dispersive equations, *Osaka J. Math.* **39** (2002), 123–147.

[2] F.A. Berezin, Wick and anti-Wick symbols of operators, *Mat. Sb. (N.S.)* **86**(128) (1971), 578–610.

[3] J. Bergh and J.Löfström, *Interpolation Spaces: An Introduction*, Springer-Verlag, Berlin, 1976.

[4] P. Boggiatto, E. Cordero and K. Gröchenig, Generalized anti-Wick operators with symbols in distributional Sobolev spaces, *Integral Equations Operator Theory*, to appear.

[5] E. Cordero and K. Gröchenig, Symbolic calculus for localization operators in modulation spaces, preprint.

[6] E. Cordero and K. Gröchenig, Time-frequency analysis of localization operators, *J. Funct. Anal.* **205** (2003), 107–131.

[7] E. Cordero, F. Nicola and L. Rodino, Microlocal analysis and applications, in *Advances in Pseudo-Differential Operators*, Editors: R. Ashino, P. Boggiatto and M. W. Wong, Birkhäuser, Basel, 2004, 1–17.

[8] E. Cordero and L. Rodino, Wick calculus: a time-frequency approach, *Osaka J. Math.*, to appear.

[9] A. Córdoba and C. Fefferman, Wave packets and Fourier integral operators, *Comm. Partial Differential Equations* **3** (1978), 979–1005. 1978.

[10] I. Daubechies, Time-frequency localization operators: a geometric phase space approach, *IEEE Trans. Inform. Theory* **34** (1988), 605–612.

[11] F. De Mari, H.G. Feichtinger and K. Nowak, Uniform eigenvalue estimates for time-frequency localization operators, *J. London Math. Soc.* (2) **65** (2002), 720–732.

[12] F. De Mari and K. Nowak, Localization type Berezin-Toeplitz operators on bounded symmetric domains, *J. Geom. Anal.* **12** (2002), 9–27.

[13] J. Du and M.W. Wong, A product formula for localization operators, *Bull. Korean Math. Soc.* **37** (2000), 77–84.

[14] H.G. Feichtinger, Banach convolution algebras of Wiener type, in *Functions, Series, Operators*, Vol. I, II, North-Holland, Amsterdam, 1983, 509–524..

[15] H.G. Feichtinger, Generalized amalgams, with applications to Fourier transform, *Canad. J. Math.* **42** (1990), 395–409.

[16] H.G. Feichtinger and K. Nowak, A first survey of Gabor multipliers, in *Advances in Gabor Analysis*, Editors: H. G. Feichtinger and T. Strohmer, Birkhäuser, Boston, 2003, 99–128.

[17] H.G. Feichtinger and G. Zimmermann, A Banach space of test functions for Gabor analysis, in *Gabor Analysis and Algorithms*, Editors: H. G. Feichtinger and T. Strohmer, Birkhäuser, Boston, 1998, 123–170.

[18] G.B. Folland, *Harmonic Analysis in Phase Space*, Princeton Univ. Press, Princeton, NJ, 1989.

[19] J.J.F. Fournier and J. Stewart, Amalgams of L^p and l^q, *Bull. Amer. Math. Soc.* **13** (1985), 1–21.

[20] K. Gröchenig, *Foundations of Time-Frequency Analysis*, Birkhäuser, Boston, 2001.

[21] K. Gröchenig and C. Heil, Modulation spaces and pseudo-differential operators, *Integral Equations Operator Theory* **34** (1999), 439–457.

[22] K. Gröchenig and G. Zimmermann, Hardy's theorem and the short-time Fourier transform of Schwartz functions, *J. London Math. Soc.* **63** (2001), 205–214, 2001.

[23] L. Hörmander, *The Analysis of Linear Partial Differential Operators III*, Springer-Verlag, Berlin, 1985.

[24] R. Howe, Quantum mechanics and partial differential equations, *J. Funct. Anal.* **38** (1980), 188–254.

[25] Y. Katznelson, *An Introduction to Harmonic Analysis*, John Wiley & Sons, New York, 1968.

[26] N. Lerner, The Wick calculus of pseudo-differential operators and energy estimates, in *New Trends in Microlocal Analysis*, Springer, Tokyo, 1997, 23–37.

[27] J. Ramanathan and P. Topiwala, Time-frequency localization via the Weyl correspondence, *SIAM J. Math. Anal.* **24** (1993), 1378–1393.

[28] M.A. Shubin, *Pseudo-Differential Operators and Spectral Theory*, Second Edition, Springer-Verlag, Berlin, 2001.

[29] E.M. Stein, *Singular Integrals and Differentiability Properties of Functions*, Princeton Univ. Press, Princeton, NJ, 1970.

[30] J. Toft, Continuity properties for modulation spaces with applications in pseudo-differential calculus I, *J. Funct. Anal.*, to appear.

[31] F. Trèves, *Topological Vector spaces, Distributions and Kernels*, Academic Press, New York, 1967.

[32] M.W. Wong, *Weyl Transforms*, Springer-Verlag, New York, 1998.

[33] M.W. Wong, *Localization Operators*, Seoul National University, Research Institute of Mathematics, Global Analysis Research Center, Seoul, 1999.

[34] M.W. Wong, *Wavelet Transforms and Localization Operators*, Birkhäuser, Basel, 2002.

E. Cordero and A. Tabacco
Department of Mathematics
Politecnico of Torino
Corso Duca degli Abruzzi 24
10129 Torino
Italy
e-mail: cordero@dm.unito.it, cordero@calvino.polito.it
anita.tabacco@polito.it

Operator Theory:
Advances and Applications, Vol. 155, 149–163

Localization Operators with L^p Symbols on Modulation Spaces

Paolo Boggiatto

Abstract. Boundedness of localization operators on modulation spaces is studied obtaining results for operators with symbol in $L^p(\mathbb{R}^{2n})$, $1 \leq p \leq \infty$. Compactness properties are then proved in the case $p < \infty$ and for bounded symbols vanishing at infinity. Finally the results obtained for localization operator are shown to extend to Weyl operators on Sobolev spaces in the case where the symbol belongs to the Shubin classes $\Gamma^k_\rho(\mathbb{R}^{2n})$. In this context the results presented here generalize well-known properties of boundedness and compactness.

Mathematics Subject Classification (2000). Primary 46E35, 42B35; Secondary 46B50.

Keywords. Localization operators, Weyl operators, Sobolev spaces, modulation spaces.

1. An Overview of the Subject

Localization operators, in some cases known as *Toeplitz operators* or *anti-Wick operators*, arise in pure and applied mathematics in connection with quite different areas of research. On one hand if $u(t)$ represents a signal, i.e. a function of the time t, then its Fourier transform $\hat{u}(\omega)$ represents the distribution of the frequencies ω contained in the signal. However frequencies in real signals vary generally in time and each frequency occupies only a bounded interval of time, such as in a piece of music every note has a finite duration. The knowledge of the frequencies contained in the whole signal, without information about the time at which they are present, is therefore not very useful for the analysis of general time varying signals. In order to overcome this problem a very basic idea, originally due to Gabor, is to focus the Fourier transform on small intervals in time and analyze the frequencies present in these intervals. This can be done by multiplying the function $u(t)$ by a cut-off, or *window*, function $\phi(t)$ before taking the Fourier transform. The analysis of the frequencies concentrated in a surrounding of a fixed time x amounts then to the calculation of the integral

$$V_\phi u(x,\omega) = \int_{\mathbb{R}^n} e^{-2\pi it\omega}\phi(t-x)u(t)dt.$$

The function $V_\phi u(x,\omega)$, called *short-time Fourier transform* (briefly *STFT*) of u, expresses the joint time-frequency content of the signal. The variables x and ξ are respectively *time* and *frequency* variables and the space $\mathbb{R}^n_x \times \mathbb{R}^n_\omega$ is called *time-frequency plane.*
Due to the Heisenberg uncertainty principle however the size of the support of the window function can not shrink arbitrarily without a loss of significance in the information obtained about the frequencies, so that an "instantaneous frequency" analysis makes in reality no sense.

The signal $u(t)$ can be reconstructed (*synthesis process*) from its STFT by the *inversion* (or *reconstruction*) formula:

$$u(s) = \frac{1}{\|\phi\|_{L^2}} \int_{\mathbb{R}^{2n}} V_\phi u(x,\omega)e^{2\pi is\omega}\phi(s-x)\,dx\,d\omega \tag{1.1}$$

The literature on this topic is considerable, we refer e.g. to [15], [21] for a more extended description of details and results. Before being reconstructed, the signal undergoes in general a *processing* or *filtering* that consists in a modification of its STFT $V_\phi u$. This can be achieved through a multiplication by a function $F(x,\omega)$ that acts as a *filter* and amplifies or annihilates different parts of the signal situated in different regions of the time-frequency plane. This procedure leads to consider operators of the type:

$$L^F_\phi u(s) = \int_{\mathbb{R}^{2n}} F(x,\omega)V_\phi u(x,\omega)e^{2\pi is\omega}\phi(s-x)\,dx\,d\omega \tag{1.2}$$

that represent the synthetised signal.
Operators of type (1.2) are called *localization operators*; a fundamental reference on this topics is [10]. A discrete version of localization operators in the context of signal analysis is studied in [14].

On the other hand, it is a remarkable fact that, independently of signal analysis, the same mathematical structures also underlie some of the basic constructions of Quantum Mechanics. Namely a state of a physical system is represented by a function $u(t) \in L^2(\mathbb{R}^n)$ and in this context the reconstruction formula (1.1) gives a decomposition of the state $u(t)$ into the *(generalized) coherent states*: $e^{2\pi is\omega}\phi(s-x)$.
These are realized as translations and modulations, i.e. multiplication by the characters $e^{i2\pi tx}$, of a fixed state function $\phi(t)$.

Operators of type (1.2) correspond in this context to the so-called *anti-Wick* quantization rule, see Berezin [3]. The formalization of this quantization in the frame of a suitable pseudodifferential calculus is developed in Shubin [26]. Anti-Wick operators have revealed to be a very useful tool in many questions in the theory of pseudodifferential operators, for reference see e.g. [4], [7], [24].

Translations and modulations, called *time-frequency shifts* from the language of signal analysis, are essentially the action of the Weyl-Heisenberg group on the Hilbert space $L^2(\mathbb{R}^n)$, so operators of type (1.2) could be said to be associated in a natural way with the Weyl-Heisenberg group. The important rôle played by group theory in the enlightening of the basic structures of localization operators is very well presented for instance in [2], [15], [20]. A general theory of localization operators associated with square integrable representations is developed in [18], [31].

For what concerns the spaces on which the operators act, they were traditionally $L^2(\mathbb{R}^n)$ or some types of Sobolev spaces.

A remarkable class of Banach spaces, originally defined to measure the time-frequency content of functions and temperate distributions, are the *modulation spaces* introduced by Feichtinger in [11] and generalized by Feichtinger and Gröchenig in [12], [13]. Their connection to localization operators is analogous to that of the Besov spaces to the wavelet transform and they contain as particular cases most of the usual Sobolev spaces. Modulation spaces has revealed an ideal setting where a considerable amount of research in time-frequency analysis and Weyl operators in the last two decades has been developed, see for references [17], [19], [22], [23], [25], [27], [28].

As symbols for Weyl operators, modulation spaces have been considered in [16] and for localization operators in [6] and [9], where various boundedness and Schatten-von Neumann properties are proved.

The study of localization operators has concentrated principally on the action of the operators on $L^2(\mathbb{R}^n)$, whereas not many results about localization operators acting on other types of Banach spaces are known. Some results in this direction are contained in [8], where L^p-boundedness and compactness are studied.

Results for Weyl operators on modulation spaces with symbols in modulation spaces are presented in [29] and [30], including also results on localization operators.

As modulation spaces appear natural to "measure" the time-frequency content of signals, whereas most natural filters belongs L^p spaces, we want to study in this paper localization operators with L^p symbols acting on modulation spaces.

Section 2 is a collection of basic properties that will be used later on. We study boundedness for localization operators in section 3 making use of time-frequency analysis techniques and interpolation theory. In section 4 compactness for localization operators is studied after we have proved some general compactness results of regularizing operators. Finally in section 5 we prove some consequences of the results of the previous two sections for Weyl operators on Sobolev spaces extending known boundedness and compactness results.

The author is thankful to J. Toft and K.-H. Gröchenig for many helpful comments and suggestions.

2. Modulation Spaces and Localization Operators

Let $z \in \mathbb{R}^d$, we use the constants convention by which we assume that in every inequality of the type $f(z) \leq Cg(z)$ the positive constant C can be different.
A weight function on $\mathbb{R}^d$ is a sub-multiplicative positive function $v(z)$, i.e. $0 < v(z_1 + z_2) \leq v(z_1)v(z_2)$ for every $z_1, z_2 \in \mathbb{R}^d$. A v-moderate weight function $m(z)$ is a function such that $0 < m(z_1 + z_2) \leq Cv(z_1)m(z_2)$. The standard example of weight function is $\langle z \rangle = \sqrt{1 + |z|^2}$.
We assume further that both $v(z)$ and $1/v(z)$ have *tempered growth* i.e. there exists $k > 0$ such that

$$v(z) + 1/v(z) \leq C\langle z \rangle^k. \tag{2.1}$$

We shall be concerned with the case $d = 2n$, $z = (x, \omega) \in \mathbb{R}^{2n}$. In this case *time-frequency shifts* of a function $u(t)$ on $\mathbb{R}^n$ are defined as $u_z(t) = M_\omega T_x u(t)$ with $T_x u(t) = u(t - x)$, $M_\omega u(t) = e^{i2\pi t\omega} u(t)$.

We summarize next some basic definitions and properties, cf.. [15].

Definition 2.1. Let g be a fixed function (*window function*), then, whenever this makes sense, the *short-time Fourier transform (STFT)* of the function u is defined as $V_g u(z) = (u, g_z)_{L^2} = \int_{\mathbb{R}^n} e^{-2\pi it\omega} u(t)\overline{g(t-x)}dt$.

Proposition 2.2. *If* $u, v, g, f \in L^2(\mathbb{R}^n)$ *then* $V_g u$ *and* $V_f v$ *are in* $L^2(\mathbb{R}^{2n})$ *and*

$$(V_g u, V_f v)_{L^2} = (u, v)_{L^2}(g, f)_{L^2}, \tag{2.2}$$

in particular $V_g : L^2(\mathbb{R}^n) \to L^2(\mathbb{R}^{2n})$ *is an isometry if* $\|g\|_{L^2} = 1$.
Furthermore the application $(g, u) \to V_g u$ *is a continuous map from* $S(\mathbb{R}^n) \times S(\mathbb{R}^n)$ *to* $S(\mathbb{R}^n)$ *extendible to a map from* $S'(\mathbb{R}^n) \times S'(\mathbb{R}^n)$ *to* $S'(\mathbb{R}^n)$.

If $m(x, \omega)$ is a fixed weight function on $\mathbb{R}^{2n}$ and $p, q \in [1, +\infty]$, let $L^{p,q}_m(\mathbb{R}^{2n})$ denote the *weighted mixed-norm space* of the measurable functions φ on $\mathbb{R}^{2n}$ such that $\|\varphi\|_{L^{p,q}_m} = (\int_\omega (\int_x m(x,\omega)^p |\varphi(x,\omega)|^p dx)^{q/p} d\omega)^{1/q} < +\infty$.

Definition 2.3. Let $m(x, \omega)$ be a $v-$moderate weight function, $p, q \in [1, +\infty]$ and $0 \neq g \in S(\mathbb{R}^n)$. The modulation space $M^{p,q}_m(\mathbb{R}^n)$ is defined as

$$M^{p,q}_m(\mathbb{R}^n) = \{u \in S'(\mathbb{R}^n) : V_g u \in L^{p,q}_m(\mathbb{R}^{2n})\} \tag{2.3}$$

The space $M^{p,q}_m(\mathbb{R}^n)$ is independent of $0 \neq g \in S(\mathbb{R}^n)$ and even more generally one can suppose $g \in M^1_v(\mathbb{R}^n)$.
The abbreviations $M^{p,q}_m(\mathbb{R}^n) = M^{p,q}_m$, $M^{p,p}_m = M^p_m$ and $M^{p,q}_1 = M^{p,q}$ are commonly used.
If $p \in [1, +\infty]$, we indicate with p' the *conjugate* of p, i.e. the real number defined by $\frac{1}{p} + \frac{1}{p'} = 1$; the same holds for q, α, β.

We summarize in the following proposition the properties of modulation spaces that we shall need.

Proposition 2.4. *Let* $g \in M^1_v(\mathbb{R}^n)$ *and let* m *be a* $v-$*moderate weight function on* $\mathbb{R}^{2n}$. *Then the following assertions hold:*

(a) *$M^{p,q}_m(\mathbb{R}^n)$ is a Banach space with respect to the norm $\|.\|_{M^{p,q}_m} := \|V_g(.)\|_{L^{p,q}_m}$ and different functions $0 \neq g \in M^1_v(\mathbb{R}^n)$ give rise to equivalent norms.*

(b) *$M^{2,2}_m(\mathbb{R}^n)$ is a Hilbert space with respect to the inner product $(.,.)_{M^{2,2}_m} := (V_g(.), V_g(.))_{L^2_m(\mathbb{R}^{2n})}$.*

(c) *If $p, q \in [1, +\infty]$ then, for $u \in M^{p,q}_m(\mathbb{R}^n)$, $v \in M^{p',q'}_{1/m}(\mathbb{R}^n)$ we have $|(V_g u, V_g v)_{L^2}| \leq \|u\|_{M^{p,q}_m} \|v\|_{M^{p',q'}_{1/m}}$, and for $1 \leq p, q < \infty$, the pairing $(.,.)_{M^{p,q}_m, M^{p',q'}_{1/m}} := (V_g(.), V_g(.))_{L^2}$ gives a standard identification of the dual of $M^{p,q}_m(\mathbb{R}^n)$ with $M^{p',q'}_{1/m}(\mathbb{R}^n)$.*

(d) *If $p_1 \leq p_2$, $q_1 \leq q_2$, with $p_1, p_2, q_1, q_2 \in [1, \infty]$, and $m_1(z) \geq m_2(z)$, we have the continuous embedding $M^{p_1,q_1}_{m_1}(\mathbb{R}^n) \hookrightarrow M^{p_2,q_2}_{m_2}(\mathbb{R}^n)$.*

(e) *$\bigcap_{s\in\mathbb{R}} M^{1,1}_{\langle . \rangle^s}(\mathbb{R}^n) = S(\mathbb{R}^n)$; $\bigcup_{s\in\mathbb{R}} M^{\infty,\infty}_{\langle . \rangle^s}(\mathbb{R}^n) = S'(\mathbb{R}^n)$ (with $\langle\ .\ \rangle = \sqrt{1 + |\ .\ |^2}$).*

(g) *We have, for $p, q \in [1, \infty]$, continuous embeddings $S(\mathbb{R}^n) \hookrightarrow M^{p,q}_m(\mathbb{R}^n) \hookrightarrow S'(\mathbb{R}^n)$ (this is the essential reason of the requirement (2.1)). Furthermore, for $1 \leq p, q < \infty$, $S(\mathbb{R}^n)$ is dense in $M^{p,q}_m$.*

The modulation space $M^{1,1}_1(\mathbb{R}^n)$ is an algebra, called *Feichtinger Algebra*. The Feichtinger algebra and its dual space $M^{\infty,\infty}_1(\mathbb{R}^n)$, contained in the spaces of tempered distributions, play in the context of modulation spaces a similar rôle as the Schwarz space $S(\mathbb{R}^n)$ and its dual $S'(\mathbb{R}^n)$ in the usual distribution theory.

If $g \in L^2(\mathbb{R}^n)$ with $\|g\|_{L^2} = 1$ then, from (2.2), we see that the product in the spaces $M^{2,2}_m$ as well as the pairing in Proposition 2.4 (c) are just a restriction/extension of the L^2 product. With abuse of notation we shall write therefore (u, v) instead of $(u, v)_{M^{2,2}_m}$ and $(u, v)_{M^{p,q}_m, M^{p',q'}_{1/m}}$.

Modulation spaces include many Sobolev-types spaces, in particular let us consider the pseudodifferential operators.

$$P_s u(x) = \int_{\mathbb{R}^{2n}} e^{i(x-y)\omega} \langle\omega\rangle^s u(y)\, dy\, d\omega,$$

$$\Lambda_s u(x) = \int_{\mathbb{R}^{2n}} e^{i(x-y)\omega} \langle(x,\omega)\rangle^s u(y)\, dy\, d\omega.$$

Then *Bessel potential spaces* and *Shubin-Sobolev spaces* (see [26], Ch. IV) are defined respectively as:

$$W^p_s(\mathbb{R}^n) = \{u \in S'(\mathbb{R}^n) : P_s u \in L^p(\mathbb{R}^n)\}$$

$$Q_s(\mathbb{R}^n) = \{u \in S'(\mathbb{R}^n) : \Lambda_s u \in L^2(\mathbb{R}^n)\}.$$

We have that the following identifications hold (cf.. [6]).

Proposition 2.5. *$M^2_{\langle\omega\rangle^s}(\mathbb{R}^n) = W^2_s(\mathbb{R}^n)$ and $M^2_{\langle z\rangle^s}(\mathbb{R}^n) = Q_s(\mathbb{R}^n)$, for every $s \in \mathbb{R}$.*

Definition 2.6. Let $\phi, \psi \in S(\mathbb{R}^n)$, $F \in S'(\mathbb{R}^n)$. The linear continuous operator from $S(\mathbb{R}^n)$ to $S'(\mathbb{R}^n)$ defined by

$$(L^F_{\phi,\psi} u, v) = (F, V_\phi u \overline{V_\psi v}), \qquad u, v \in S(\mathbb{R}^n), \tag{2.4}$$

is called *localization operator* with symbol F and window functions ϕ, ψ.

More generally the definition of $L^F_{\phi,\psi}$ can be extended to windows functions $\phi, \psi \in M^1_m(\mathbb{R}^n)$, see [9].

Naturally it is also convenient to rewrite (2.4) more explicitly and accordingly to (1.2) we have:

$$L^F_{\phi,\psi}u(t) = \int_{\mathbb{R}^{2n}} F(z)(u,\phi_z)\psi_z(t)dz, \qquad u \in S(\mathbb{R}^n), \tag{2.5}$$

whenever the integral exists or can be interpreted in a week sense. As can easily be verified, we also have the factorization

$$L^F_{\phi,\psi}u = V^*_\psi F V_\phi u \tag{2.6}$$

where V^*_ψ is the adjoint of $V_\psi : L^2(\mathbb{R}^n) \to L^2(\mathbb{R}^{2n})$.
If $\psi = \phi = g$ with $g(z)$ $L^2(\mathbb{R}^n)$–normalized Gaussian, then the localization operator $L^F_{g,g}$ is also called *anti-Wick* operator. These operators have been widely used in pseudo-differential calculus independently from the theory of localization operators and signal analysis, see [26] Chp. IV.24 and [4] Chp. I.6.

3. Boundedness of Localization Operators

Throughout this section let $v(z)$ be a weight function with tempered growth and $m_1(z), m_2(z)$ be v–moderate weight functions. We further assume $\phi, \psi \in M^1_v(\mathbb{R}^n)$ are window functions, F is a measurable function belonging to $S'(\mathbb{R}^n)$ and we indicate with $L^F_{\phi,\psi}$ the localization operator defined as in (2.4) (or equivalently (2.5),(2.6)).

We begin with a technical lemma about mixed-norm spaces.

Lemma 3.1. *Let $\alpha, \beta, \gamma, \delta \in [1, +\infty]$ and $F \in L^{\alpha\gamma,\beta\delta}(\mathbb{R}^n), G \in L^{\alpha'\gamma,\beta'\delta}(\mathbb{R}^n)$. Then $FG \in L^{\gamma,\delta}(\mathbb{R}^n)$ and*

$$\|FG\|_{L^{\gamma,\delta}(\mathbb{R}^n)} \leq \|F\|_{L^{\alpha\gamma,\beta\delta}(\mathbb{R}^n)} \|G\|_{L^{\alpha'\gamma,\beta'\delta}(\mathbb{R}^n)} \tag{3.1}$$

Proof. It is just a straightforward computation based on the Hölder inequality, generalized to mixed-norm spaces: $\int_{\mathbb{R}^{2n}} |F(z)H(z)|dz \leq \|F\|_{L^{p,q}_m} \|H\|_{L^{p',q'}_{1/m}}$ □

As the topology of modulation spaces is not given by a fixed standard norm but by a family of equivalent norms, we shall have boundedness results where operators norm estimates are modulo a multiplicative constant.

Proposition 3.2. *Let $\alpha, \beta, p, q \in [1, \infty]$ and F, m_1, m_2 satisfy the condition*

$$\frac{F(z)m_2(z)}{m_1(z)} \in L^{\alpha'p,\beta'q}(\mathbb{R}^{2n}). \tag{3.2}$$

Then the localization operator $L^F_{\phi,\psi} : M^{\alpha p,\beta q}_{m_1}(\mathbb{R}^n) \to M^{p,q}_{m_2}(\mathbb{R}^n)$ is bounded with norm estimate

$$\|L^F_{\phi,\psi}\| \leq C\|\phi\|_{M^1_v}\|\psi\|_{M^1_v} \left\| \frac{F(z)m_2(z)}{m_1(z)} \right\|_{L^{\alpha'p,\beta'q}}. \tag{3.3}$$

Proof. Suppose that $g(z)$ is a $L^2(\mathbb{R}^n)-$normalized Gaussian, then from (2.6) and the definition of the STFT, and using the fact that m_2 is $v-$moderate and $|V_\psi g_z(w)| = |T_z V_\psi g(w)|$, $w \in \mathbb{R}^{2n}$, we can write

$$\begin{aligned} m_2(z)\left|(V_g L^F_{\phi,\psi}u)(z)\right| &= m_2(z)\left|(L^F_{\phi,\psi}u, g_z)\right| \\ &= m_2(z)\left|(FV_\phi u, V_\psi g_z)\right| \\ &\leq C\int m_2(w)\ |F(w)V_\phi u(w)|\ \ v(z-w)\ |V_\psi g_z(w)|dw \\ &= \int m_2(w)\ |F(w)V_\phi u(w)|\ \ v(z-w)\ |T_z V_\psi g(w)|dw \\ &= \left(m_2|FV_\phi u| * v|\check{V_\psi g}|\right)(z) \end{aligned}$$

where we have denoted $\check{V_\psi g}(w) = V_g\psi(-w)$. Using this estimate, the mixed-norm convolution estimates $\|F * G\|_{L^{p,q}} \leq C\|F\|_{L^{p,q}}\|G\|_{L^1}$ and Lemma 3.1 with $\gamma = p, \delta = q$ we have

$$\begin{aligned} \|L^F_{\phi,\psi}u\|_{M^{p,q}_{m_2}} &= \|m_2 V_g L^F_{\phi,\psi}u\|_{L^{p,q}} \\ &\leq \|m_2|FV_\phi u| * v|\check{V_\psi g}|\|_{L^{p,q}} \\ &\leq \|m_2 FV_\phi u\|_{L^{p,q}}\|v\check{V_\psi g}\|_{L^1} \\ &\leq \left\|\frac{m_2}{m_1}F\right\|_{L^{\alpha' p,\beta' q}} \|m_1 V_\phi u\|_{L^{\alpha p,\beta q}}\|v\check{V_\psi g}\|_{L^1} \end{aligned} \tag{3.4}$$

A simple computation shows that $\check{V_\psi g}(z) = e^{-2\pi i x\omega}V_g\psi(z)$, for every $z = (x,\omega) \in \mathbb{R}^{2n}$, so that $|\check{V_\psi g}(z)| = |V_g\psi(z)|$ and we have

$$\|v\check{V_\psi g}\|_{L^1} = \|vV_g\psi\|_{L^1} = \|\psi\|_{M^1_v} \tag{3.5}$$

On the other hand, supposing g is not orthogonal to ϕ, the pointwise estimate holds (see e.g. [15]), Lemma 11.3.3):

$$|V_\phi u(z)| \leq \frac{1}{|(g,\phi)|}(|V_g u| * |V_\phi g|)(z).$$

Using this estimate, the mixed-norm convolution estimate
$\|F * G\|_{L^{p,q}_{m_1}} \leq C\|F\|_{L^{p,q}_{m_1}}\|G\|_{L^1_v}$ and (3.5), we have

$$\begin{aligned} \|m_1 V_\phi u\|_{L^{\alpha p,\beta q}} = \quad \|V_\phi u\|_{L^{\alpha p,\beta q}_{m_1}} &\leq C\|(|V_g u| * |V_\phi g|)\ \|_{L^{\alpha p,\beta q}_{m_1}} \\ &\leq C\|V_g u\|_{L^{\alpha p,\beta q}_{m_1}}\|V_\phi g\|_{L^1_v} \\ &= C\|V_g u\|_{L^{\alpha p,\beta q}_{m_1}}\|V_g \phi\|_{L^1_v} \\ &= C\|u\|_{M^{\alpha p,\beta q}_{m_1}}\|\phi\|_{M^1_v} \end{aligned} \tag{3.6}$$

From (3.4), (3.5), (3.6) we obtain

$$\|L^F_{\phi,\psi}u\|_{M^{p,q}_{m_2}} \leq C\|u\|_{M^{\alpha p,\beta q}_{m_1}}\|\phi\|_{M^1_v}\|\psi\|_{M^1_v}\left\|\frac{m_2}{m_1}F\right\|_{L^{\alpha' p,\beta' q}} \tag{3.7}$$

and therefore the thesis is proved. □

If we let $\alpha p = p_1$, $\beta q = q_1$, $\alpha' p = p_0$, $\beta' q = q_0$, $p = p_2$, $q = q_2$, in the previous proposition we get the following reformulation.

Proposition 3.3. *Assume that* $p_0, p_1, p_2, q_0, q_1, q_2 \in [1, \infty]$, *and* F, m_1 *and* m_2 *satisfy*

$$\frac{1}{p_0} = \frac{1}{p_2} - \frac{1}{p_1}, \qquad \frac{1}{q_0} = \frac{1}{q_2} - \frac{1}{q_1}, \qquad \frac{F(z)m_2(z)}{m_1(z)} \in L^{p_0,q_0}(\mathbb{R}^{2n}). \tag{3.8}$$

Then the localization operator $L^F_{\phi,\psi} : M^{p_1,q_1}_{m_1} \longrightarrow M^{p_2,q_2}_{m_2}$ *is bounded with norm estimate* $\|L^F_{\phi,\psi}\| \leq C\|\phi\|_{M^1_v}\|\psi\|_{M^1_v}\|Fm_2/m_1\|_{L^{p_0,q_0}}$.

Remark 3.4. We remark that necessarily we have $p_2 \leq p_1$ and $q_2 \leq q_1$ (strictly for $p_0 < \infty$, $q_0 < \infty$) and therefore $M^{p_2,q_2}_{m_2} \hookrightarrow M^{p_1,q_1}_{m_1}$ so that we observe a certain regularizing effect of the corresponding localization operators with respect to the p, q indices of modulation spaces.

A corollary of proposition 3.3 is the following.

Corollary 3.5. *Assume that* $p_0, p_1, q_0, q_1, \in [1, \infty]$ *satisfy the condition*

$$\frac{1}{p_0} + \frac{1}{p_1} \leq 1, \quad \frac{1}{q_0} + \frac{1}{q_1} \leq 1, \tag{3.9}$$

$F \in L^{p_0,q_0}(\mathbb{R}^{2n})$ *and* m *is* v*-moderate. Then the localization operator* $L^F_{\phi,\psi}$ *is bounded on* M^{p_1,q_1}_m *with norm estimate* $\|L^F_{\phi,\psi}\| \leq C\|\phi\|_{M^1_v}\|\psi\|_{M^1_v}\|F\|_{L^{p_0,q_0}}$.

Proof. Under condition (3.9) there exists $p_2 \in [1, \infty]$, $q_2 \in [1, \infty]$ satisfying (3.8) and therefore Proposition 3.3, in the case $m_1 = m_2 = m$, shows that $L^F_{\phi,\psi}$ is bounded from M^{p_1,q_1}_m to M^{p_2,q_2}_m. However, condition (3.8) implies $p_2 \leq p_1$ and $q_2 \leq q_1$, so that the continuous inclusion $M^{p_2,q_2}_m \hookrightarrow M^{p_1,q_1}_m$ shows that $L^F_{\phi,\psi}$ is bounded on M^{p_1,q_1}_m. □

We shall see in Proposition 3.8 that condition (3.9) can be omitted at least in the case $p_0 = q_0$.

In view of an application of interpolation theory, we state now explicitly two particular cases of Proposition 3.3. Suppose m and σ are $v-$moderate weight functions and consider the case $m_1 = m$, $m_2 = m\sigma$. Proposition 3.3, for $p_0 = q_0 = \infty$, gives then the following result.

Proposition 3.6. *Let* $F \in L^\infty_\sigma(\mathbb{R}^{2n})$. *Then the operator* $L^F_{\phi,\psi}$ *is bounded from* $M^{p,q}_m(\mathbb{R}^n)$ *to* $M^{p,q}_{m\sigma}(\mathbb{R}^n)$ *for every* $1 \leq p, q \leq +\infty$ *with norm estimate:* $\|L^F_{\phi,\psi}\| \leq C\|\phi\|_{M^1_v}\|\psi\|_{M^1_v}\|F\|_{L^\infty_\sigma}$.

We observe that Proposition 3.6 contains Remark 2, sec.3 of [6].
The other particular case is $p_0 = q_0 = 1$ where, from Proposition 3.3 again with $m_1 = m$, $m_2 = m\sigma$, follows the following stronger boundedness result.

Proposition 3.7. *Let $F \in L^1_\sigma(\mathbb{R}^{2n})$. Then, for every $p_1, q_1, p_2, q_2 \in [1, +\infty]$, the operator $L^F_{\phi,\psi}$ is bounded from $M^{p_1,q_1}_m(\mathbb{R}^n)$ to $M^{p_2,q_2}_{m\sigma}(\mathbb{R}^n)$ and $\|L^F_{\phi,\psi}\| \leq C\|\phi\|_{M^1_v}\|\psi\|_{M^1_v}\|F\|_{L^1_\sigma}$.*

Proof. Proposition 3.3 yields in this case the boundedness from $M^{\infty,\infty}_m(\mathbb{R}^n)$ to $M^{1,1}_{m\sigma}(\mathbb{R}^n)$. Recalling the continuous immersions between modulation spaces (Proposition 2.4, part (d)) we obtain the following bounded sequence of operators

$$M^{p_1,q_1}_m(\mathbb{R}^n) \overset{id}{\hookrightarrow} M^{\infty,\infty}_m(\mathbb{R}^n) \overset{L^F_{\phi,\psi}}{\longrightarrow} M^{1,1}_{m\sigma}(\mathbb{R}^n) \overset{id}{\hookrightarrow} M^{p_2,q_2}_{m\sigma}(\mathbb{R}^n).$$

The norm estimate follows immediately from (3.3). □

Consider now the space $B(M^{p,q}_m, M^{p,q}_{m\sigma})$ of bounded linear operators from $M^{p,q}_m$ to $M^{p,q}_{m\sigma}$. From Proposition 3.6 we have that the linear map $F \to L^F_{\phi,\psi}$ is bounded from $L^\infty_\sigma(\mathbb{R}^{2n})$ to $B(M^{p,q}_m, M^{p,q}_{m\sigma})$ and from Proposition 3.7 with $p_1 = p_2 = p$, $q_1 = q_2 = q$, the same map is also bounded from $L^1_\sigma(\mathbb{R}^{2n})$ to $B(M^{p,q}_m, M^{p,q}_{m\sigma})$. Then using interpolation (see e.g. Thm 2.10 of [31]) we have proved the following proposition.

Proposition 3.8. *Let $F \in L^r_\sigma(\mathbb{R}^{2n})$, $r \in [1, \infty]$, and m, σ be v-moderate weight functions. Then for every $p, q, \in [1, +\infty]$ we have a bounded localization operator*

$$L^F_{\phi,\psi} : M^{p,q}_m(\mathbb{R}^n) \to M^{p,q}_{m\sigma}(\mathbb{R}^n).$$

with norm estimate $\|L^F_{\phi,\psi}\| \leq C\|\phi\|_{M^1_v}\|\psi\|_{M^1_v}\|F\|_{L^r_\sigma}$.

We mention also the following other result obtained by interpolation.

Proposition 3.9. *Under the same hypothesis of Proposition 3.8, $L^F_{\phi,\psi}$ is a bounded localization operator between the following modulation spaces:*

$$\begin{array}{ll} a) & L^F_{\phi,\psi} : M^{\infty,\infty}_m \to M^{r,r}_m \\ b) & L^F_{\phi,\psi} : M^{r',r'}_m \to M^{1,1}_m \end{array} \tag{3.10}$$

with consequent norm estimates.

Proof. In case (a) we interpolate between $B(M^{\infty,\infty}_m)$ and $B(M^{\infty,\infty}_m, M^{1,1}_m)$. In case (b) we interpolate between $B(M^{1,1}_m)$ and $B(M^{\infty,\infty}_m, M^{1,1}_m)$, (for interpolation between modulation spaces see e.g. [11], Thm. 6.1 (d)). □

4. Compact Localization Operators

Let E_1 and E_2 be Banach spaces. We denote with $B(E_1, E_2)$ the space of bounded linear operators from E_1 to E_2 and with $K(E_1, E_2)$ the subspace of compact operators (if $E_1 = E_2$ we write simply $B(E_1)$ and $K(E_1)$).

Lemma 4.1. *Let S be a dense subset in E_1 and $T \in B(E_1, E_2)$. Then the following assertions are equivalent.*

i) *For every sequence $\{u_j\} \subset S$, weekly convergent in E_1, the sequence Tu_j is strongly convergent in E_2.*
ii) *For every sequence $\{u_j\} \subset E_1$, weekly convergent in E_1, the sequence Tu_j is strongly convergent in E_2.*

Proof. We show that i) implies ii). Let $u_j \in E_1$ be weekly convergent to zero. For Every $u_j \in E_1$ there exists a sequence $\{v_j^k\}_{k=1}^{\infty} \in S$ such that $\|u_j - v_j^{(k)}\|_{E_1} \leq 1/k$. By Cantor diagonal method $v_j^{(j)}$ converges weekly to zero in E_1 as $j \to \infty$, then $Tv_j^{(j)}$ converge strongly to zero in E_2. We have
$\|Tu_j\| \leq \|Tu_j - Tv_j^{(j)}\|_{E_2} + \|Tv_j^j\|_{E_2} \leq \|T\|_{B(E_1,E_2)} \ \|u_j - v_j^{(j)}\|_{E_1} + \|Tv_j^{(j)}\|_{E_2}$, which tends to zero. □

If E_1 is reflexive then the compactness of T is equivalent to condition ii) of the previous proposition (cf.. [1]) and we have therefore:

Proposition 4.2. *In the same hypothesis of the previous lemma, if we assume further that E_1 is reflexive, then* i) *is equivalent to $T \in K(E_1, E_2)$.*

With the following proposition we formalize a situation that occurs in many practical realizations of the Banach spaces E_1 and E_2.

Proposition 4.3. *Suppose that T is a linear sequentially continuous operator from $S'(\mathbb{R}^{2n})$ to $S(\mathbb{R}^{2n})$ (i.e. it takes pointwise convergent sequences of tempered distributions into sequences converging in $S(\mathbb{R}^{2n})$). Let E_1 be a reflexive Banach space and E_2 a Banach space satisfying the following conditions:*

i) *$S(\mathbb{R}^{2n})$ is dense in E_1; $S(\mathbb{R}^{2n}) \subset E_1{}'$.*
ii) *$S(\mathbb{R}^{2n}) \hookrightarrow E_2$ is a continuous embedding.*
iii) *If $\alpha \in E_1{}'$, $u \in E_1$ then $\alpha(u) = \int_{\mathbb{R}^{2n}} \alpha u$ whenever $\alpha \in S(\mathbb{R}^{2n})$ and $u \in S(\mathbb{R}^{2n})$ (note that this makes sense from* i)*).*

Then $T \in K(E_1, E_2)$.

Proof. Suppose $u_j \in S(\mathbb{R}^{2n})$ converges weekly to zero in E_1 as $j \to \infty$, according to proposition 4.2 we just need to show that Tu_j converge strongly to zero in E_2. For every functional $\alpha \in E_1{}'$ we have that $\alpha(u_j)$ converges to zero. As $S(\mathbb{R}^{2n}) \subset E_1{}'$, this is true in particular for every $\alpha \in S(\mathbb{R}^{2n})$. If we now consider u_j as tempered distributions applied to the Schwarz function α, then $u_j(\alpha) = \int \alpha u_j = \alpha(u_j) \to 0$, which means that u_j converge to zero in $S'(\mathbb{R}^{2n})$. From the sequential continuity of T, then Tu_j converge to zero in $S(\mathbb{R}^{2n})$ and from the continuous embedding $S(\mathbb{R}^{2n}) \hookrightarrow E_1$ we have that Tu_j converge strongly to zero in E_2. □

We recall that every localization operator $L^F_{\phi,\psi}$ can be written as Weyl operator W^σ with symbols $\sigma = F * Wig(\phi, \psi)$, (cf.. [6]), where $Wig(\phi, \psi)$ is the Wigner transform of the window functions ϕ and ψ defined by:

$$Wig(\phi,\psi)(x,\omega) = \int_{\mathbb{R}^n} e^{-2\pi it\omega}\phi(x+t/2)\overline{\psi(x-t/2)}\,dt.$$

Let us consider now localization operators with compact supported symbols and Schwarz window functions.

Proposition 4.4. *Let E_1 and E_2 be Banach spaces satisfying the conditions of proposition 4.3. Suppose $F \in \mathcal{E}'(\mathbb{R}^{2n})$ (the space of distributions with compact support) and $\phi,\psi \in S(\mathbb{R}^{2n})$, then $L^F_{\phi,\psi} \in K(E_1,E_2)$.*

Proof. We recall that a Weyl operator W^σ is regularizing i.e. sequentially continuous from $S'(\mathbb{R}^{2n})$ to $S(\mathbb{R}^{2n})$ if and only if its Weyl symbol σ is in $S(\mathbb{R}^{2n})$ (cf.. [4]). We have here $L^F_{\phi,\psi} = W^\sigma$ with $\sigma = F * Wig(\phi,\psi)$, where $Wig(\phi,\psi) \in S(\mathbb{R}^{2n})$ thanks to the mapping property $Wig : S(\mathbb{R}^{2n}) \times S(\mathbb{R}^{2n}) \to S(\mathbb{R}^{2n})$ of the Wigner transform. If $\sigma \in S(\mathbb{R}^{2n})$, then the operator $L^F_{\phi,\psi}$ is regularizing and proposition 4.3 implies that it is compact between E_1 and E_2.
We have therefore only to show that the convolution between a compact supported distribution $F \in \mathcal{E}'(\mathbb{R}^{2n})$ and a Schwarz function $g \in S(\mathbb{R}^{2n})$ belongs to $S(\mathbb{R}^{2n})$. Actually from the Paley-Wiener Theorem $\hat{F}$ is an entire function with at most polynomial growth so that $\hat{F}\hat{g} \in S(\mathbb{R}^{2n})$ and therefore $F * g = \mathcal{F}^{-1}(\hat{F}\hat{g}) \in S(\mathbb{R}^{2n})$. □

Most of the boundedness results of the previous sections can now be improved to compactness results by approximation with operators with compact supported symbols and Schwarz window functions. We show this with the following two propositions.

Let as usual m and σ be $v-$moderate weights on $\mathbb{R}^{2n}$.

Proposition 4.5. *Assume that $\phi,\psi \in M^1_v(\mathbb{R}^n)$, $p_0,q_0 \in [1,\infty)$, $p_1,q_1 \in (1,\infty)$, $p_2,q_2 \in [1,\infty]$, and suppose that:*

$$\frac{1}{p_0} = \frac{1}{p_2} - \frac{1}{p_1}, \qquad \frac{1}{q_0} = \frac{1}{q_2} - \frac{1}{q_1}; \qquad F \in L^{p_0,q_0}_\sigma(\mathbb{R}^{2n}). \tag{4.1}$$

Then $L^F_{\phi,\psi} \in K(M^{p_1,q_1}_m, M^{p_2,q_2}_{m\sigma})$.

Proof. Suppose first $F \in \mathcal{E}'(\mathbb{R}^{2n})$, $\phi,\psi \in S(\mathbb{R}^{2n})$. As $p_1,q_1 \in (1,\infty)$, the modulation space M^{p_1,q_1}_m is reflexive and we easily see that all other conditions required by proposition 4.4 are satisfied with $E_1 = M^{p_1,q_1}_m$ and $E_2 = M^{p_2,q_2}_{m\sigma}$, so the thesis holds in this case.
Suppose now that $F\sigma \in L^{p_0,q_0}(\mathbb{R}^{2n})$ and $\phi,\psi \in M^1_v(\mathbb{R}^n)$. $S(\mathbb{R}^{2n})$ is dense in $M^1_v(\mathbb{R}^n)$. On the other hand, as $p_0,q_0 \in [1,\infty)$, the space $\mathcal{E}'(\mathbb{R}^{2n}) \cap L^{p_0,q_0}(\mathbb{R}^{2n})$ is dense in $L^{p_0,q_0}(\mathbb{R}^{2n})$ and, if χ_j is the characteristic function of the ball $B^j_0 = \{z \in \mathbb{R}^{2n} : |z| \le j\}$, then $\chi_j F\sigma \to F\sigma$ in $L^{p_0,q_0}(\mathbb{R}^{2n})$ as $j \to \infty$. Applying proposition 3.3 with $m_1 = m$, $m_2 = m\sigma$ it follows then that the operator $L^F_{\phi,\psi}$ is a limit of compact operators with symbols $F_j = \chi_j F \in \mathcal{E}'(\mathbb{R}^{2n})$ and windows in $S(\mathbb{R}^{2n})$ and is therefore compact. □

In analogous way but relying on Proposition 3.8 instead of 3.3 we have the following proposition.

Proposition 4.6. *Suppose m and σ, are $v-$ moderate weight functions. For $\phi, \psi \in M^1_v(\mathbb{R}^n)$, $p, q \in (1,\infty)$, if $F \in L^r_\sigma(\mathbb{R}^{2n})$, with $r \in [1,\infty)$, then $L^F_{\phi,\psi} \in K(M^{p,q}_m, M^{p,q}_{m\sigma})$.*

We have excluded the case $r = \infty$ because $\mathcal{E}'(\mathbb{R}^{2n}) \cap L^\infty(\mathbb{R}^{2n})$ is not dense in $L^\infty(\mathbb{R}^{2n})$. In this case however $F\sigma$ can still be approximated by compact supported functions if it vanishes at infinity so that we have

Proposition 4.7. *For $\phi, \psi \in M^1_v(\mathbb{R}^n)$, $p, q \in (1,\infty)$, if $F \in L^\infty_\sigma(\mathbb{R}^{2n})$ and $\lim_{R\to\infty} (\text{ess}\sup_{|x|\geq R} |F\sigma(x)|) = 0$ then $L^F_{\phi,\psi} \in K(M^{p,q}_m, M^{p,q}_{m\sigma})$.*

5. Weyl Operators and Sobolev Spaces

In this final section we point out how some of the boundedness and compactness results of the previous sections actually contain and generalize well-known properties about boundedness and compactness of localization and Weyl operators on Sobolev spaces.

We begin by recalling that modulations spaces $M^{p,q}_m$, for $p = q = 2$, coincide with widely used Sobolev spaces. Namely, according to proposition 2.5, for $\sigma(x,\omega) = \langle\omega\rangle^s$, $(\omega \in \mathbb{R}^n)$, we have the Bessel potential spaces $W^2_s(\mathbb{R}^n)$ (more usually denoted by $H_s(\mathbb{R}^n)$) whereas for $\sigma(z) = \langle z\rangle^s$, $(z = (x,\omega) \in \mathbb{R}^{2n})$, we obtain the Shubin-Sobolev spaces $Q_s(\mathbb{R}^n)$. Proposition 3.3, 3.8 can therefore be viewed as natural extensions to modulation spaces of known boundedness results for localization operators on these types of Sobolev spaces, as well as on $L^2(\mathbb{R}^n)$ (cf.. [31], [5]).

We focus now our attention to the case where symbol classes of pseudo-differential calculus are considered. We consider here the Shubin classes $\Gamma^m_\rho(\mathbb{R}^{2n})$ but the same could be stated for other classes of symbols. We briefly recall from [26], Chp. IV, the definition.

Definition 5.1. If $k \in \mathbb{R}, \rho \in (0,1]$ then $\Gamma^k_\rho(\mathbb{R}^{2n})$ indicates the class of C^∞ functions on $\mathbb{R}^{2n}$ such that for every multi-index $\gamma \in \mathbb{N}^{2n}$, there exists a positive constant C_γ for which

$$|\partial^\gamma_z a(z)| \leq C_\gamma \langle z\rangle^{k-\rho|\gamma|}, \quad (z \in \mathbb{R}^{2n}). \tag{5.1}$$

In this frame one main boundedness result is the following (cf.. Shubin [26], Chp. IV, Thm 25.2)

Proposition 5.2. *If $F \in \Gamma^k_\rho(\mathbb{R}^{2n})$, $k \in \mathbb{R}, \rho \in (0,1]$, then the anti-Wick operator $L^F_{g,g}$, as well as the Weyl operator W^F, are bounded from Q_s to Q_{s-k}.*

For what concerns anti-Wick operators, as $L^F_{g,g}$, is in particular a localization operator and the hypothesis $F \in \Gamma^k_\rho(\mathbb{R}^{2n})$ implies $F(z) \in L^\infty_{\langle z\rangle^{-k}}(\mathbb{R}^{2n})$, we see

that Proposition 3.6 contains Proposition 5.2 as particular case for $m(z) = \langle z \rangle$, $\sigma(z) = \langle z \rangle^{-k}$, $p = q = 2$.
Furthermore it is clear that only the estimate on the symbol $a(z)$ and not on its derivative is required for the boundedness. The much stronger requirement $F \in \Gamma_\rho^k(\mathbb{R}^{2n})$ plays a key rôle in the connection with Weyl operators.
Actually whenever we have a calculus that allows an "approximation" of Weyl operators with anti-Wick operators one can transfer results from the latter to the former operators. We show this in the following proposition that extends Proposition 5.2 also for what concerns Weyl operators.

Proposition 5.3. *Suppose $F \in \Gamma_\rho^k(\mathbb{R}^{2n})$, $\rho \in (0,1]$ and $k \in \mathbb{R}$, then the Weyl operator W^F is bounded from $M_m^{p,q}$ to $M_{m/\langle . \rangle^k}^{p,q}$, for every $1 \leq p, q \leq +\infty$, and in particular is bounded from Q_s to Q_{s-k}.*

Proof. According to a basic result of the Weyl calculus for the Shubin symbol classes (see [26], Chp. IV, Thm. 24.2) we can write $W^F = L_{g,g}^{\tilde{F}} + R$ where $L_{\phi,\phi}^{\tilde{F}}$ is an anti-Wick operator with symbol $\tilde{F}$ belonging to the same symbol class $\Gamma_\rho^k(\mathbb{R}^{2n})$ as F, whereas R is a regularizing operator from $S'(\mathbb{R}^n)$ to $S(\mathbb{R}^n)$.
From Proposition 2.4 (g), for every pair of modulation spaces $M_{m_1}^{p_1,q_1}, M_{m_2}^{p_2,q_2}$, as long as the weights m_1, m_2 have tempered growth, we have the sequence of continuous operators $M_{m_1}^{p_1,q_1}(\mathbb{R}^n) \overset{Id}{\hookrightarrow} S'(\mathbb{R}^n) \overset{R}{\to} S(\mathbb{R}^n) \overset{Id}{\hookrightarrow} M_{m_2}^{p_2,q_2}(\mathbb{R}^n)$ that shows R is a bounded map between modulation spaces. $\tilde{F} \in \Gamma_\rho^k$ implies $\tilde{F} \in L_{\langle z \rangle^{-k}}^\infty(\mathbb{R}^{2n})$ so again the thesis follows from Proposition 3.6. □

A part from the case of Shubin classes we remark that Proposition 3.3 extends obviously to Weyl operators whose symbol $G(z)$ can be written in the form $G = F * Wig(\phi, \psi)$ for some functions F, ϕ, ψ and such that condition (3.8) is satisfied because in this case the Weyl operator W^G actually coincides with the localization operator $L_{\phi,\psi}^F$, see [6].

We conclude by considering compact operators. One central result of compactness for pseudodifferential operators in the frame of Shubin-Sobolev spaces is the following.

Proposition 5.4. *If $F \in \Gamma_\rho^k(\mathbb{R}^{2n})$, $k \in \mathbb{R}, \rho \in (0,1]$, then the anti-Wick operator $L_{g,g}^F$, as well as the Weyl operator W^F, are compact from Q_s to $Q_{s-k-\epsilon}$ for every $\epsilon > 0$.*

This result is proved in [26] (Chp. IV, Prop. 25.4) by means of a global version of the Rellich Lemma on compact immersion between Shubin-Sobolev spaces. With the following proposition we show that it can also be deduced as particular case from the compactness results of section 4.

Proposition 5.5. *Suppose $F \in \Gamma_\rho^k(\mathbb{R}^{2n})$, $\rho \in (0,1]$ and $k \in \mathbb{R}$, then the Weyl operator W^F is compact from $M_m^{p,q}$ to $M_{m \langle . \rangle^{-k-\epsilon}}^{p,q}$, for every $\epsilon > 0$ and $p, q \in (1, \infty)$. In particular it is compact from $Q_s(\mathbb{R}^n)$ to $Q_{s-k-\epsilon}(\mathbb{R}^n)$.*

Proof. As in proposition 5.3 we can write $W^F = L^{\tilde{F}}_{g,g} + R$ where $\tilde{F} \in \Gamma^k_\rho(\mathbb{R}^{2n})$, R is a regularizing operator and g the $L^2(\mathbb{R}^{2n})$–normalized gaussian function. As $\tilde{F} \in \Gamma^k_\rho(\mathbb{R}^{2n})$, we have that for every $\epsilon > 0$, $\tilde{F}(z)\langle z\rangle^{-k-\epsilon} \in L^\infty(\mathbb{R}^{2n})$ and vanishes at infinity. It follows from proposition 4.7 that $L^{\tilde{F}}_{g,g}$ is compact from $Q_s(\mathbb{R}^n)$ to $Q_{s-k-\epsilon}(\mathbb{R}^n)$. According to Proposition 4.3 also R is compact so the thesis is proved. □

References

[1] H.W. Alt, *Linear Functional Analysis*, Springer-Verlag, New York, 1993.

[2] S.T. Ali, J.-P. Antoine and J.-P. Gazeau, *Coherent States, Wavelets and Their Generalizations*, Springer-Verlag, New York, 2000.

[3] F.A. Berezin, Wick and anti-Wick symbols of operators, *Math. Sb. (N.S.)*, **86**(128) (1971), 578–610.

[4] P. Boggiatto, E. Buzano and L. Rodino, *Hypoellipticity and Spectral Theory*, Akademie Verlag, Berlin, 1996.

[5] P. Boggiatto and E. Cordero, Anti-Wick quantization of tempered distributions, in *Progress in Analysis*, Editors: H. G. W. Begehr, R. P. Gilbert and M. W. Wong, World Scientific, 2003, 665–662.

[6] P. Boggiatto, E. Cordero and K. H. Gröchenig, Generalized anti-Wick operators with symbols in distributional Sobolev spaces, *Integral Equations Operator Theory*, to appear.

[7] P. Boggiatto and L. Rodino, Quantization and pseudo-differential operators, *Cubo Mat. Ed.* **5**, 2003.

[8] P. Boggiatto and M.W. Wong, Two-wavelet localization operators on $L^p(\mathbb{R}^n)$ for the Weyl-Heisenberg group, *Integral Equations Operator Theory* **49** (2004), 1–10.

[9] E. Cordero and K.H. Gröchenig, Time-frequency analysis of localization operators, *J. Funct. Anal.* **205** (2003), 107–131.

[10] I. Daubechies, Time-frequency localization operators: a geometric phase space approach, *IEEE Trans. Inform. Theory* **34** (1988), 605–612.

[11] H.G. Feichtinger, Modulation spaces on locally lompact abelian groups, in *Wavelets and Their Applications*, Editors: M. Krishna, R. Radha and S. Thangavelu, Allied Publishers, 2003, 99–140.

[12] H.G. Feichtinger and K.H. Gröchenig, Banach spaces related to integral group representations and their atomic decompositions, I, *J. Funct. Anal.* **86** (1989), 307–340.

[13] H.G. Feichtinger and K.H. Gröchenig, Banach spaces related to integral group representations and their atomic decompositions, II, *Monatsh. Math.* **108** (1989), 129–148.

[14] H.G. Feichtinger and K. Nowak, A first survey of Gabor multipliers, in *Advances in Gabor Analysis*, Editors: H.G. Feichtinger, T. Strohmer, Birkhäuser Boston, 2003, 99–128.

[15] K.H. Gröchenig, *Foundations of Time-Frequency Analysis*, Birkhäuser, Boston, 2001.

[16] K.H. Gröchenig and C. Heil, Modulation spaces as symbol classes for pseudo-differential operators, in *Wavelets and Their Applications*, Editors: M. Krishna, R. Radha and S. Thangavelu, Allied Publishers, 2003, 151–169.

[17] K.H. Gröchenig and C. Heil, Modulation spaces and pseudo-differential operators, *Integral Equations Operator Theory* **34** (1999), 439–457.

[18] Z. He and M.W. Wong, Localization operators associated to square integrable group representations, *Panamer. Math. J.* **6** (1996), 93–104.

[19] C. Heil, J. Ramanathan and P. Topiwala, Singular values of compact pseudo-differential operators, *J. Funct. Anal.* **150** (1997), 426–452.

[20] C. Heil and D. Walnut, Continuous and discrete wavelet tranforms, *SIAM Rev.* **4** (1989), 628–666.

[21] G. Kaiser, *A Friendly Guide to Wavelets*, Birkhäser, Boston, 1994.

[22] D. Labate, Pseudo-differential operators on modulation spaces, *J. Math. Anal. Appl.* **262** (2001), 242–255.

[23] D. Labate, Time-frequency analysis of pseudo-differential operators, *Monatsh. Math.* **133** (2001), 143–156.

[24] N. Lerner, The Wick calculus of pseudo-differential operators and some of its applications, *Cubo Mat. Ed.* **5** (2003), 213–236.

[25] S. Pilipovic and N. Teofanov, Pseudo-differential operators and ultramodulation spaces, *J. Funct. Anal.*, to appear.

[26] M.A. Shubin, *Pseudo-Differential Operators and Spectral Theory*, Second Edition, Springer-Verlag, Berlin, 2001.

[27] K. Tachizawa, The boundedness of pseudo-differential operators on modulation spaces, *Math. Nachr.* **168** (1994), 263–277.

[28] K. Tachizawa, Pseudo-differential operators and Wilson bases, *J. Math. Pures Appl.* **75** (1996), 509–529.

[29] J. Toft, Continuity properties for modulation spaces with applications to pseudo-differential calculus I, *J. Funct. Anal.*, to appear.

[30] J. Toft, Weighted modulation spaces and pseudo-differential operators, Research Report 2003:05 (2003), Blekinge Institute of Technology, Karlskrona.

[31] M.W. Wong, *Wavelet Transform and Localization Operators*, Birkhäuser-Verlag, Basel, 2002.

Paolo Boggiatto
Dip. Matematica, Università di Torino
Via C. Alberto 10
10123 Torino
Italy
e-mail: boggiatto@dm.unito.it

Operator Theory:
Advances and Applications, Vol. 155, 165–186

Convolutions and Embeddings for Weighted Modulation Spaces

Joachim Toft

Abstract. Let $M^{p,q}_{(\omega)}$ be the modulation space with parameters $p, q \in [1, \infty]$ and weight function ω. Also let $M^{p,q} = M^{p,q}_{(\omega_0)}$, $\omega_0 = 1$. We prove that for certain ω, there is a canonical homeomorphism $M^{p,q}_{(\omega)} \to M^{p,q}$, and use this result to extend well-known embeddings for $M^{p,q}$-spaces to embeddings between certain $M^{p,q}_{(\omega)}$-spaces and Sobolev-Besov spaces. We also give a convenient definition for convolutions between elements in $M^{p,q}_{(\omega)}$-spaces, and prove certain Hölder-Young properties.

Mathematics Subject Classification (2000). Primary 42B35, 46E35, 44A35; Secondary 35S05, 47B37.

Keywords. Modulation spaces, embeddings, convolutions, Besov spaces, Sobolev spaces.

0. Introduction

The aim of the paper is to extend certain continuity properties in [16] and [18] concerning standard modulation spaces to weighted modulation spaces. In particular convolution and embedding properties between modulation spaces and Sobolev/Besov spaces are discussed. The investigations are based on a result which concerns relations between modulation spaces with different weights, which in many cases reduces the problems so that only standard modulation spaces occur.

During the period 1980–1983, H. Feichtinger introduced in [5] and [7] modulation spaces, as appropriate families of Banach spaces of functions and distributions to have in background when discussing certain problems within time-frequency analysis. For an appropriate weight function ω, the modulation space $M^{p,q}_{(\omega)}$ is obtained by imposing a mixed $L^{p,q}_{(\omega)}$-norm on the short time Fourier transform (or alternatively certain ambiguity functions), and when having such spaces in background, certain decay and propagation properties at infinity as well as certain localization properties for distributions may be considered. If $\omega = 1$, then the standard modulation space $M^{p,q}$ is obtained.

During the last ten years, modulation spaces have also been used in the theory of pseudo-differential operators and supply this field with new types of symbol classes as well as alternative continuity considerations for such operators. (For more information, see [16]–[18] and their references.)

In order to describe the results in more details, the definition of modulation spaces is recalled. Let $p, q \in [1, \infty]$, $\chi \in \mathscr{S}(\mathbf{R}^m) \setminus 0$ be fixed, $\mathscr{F}$ be a Fourier transform, and set $\tau_x \chi = \chi(\cdot - x)$. Let also $0 < \omega \in L^\infty_{loc}(\mathbf{R}^{2m})$ be an appropriate function. For any measurable function f on $\mathbf{R}^{2m}$, and $a \in \mathscr{S}'(\mathbf{R}^m)$ set

$$\begin{aligned} \|f\|_{L^{p,q}_{(\omega)}} &\equiv \Big(\int \Big(\int |f(x,\xi)\omega(x,\xi)|^p \, dx \Big)^{q/p} d\xi \Big)^{1/q}, \\ f_a(x,\xi) &= \mathscr{F}(a\,\tau_x\chi)(\xi), \quad \|a\|_{M^{p,q}_{(\omega)}} \equiv \|f_a\|_{L^{p,q}_{(\omega)}}, \end{aligned} \tag{0.1}$$

with obvious modifications when $p = \infty$ and/or $q = \infty$. (We use the same notations for the usual function and distribution spaces, e. g. in [11].) Then $L^{p,q}_{(\omega)}(\mathbf{R}^{2m})$ and the modulation space $M^{p,q}_{(\omega)}(\mathbf{R}^m)$ consist of all measurable f on $\mathbf{R}^{2m}$ and all $a \in \mathscr{S}'(\mathbf{R}^m)$ respectively such that $\|f\|_{L^{p,q}_{(\omega)}}$ and $\|a\|_{M^{p,q}_{(\omega)}}$ are finite. An important case appears when $\omega(x,\xi) = (1 + |\xi|^2)^{s/2}$, $s \in \mathbf{R}$, and then the notation $M^{p,q}_{s,0}$ is used instead of $M^{p,q}_{(\omega)}$.

In Section 3 a classification result for modulation spaces is presented, which in particular generalizes Theorem 6.1 in [7], which asserts that

$$a \in M^{p,q}_{s,0}, \quad \text{if and only if} \quad (1 + |D|^2)^{s/2} a \in M^{p,q}. \tag{0.2}$$

Here $\varphi(D)$ is the pseudo-differential operator which means a multiplication by the function φ on the Fourier transform side.

Such properties are then used in Section 4 in order to generalize embedding results in [16] or [18]. For example, it is proved that if $s \in \mathbf{R}$ and

$$\begin{aligned} \theta_1(p,q) &= \max(0, q^{-1} - \min(p^{-1}, p'^{-1})), \\ \theta_2(p,q) &= \min(0, q^{-1} - \max(p^{-1}, p'^{-1})), \qquad p, q \in [1, \infty], \end{aligned} \tag{0.3}$$

then

$$H^p_{s+\mu m\theta_1(p,q)}(\mathbf{R}^m) \subset M^{p,q}_{s,0}(\mathbf{R}^m) \subset H^p_{s+\mu m\theta_2(p,q)}(\mathbf{R}^m), \quad \mu > 1. \tag{0.4}$$

Here p' denotes the conjugate exponent of p, i. e. p and p' satisfy $1/p + 1/p' = 1$, and H^p_s is the Sobolev space of distributions with s derivatives in L^p. In Section 4, a refinement of (0.4) is proved, where the Sobolev spaces are replaced by Besov spaces. In particular, an improvement of Theorem F.4 in [9], by P. Gröbner is obtained. Results related to (0.4) may also be found in [12].

In Section 5, certain convolution properties for modulation spaces are considered. Here a convenient definition of convolutions between elements in modulation spaces is presented, which also fit when $\mathscr{S}$ is not dense in some of the modulation spaces. It is proved that if ω_0, ω_1 and ω_2 are appropriate weight functions and $a_j \in M^{p_j,q_j}_{(\omega_j)}$ for $j = 1, 2$, then

$$\|a_1 * a_2\|_{M^{p_0,q_0}_{(\omega_0)}} \le C\|a_1\|_{M^{p_1,q_1}_{(\omega_1)}}\|a_2\|_{M^{p_2,q_2}_{(\omega_2)}}, \tag{0.5}$$

provided $1/p_1 + 1/p_2 = 1 + 1/p_0$ and $1/q_0 = 1/q_1 + 1/q_2$.

The convolution results in Section 5 open up for using modulation space theory in different fields, for example when dealing with fundamental solutions for linear partial differential equations. They can also be used in other problem areas, for example when dealing with convolution operators. In forthcoming papers by the author, the convolution properties are used for analyzing Toeplitz operators (or localization operators) in terms of modulation spaces and pseudo-differential operators. (Cf. [2], [4] or [16]–[18].)

1. Preliminaries

In this section, some basic and well-known properties for modulation spaces are presented. A lot of the proofs are omitted since they are given in [7] or [10], where more complete overviews of the subject are presented.

We start by discussing general properties for the weight function ω in (0.1). It is usually required that ω should be *v-moderate* for some appropriate function $v \in L^\infty_{loc}(\mathbf{R}^m)$. This means that for some constant $C > 0$, then

$$\omega(x_1 + x_2) \le C\omega(x_1)v(x_2), \quad x_1, x_2 \in \mathbf{R}^m. \tag{1.1}$$

The function v is then said to *moderate* ω.

Definition 1.1. The cone $\mathscr{P}(\mathbf{R}^m)$ consists of all $0 < \omega \in L^\infty_{loc}(\mathbf{R}^m)$ such that ω is v-moderate, for some polynomial v on $\mathbf{R}^m$. The set $\mathscr{P}_0(\mathbf{R}^m)$ consists of all smooth $\omega \in \mathscr{P}(\mathbf{R}^m)$ such that $(\partial^\alpha \omega)/\omega$ is bounded for every α.

Note that if $\omega \in \mathscr{P}$, then $\omega(x) + \omega(x)^{-1} \le v(x)$, $x \in \mathbf{R}^m$ for some polynomial v on $\mathbf{R}^m$.

If ω_1, ω_2 are positive functions, then $\omega_2 \prec \omega_1$ means that $\omega_2 \le C\omega_1$ for some constant $C > 0$, and they are equivalent, which is written $\omega_1 \sim \omega_2$, when $\omega_1 \prec \omega_2 \prec \omega_1$.

For any $\omega \in \mathscr{P}(\mathbf{R}^m)$, let $v_\omega(x) \equiv \sup_y(\omega(x + y)/\omega(y))$. Then ω is v_ω-moderate. The following lemma shows that v_ω may be considered as a minimal element among the functions which moderate ω.

Lemma 1.2. *Assume that $\omega \in \mathscr{P}(\mathbf{R}^m)$. Then the following are true:*

(1) *ω is v_ω-moderate;*

(2) *if $x, x_1, x_2 \in \mathbf{R}^m$, then $v_\omega \in \mathscr{P}(\mathbf{R}^m)$ and satisfies $v_\omega(x_1+x_2) \le v_\omega(x_1)v_\omega(x_2)$ and $v_\omega(x) \ge 1$ with equality when $x = 0$;*

(3) *if $v > 0$ moderates ω, then $v_\omega \prec v$;*

(4) *there is a function $\omega_0 \in \mathscr{P}_0(\mathbf{R}^m)$ such that $\omega_0 \sim \omega$;*

(5) *if in addition* $\omega \in C(\mathbf{R}^m)$, *then* $v_\omega \in C(\mathbf{R}^m)$.

Proof. The assertion (4) follows by letting $\omega_0 = \omega * \varphi$ for some $0 \le \varphi \in \mathscr{S}(\mathbf{R}^m)\backslash 0$. The other assertions follow by straight-forward computations and are left for the reader. □

So far, the Fourier transform has not been specified. From now on it is assumed that the Fourier transform on $\mathscr{S}'(\mathbf{R}^m)$ has the form

$$\mathscr{F} f(\xi) = \widehat{f}(\xi) \equiv (2\pi)^{-m/2} \int f(x) e^{-i\langle x, \xi \rangle}\, dx,$$

when $f \in \mathscr{S}(\mathbf{R}^m)$. In particular, $\mathscr{F}$ is normalized in the sense that it is unitary on $L^2(\mathbf{R}^m)$.

Next some notations are introduced. Assume that $\mathscr{B}_1$ and $\mathscr{B}_2$ are Banach spaces. Then $\mathscr{B}_1 \hookrightarrow \mathscr{B}_2$ means that $\mathscr{B}_1$ is continuously embedded in $\mathscr{B}_2$, i. e. $\mathscr{B}_1 \subset \mathscr{B}_2$ and $\|x\|_{\mathscr{B}_2} \le C\|x\|_{\mathscr{B}_1}$, for some constant $C > 0$ which is independent of $x \in \mathscr{B}_1$.

The dual form between function or distribution spaces and their duals is denoted by $\langle \cdot, \cdot \rangle$. For admissible a and b in $\mathscr{S}'(\mathbf{R}^m)$, let $(a, b) = \langle a, \overline{b} \rangle$. Then $(\cdot, \cdot)$ on L^2 is the usual scalar product.

Assume next that $v(x_1, x_2) = (v_0 \otimes 1)(x_1, x_2)$, where $(x_1, x_2) \in V_1 \oplus V_2 = \mathbf{R}^m$, $V_1 \perp V_2$ and $v_0 \in \mathscr{S}'(V_1)$. Then $v(x_1, x_2)$ is identified with $v_0(x_1)$, and we set $v(x_1, x_2) = v(x_1)$.

Definition 1.3. Assume that $\omega \in \mathscr{P}(\mathbf{R}^{2m})$, and that $p, q \in [1, \infty]$. Then the space $M^{p,q}_{(\omega)}(\mathbf{R}^m)$ is called a *modulation space.* If in addition $\omega = 1$ everywhere, then $M^{p,q}_{(\omega)}(\mathbf{R}^m)$ is called a *standard* modulation space, and is denoted by $M^{p,q}(\mathbf{R}^m)$ instead of $M^{p,q}_{(\omega)}(\mathbf{R}^m)$. In the case $p = q$, the notations $M^p_{(\omega)}(\mathbf{R}^m)$ and $M^p(\mathbf{R}^m)$ are used instead of $M^{p,p}_{(\omega)}(\mathbf{R}^m)$ and $M^{p,p}(\mathbf{R}^m)$ respectively.

A reason not to index $M^{p,q}_{(\omega)}$ by the fixed function $\chi \in \mathscr{S} \setminus 0$ is given by Proposition 1.4 below. The convention of indexing weight functions with parenthesis is used also in other situations. For example, if $\omega \in \mathscr{P}(\mathbf{R}^m)$, then $L^p_{(\omega)}(\mathbf{R}^m)$ is the set of all measurable functions f on $\mathbf{R}^m$ such that $f\omega \in L^p(\mathbf{R}^m)$, i. e. $\|f\|_{L^p_{(\omega)}} \equiv \|f\omega\|_{L^p}$ is finite.

Some basic properties for modulation spaces are given in the following proposition. Here recall that p' denotes the conjugate exponent of p, i. e. $1/p + 1/p' = 1$.

Proposition 1.4. *Assume that* $\omega, \omega_1, \omega_2 \in \mathscr{P}(\mathbf{R}^{2m})$ *and* $p, q \in [1, \infty]$. *Then the following are true:*

(1) $M^{p,q}_{(\omega)}(\mathbf{R}^m)$ *is a Banach space which is independent of the choice of* $\chi \in \mathscr{S}(\mathbf{R}^m) \setminus 0$ *in* (0.1). *Moreover, different choices of* χ *give rise to equivalent norms;*

(2) *if* $p_1, p_2, q_1, q_2 \in [1, \infty]$ *such that* $p_1 \le p_2$, $q_1 \le q_2$ *and* $\omega_2 \prec \omega_1$, *then*

$$\mathscr{S}(\mathbf{R}^m) \hookrightarrow M^{p_1,q_1}_{(\omega_1)}(\mathbf{R}^m) \hookrightarrow M^{p_2,q_2}_{(\omega_2)}(\mathbf{R}^m) \hookrightarrow \mathscr{S}'(\mathbf{R}^m);$$

(3) *the scalar product* $(\,\cdot\,,\,\cdot\,)$ *on* L^2 *extends to a continuous bilinear map from* $M^{p,q}_{(\omega)}(\mathbf{R}^m) \times M^{p',q'}_{(1/\omega)}(\mathbf{R}^m)$ *to* $\mathbf{C}$. *On the other hand, if* $\|a\| = \sup |(a,b)|$, *where the supremum is taken over all* $b \in M^{p',q'}_{(1/\omega)}(\mathbf{R}^m)$ *such that* $\|b\|_{M^{p',q'}_{(1/\omega)}} \le 1$, *then* $\|\cdot\|$ *and* $\|\cdot\|_{M^{p,q}_{(\omega)}}$ *are equivalent norms;*

(4) *if* $p, q < \infty$, *then* $\mathscr{S}(\mathbf{R}^m)$ *is dense in* $M^{p,q}_{(\omega)}(\mathbf{R}^m)$, *and the dual space for* $M^{p,q}_{(\omega)}(\mathbf{R}^m)$ *is identified with* $M^{p',q'}_{(1/\omega)}(\mathbf{R}^m)$, *through the form* $(\cdot,\cdot)$. *Moreover,* $\mathscr{S}(\mathbf{R}^m)$ *is weakly dense in* $M^{\infty}_{(\omega)}(\mathbf{R}^m)$.

Remark 1.5. Many other properties for modulation spaces are presented in [10], [5]–[7], [8], [13], [16] or [18]. For example, if $p \in [1,\infty]$, $q_1 = \min(p,p')$ and $q_2 = \max(p,p')$, then $M^{p,q_1} \subset L^p \subset M^{p,q_2}$. In particular, $M^2 = L^2$. The space $M^{\infty,1}$ contains all smooth functions which are bounded together with all their derivatives. The space $M^{1,\infty}$ contains C'_B, the set of all measures on $\mathbf{R}^m$ with bounded mass, and if $p_0 \le p$, then $M^{p_0,1}(\mathbf{R}^m) \subset C(\mathbf{R}^m) \cap L^p$.

Proposition 1.4 (1) permits us to be rather vague according to the choice of $\chi \in \mathscr{S} \setminus 0$ in (0.1). For example, if $C > 0$ is a constant and Ω is a subset of $\mathscr{S}'$, then $\|a\|_{M^{p,q}_{(\omega)}} \le C$ for every $a \in \Omega$, means that the inequality holds for some choice of $\chi \in \mathscr{S} \setminus 0$ and every $a \in \Omega$. Evidently, for any other choice of $\chi \in \mathscr{S} \setminus 0$, a similar inequality is true provided C has been replaced by a larger constant, if necessary.

For $\mathscr{S}$ not to be dense in $M^{p,q}_{(\omega)}$, it suffices that $p = \infty$ and/or $q = \infty$. This possibility causes problems in uniqueness properties when one extends certain properties of $\mathscr{S}$ to modulation spaces. In these cases, we pay some attentions to find appropriate constructions and definitions.

Remark 1.6. Assume that $\omega \in \mathscr{P}(\mathbf{R}^{2m})$, $a \in \mathscr{S}'(\mathbf{R}^m)$ and $\chi \in \mathscr{S}(\mathbf{R}^m)$. Then Parseval's formula gives

$$\mathscr{F}(a\,\tau_x\widehat{\chi})(\xi) = e^{-i\langle x,\xi\rangle} H_a(-x,\xi), \quad \text{where} \quad H_a(x,\xi) = \mathscr{F}(\widehat{a}\,\tau_\xi\chi)(x).$$

Hence if $\omega_0(x,\xi) = \omega(-x,\xi)$, then $\|a\| = \|H_a\|_{L^{p,q}_{(\omega_0)}}$ is a norm which is equivalent to $\|a\|_{M^{p,q}_{(\omega)}}$, by Proposition 1.4 (1).

In some considerations, Minkowski's inequality are used in somewhat general form. Recalling that for a $d\nu$-measurable function f with values in the Banach space $\mathscr{B}$ with norm $\|\cdot\|$, Minkowski's inequality asserts that $\|\int f\,d\nu\| \le \int \|f\|\,d\nu$. In our applications, $\mathscr{B}$ is equal to $L^p(d\mu)$, for some $p \in [1,\infty]$, and Minkowski's inequality takes the form

$$\Big(\int \Big|\int f(x,y)\,d\nu(y)\Big|^p\,d\mu(x)\Big)^{1/p} \le \int \Big(\int |f(x,y)|^p\,d\mu(x)\Big)^{1/p}\,d\nu(y).$$

Remark 1.7. Assume that $\omega \in \mathscr{P}(\mathbf{R}^{2m})$ and $p,q \in [1,\infty]$, and let $\mathcal{M}^{p,q}_{(\omega)}(\mathbf{R}^m)$ be the completions of $\mathscr{S}(\mathbf{R}^m)$ under the norms $\|\cdot\|_{M^{p,q}_{(\omega)}}$. Then it easily follows that the most of the properties which are valid for $M^{p,q}_{(\omega)}$ also hold for $\mathcal{M}^{p,q}_{(\omega)}$.

We finish this section by discussing weight functions which are of particular interests. For any $s,t \in \mathbf{R}$ set

$$\begin{aligned} \sigma_t(x) &= \langle x\rangle^t, \quad \sigma_{s,t}(x,\xi) = \langle \xi\rangle^s \langle x\rangle^t, \\ v_t(x) &= \langle x\rangle^{|t|}, \quad v_{s,t}(x,\xi) = \langle \xi\rangle^{|s|}\langle x\rangle^{|t|}, \quad x,\xi \in \mathbf{R}^m. \end{aligned} \tag{1.2}$$

Here and in what follows, $\langle x\rangle \equiv (1+|x|^2)^{1/2}$. Then $\sigma_t, v_t \in \mathscr{P}_0(\mathbf{R}^m)$ and $\sigma_{s,t}, v_{s,t} \in \mathscr{P}_0(\mathbf{R}^{2m})$ for every $s,t \in \mathbf{R}$, and σ_t is v_t-moderate and $\sigma_{s,t}$ is $v_{s,t}$-moderate. Obviously, $\sigma_s(x,\xi) = (1+|x|^2+|\xi|^2)^{s/2}$, and $\sigma_{s,t} = \sigma_t \otimes \sigma_s$. Moreover, if $\omega \in \mathscr{P}(\mathbf{R}^m)$, then ω is v_t-moderate provided t is chosen large enough.

For convenience, the notations $L^p_s(\mathbf{R}^m)$, $M^{p,q}_s(\mathbf{R}^m)$ and $M^{p,q}_{s,t}(\mathbf{R}^m)$ are used instead of $L^p_{(\sigma_s)}$, $M^{p,q}_{(\sigma_s)}(\mathbf{R}^m)$ and $M^{p,q}_{(\sigma_{s,t})}(\mathbf{R}^m)$ respectively.

2. The Narrow Convergence

In this section, the narrow convergence is discussed. Proposition 1.4 does not guarantee that elements in $M^{p,q}_{(\omega)}$ may be approximated by elements in $\mathscr{S}$ when $p=\infty$ and/or $q=\infty$. The cases $(p,q)=(1,\infty)$ or $(p,q)=(\infty,1)$ are critical since $\mathscr{S}$ is not dense neither in $M^{\infty,1}_{(\omega)}$ nor in $M^{1,\infty}_{(\omega)}$, which might cause problems when using weak* convergence as well. Here it is seen that some of these problems are avoided by modifying the definition of narrow convergence in [13], [15], [16] and [18] to weighted modulation spaces.

We start by giving the definition of the narrow convergence. For any $a \in \mathscr{S}'(\mathbf{R}^m)$, $\omega \in \mathscr{P}(\mathbf{R}^{2m})$, $\chi \in \mathscr{S}(\mathbf{R}^m)$ and $p \in [1,\infty]$, let

$$H_{a,\omega,p}(\xi) = \Big(\int |\mathscr{F}(a\,\tau_x\chi)(\xi)\omega(x,\xi)|^p\,dx \Big)^{1/p}.$$

Definition 2.1. (Cf. [13], [15], [16] or [18].) Let $a, a_j \in M^{p,q}_{(\omega)}(\mathbf{R}^m)$, $j=1,2,\dots$. Then a_j is said to converge *narrowly* to a (with respect to $p,q \in [1,\infty]$, $\chi \in \mathscr{S}(\mathbf{R}^m)\setminus 0$ and $\omega \in \mathscr{P}(\mathbf{R}^{2m})$), if the following conditions are satisfied:

(1) $a_j \to a$ in $\mathscr{S}'(\mathbf{R}^m)$ as $j\to\infty$;
(2) $H_{a_j,\omega,p}(\xi) \to H_{a,\omega,p}(\xi)$ in $L^q(\mathbf{R}^m)$ as $j\to\infty$.

Remark 2.2. Assume that $a, a_1, a_2, \dots \in \mathscr{S}'(\mathbf{R}^m)$ satisfies (1) in Definition 2.1, and assume that $\xi \in \mathbf{R}^m$. Then it follows from Fatou's lemma that

$$\liminf_{j\to\infty} H_{a_j,\omega,p}(\xi) \ge H_{a,\omega,p}(\xi) \quad \text{and} \quad \liminf_{j\to\infty} \|a_j\|_{M^{p,q}_{(\omega)}} \ge \|a\|_{M^{p,q}_{(\omega)}}.$$

The following result generalizes Proposition 2.3 in [15].

Proposition 2.3. *Assume that $p, q \in [1,\infty]$ such that $q < \infty$, and that $\omega \in \mathscr{P}(\mathbf{R}^{2m})$. Then $C_0^\infty(\mathbf{R}^m)$ is dense in $M^{p,q}_{(\omega)}(\mathbf{R}^m)$ with respect to the narrow convergence.*

Proof. By an application of Cantor's diagonal principle, it suffices to prove that C_0^∞ is narrowly dense in $M^{p,q}_{(\omega)} \cap \mathscr{E}$, and that $M^{p,q}_{(\omega)} \cap \mathscr{E}$ is narrowly dense in $M^{p,q}_{(\omega)}$.

Therefore, assume that $a \in M^{p,q}_{(\omega)}(\mathbf{R}^m)$, and choose a function $\phi \in C_0^\infty(\mathbf{R}^m)$ such that ϕ and $\widehat{\phi}$ are non-negative functions, and $\phi(0) = 1$. Set $a_j = a\,\phi(\varepsilon_j\,\cdot\,)$, where $\varepsilon_j \searrow 0$ as $j \to \infty$. It shall be proved that a_j converges to a narrowly.

Clearly, $a_j \to a$ in $\mathscr{S}'$ as $j \to \infty$. We have to prove that $H_{a_j,\omega,p} \to H_{a,\omega,p}$ in L^q as $j \to \infty$. Let $v = v_\omega$. (Cf. Section 1.) Then v is bounded by polynomials, $v(0) = 1$ and $\omega(x_1+x_2, \xi_1+\xi_2) \le \omega(x_1,\xi_1)v(x_2,\xi_2)$. Hence, Minkowski's inequality gives

$$\begin{aligned} H_{a_j,\omega,p}(\xi) &= (2\pi)^{-m/2}\Big(\int\Big|\int(\mathscr{F}(\chi_x a)(\xi-\varepsilon_j\eta)\widehat{\phi}(\eta)\omega(x,\xi)\,d\eta\Big|^p dx\Big)^{1/p} \\ &\le (2\pi)^{-m/2}\int\Big(\int|(\mathscr{F}(\chi_x a)(\xi-\varepsilon_j\eta)\omega(x,\xi)|^p dx\Big)^{1/p}\widehat{\phi}(\eta)\,d\eta \\ &\le (2\pi)^{-m/2}\int H_{a,\omega,p}(\xi-\varepsilon_j\eta)(v(0,\varepsilon_j\eta)\widehat{\phi}(\eta))\,d\eta \equiv U_j(\xi). \end{aligned}$$

Since $(2\pi)^{-m/2}\int\widehat{\phi}\,d\xi = \phi(0) = 1$, v is bounded by polynomials, $v(0) = 1$, and $H_{a,\omega,p} \in L^q$, it follows that $U_j \to H_{a,\omega,p}$ in L^q as $j \to \infty$. Hence $\|U_j - U_k\|_{L^q} \to 0$ as $j,k \to \infty$, and since $H_{a_j,\omega,p} \le U_j$,

$$\limsup_{j\to\infty} H_{a_j,\omega,p}(\xi) \le H_{a,\omega,p}(\xi), \qquad \text{a. e.}$$

It now follows that $H_{a_j,\omega,p} \to H_{a,\omega,p}$ in L^q as $j \to \infty$ from Remark 2.2 and a generalization of Lebesgue's theorem which asserts that if $f_j \to f$ a. e. as $j \to \infty$ and if there exists a sequence $g_j \in L^q$ such that $|f_j| \le g_j$ and $\|g_j - g_k\|_{L^q} \to 0$ as $j,k \to \infty$, then $\|f - f_j\|_{L^q} \to 0$ as $j \to 0$. Hence $M^{p,q}_{(\omega)} \cap \mathscr{E}'$ is narrowly dense in $M^{p,q}_{(\omega)}$.

Assume next that $a \in M^{p,q}_{(\omega)} \cap \mathscr{E}'$ and let $a_j = \varepsilon_j^{-m}\phi(\,\cdot\,/\varepsilon_j) * a$, where $\phi \in C_0^\infty$ is non-negative, $\int\phi\,dx = 1$ and $\varepsilon_j \searrow 0$ as $j \to \infty$. We shall prove that $a_j \in C_0^\infty$ converges to a narrowly as $j \to \infty$.

By a straight-forward computation it follows that

$$|\mathscr{F}(a_j\,\tau_x\chi)(\xi)| = \Big|\int\mathscr{F}(a\,\tau_{x-\varepsilon_j y}\chi)(\xi)e^{-i\varepsilon_j\langle y,\xi\rangle}\phi(y)\,dy\Big|. \tag{2.1}$$

Hence Minkowski's inequality gives

$$H_{a_j,\omega,p}(\xi) \le \int\Big(\int|\mathscr{F}(a\,\tau_{x-\varepsilon_j y}\chi)(\xi)\omega(x,\xi)|^p\,dx\Big)^{1/p}\phi(y)\,dy \le U_j(\xi),$$

where

$$U_j(\xi) = \int \Big(\int |\mathscr{F}(a\,\tau_{x-\varepsilon_j y}\chi)(\xi)\omega(x-\varepsilon_j y,\xi)|^p\,dx\Big)^{1/p} v(\varepsilon_j y,0)\phi(y)\,dy$$

$$= C_j H_{a,\omega,p}(\xi)$$

and $C_j = \int v(\varepsilon_j y,0)\phi(y)\,dy$. Since $C_j \to 1$ as $j \to \infty$, the same arguments as in the first part of the proof now show that $H_{a_j,\omega,p} \to H_{a,\omega,p}$ in L^q as $j \to \infty$. The proof is complete. □

Next, an appropriate extension of the form $(\cdot,\cdot)$ to $M^{\infty,1}_{(\omega)} \times M^{1,\infty}_{(1/\omega)}$ is discussed. Assume that $a,b,\chi \in \mathscr{S}$ such that $\|\chi\|_{L^2} = 1$, and $\omega \in \mathscr{P}(\mathbf{R}^{2m})$. Then Parseval's formula gives

$$(a,b) = \iint \mathscr{F}(a\,\tau_x\chi)(\xi)\overline{\mathscr{F}(b\,\tau_x\chi)(\xi)}\,dx d\xi\,. \tag{2.2}$$

In the case $a \in M^{\infty,1}_{(\omega)}$ and $b \in M^{1,\infty}_{(1/\omega)}$ or $a \in M^{1,\infty}_{(1/\omega)}$ and $b \in M^{\infty,1}_{(\omega)}$, let (a,b) be defined as the right-hand side of (2.2). This definition makes sense, since the integrand in (2.2) belongs to L^1. The following lemma shows that this extension is unique when continuity with respect to the narrow convergence is required.

Lemma 2.4. *Assume that $\omega \in \mathscr{P}(\mathbf{R}^{2m})$, $a \in M^{\infty,1}_{(\omega)}(\mathbf{R}^m)$ and $b \in M^{1,\infty}_{(1/\omega)}(\mathbf{R}^m)$. Then the following are true:*

1. $|(a,b)| \le C\|a\|_{M^{\infty,1}_{(\omega)}}\|b\|_{M^{1,\infty}_{(1/\omega)}}$;
2. *if $a \in \mathscr{S}$ or $b \in \mathscr{S}$, then (a,b) is the usual scalar product between a distribution and a tempered function;*
3. *assume that $a_j \in M^{\infty,1}_{(\omega)}$, $j = 1,2,\dots$, converges narrowly to a as $j \to \infty$. Then $(a_j,b) \to (a,b)$ as $j \to \infty$;*
4. *the form $(\,\cdot\,,\,\cdot\,)$ on $M^{\infty,1}_{(\omega)}(\mathbf{R}^m) \times M^{1,\infty}_{(1/\omega)}(\mathbf{R}^m)$ does not depend on $\chi \in \mathscr{S}$, provided $\|\chi\|_{L^2} = 1$.*

Proof. The assertions (1) and (2) follow immediately from (2.2) together with Hölder's inequality and Parseval's formula respectively.

(3) Let $a_0 = a$. Then the narrow convergence implies for every $(x,\xi) \in \mathbf{R}^{2m}$ that $\mathscr{F}(a_j\,\tau_x\chi)(\xi) \to \mathscr{F}(a_0\,\tau_x\chi)(\xi)$ as $j \to \infty$. Moreover, if $U(x,\xi) = |\mathscr{F}(b\,\tau_x\chi)(\xi)/\omega(x,\xi)|$, then

$$|\mathscr{F}(a_j\,\tau_x\chi)(\xi)\overline{\mathscr{F}(b\,\tau_x\chi)(\xi)}| \le H_{a_j,\omega,\infty}(\xi)U(x,\xi) \in L^1(\mathbf{R}^{2m}).$$

Since $a_j \to a$ narrowly as $j \to \infty$, it follows that

$$\|(H_{a_j,\omega,\infty} - H_{a,\omega,\infty})U\|_{L^1} \le \|H_{a_j,\omega,\infty} - H_{a,\omega,\infty}\|_{L^1}\|b\|_{M^{1,\infty}_{(1/\omega)}} \to 0,$$

as $j \to \infty$. It follows now that

$$\mathscr{F}(a_j\,\tau_x\chi)(\xi)\overline{\mathscr{F}(b\,\tau_x\chi)(\xi)} \to \mathscr{F}(a\,\tau_x\chi)(\xi)\overline{\mathscr{F}(b\,\tau_x\chi)(\xi)}$$

in $L^1(\mathbf{R}^{2m})$ as $j \to \infty$, by a generalization of Lebesgue's theorem. (Cf. the proof of Proposition 2.3.) This implies that $(a_j,b) \to (a,b)$ as $j \to \infty$, and (3) follows.

(4) Assume that $\psi \in \mathscr{S}(\mathbf{R}^m)$ such that $\|\psi\|_{L^2} = 1$. Then, by Proposition 2.3 and its proof, there is a sequence $a_j \in C_0^\infty(\mathbf{R}^m)$ which converges narrowly to a with respect to χ and ψ, as $j \to \infty$. The result follows now from (2) and (3). The proof is complete. □

From now on, it is always assumed that the form $(\,\cdot\,,\,\cdot\,)$, on $M^{\infty,1}_{(\omega)} \times M^{1,\infty}_{(1/\omega)}$, is given by (2.2).

3. Relations between Modulation Spaces with Different Weights

In this section it is proved that there is a canonical homeomorphism between modulation spaces with different weight functions. In particular, the spaces $M^{p,q}_s$ and $M^{p,q}_{s,t}$ may be constructed in a quite natural way from the corresponding standard space $M^{p,q}$.

Lemma 3.1. *Assume that $\omega \in \mathscr{P}_0(\mathbf{R}^m)$, $\kappa \in C^\infty(\mathbf{R}^m)$ such that $\partial^\alpha \kappa/\omega \in L^\infty$ for every α, and that $N > m + N_0$, where $N_0 \geq 0$ is chosen such that ω is v_{N_0}-moderate. Let*

$$\Phi(x,y) = \frac{\kappa(x)}{\omega(y)\langle x-y\rangle^N}, \quad x,y \in \mathbf{R}^m,$$

and let $\Phi_2(x,\xi)$ be the partial Fourier transform of $\Phi(x,y)$ with respect to the y-variable. Then for any α and β, $\partial_x^\alpha \xi^\beta \Phi_2(x,\xi)$ is a continuous and bounded function on $\mathbf{R}^{2m}$, and

$$\int \sup_x |\partial_x^\alpha \xi^\beta \Phi_2(x,\xi)|\, d\xi < \infty. \tag{3.1}$$

Proof. We start by proving that $\Phi_2 \in C \cap L^\infty$. From the facts that $\kappa/\omega \in L^\infty$, and ω is v_{N_0}-moderate, it follows that Φ is bounded. Hence it suffices to prove that

$$R_\Phi = \sup_x \int \Phi(x,y)\, dy < \infty, \tag{3.2}$$

since $\|\Phi_2\|_{L^\infty} \leq C R_\Phi$ for some constant C.

For some $s > 0$ we have that $N = m + N_0 + s$, and it follows from the inequalities $\kappa(x) \leq C\omega(x) \leq C'\omega(y)\langle x-y\rangle^{N_0}$, for some constants C and C', that

$$\int \Phi(x,y)\, dy < C' \int \langle x-y\rangle^{-m-s}\, dy = C' \int \langle y\rangle^{-m-s}\, dy < \infty.$$

This gives (3.2).

From the assumptions it follows that for any multi-indices α and β, there is a constant $C_{\alpha,\beta}$ such that $|\partial_x^\alpha \partial_y^\beta \Phi(x,y)| \leq C_{\alpha,\beta}\Phi(x,y)$. By applying the Fourier transform on the y-variable and using (3.2), we get

$$\sup_{x,\xi} |\partial_x^\alpha \xi^\beta \Phi_2(x,\xi)| \leq C \sup_x \|\partial_x^\alpha \partial_y^\beta \Phi(x,\,\cdot\,)\|_{L^1} \leq C'_{\alpha,\beta} R_\Phi,$$

and it follows from similar arguments as those above that $\partial_x^\alpha \xi^\beta \Phi_2(x,\xi)$ is a bounded and continuous function for every α and β. Since β may be chosen arbitrary, (3.1) follows. The proof is complete. □

For any $a \in \mathscr{S}(\mathbf{R}^{2m})$, the pseudo-differential operator $a(x,D)$ on $\mathscr{S}(\mathbf{R}^m)$ is defined by

$$a(x,D)f(x) = (2\pi)^{-n} \iint a(x,\xi)f(y)e^{i\langle x-y,\xi\rangle}\,dyd\xi\,. \tag{3.3}$$

The definition extends to any $a \in \mathscr{S}'(\mathbf{R}^{2m})$, and then $a(x,D)$ is continuous from $\mathscr{S}(\mathbf{R}^m)$ to $\mathscr{S}'(\mathbf{R}^m)$. In the case $a \in \mathscr{P}_0$, then $a(x,D)$ is continuous on $\mathscr{S}(\mathbf{R}^m)$, and on $\mathscr{S}'(\mathbf{R}^m)$. (Cf. [11].)

The following result generalizes Theorem 6.1 in [7].

Theorem 3.2. *Assume that $p,q \in [1,\infty]$, $\omega_1 \in \mathscr{P}(\mathbf{R}^{2m})$ and $\omega_2 \in \mathscr{P}_0(\mathbf{R}^{2m})$ such that $\omega_2(x,\xi) = \omega_2(x)$ or $\omega_2(x,\xi) = \omega_2(\xi)$. Then $a \mapsto \omega_2(x,D)a$ is a homeomorphism from $M^{p,q}_{(\omega_1\omega_2)}(\mathbf{R}^m)$ to $M^{p,q}_{(\omega_1)}(\mathbf{R}^m)$.*

Proof. Assume first that $\omega_2(x,\xi) = \omega_2(x)$, and that N_1 is chosen such that ω_1 and ω_2 are v_{N_1}-moderate. Let Φ be as in Lemma 3.1 with $\kappa = \omega = \omega_2$, and set $\chi_0 = \langle\,\cdot\,\rangle^{N_2}\chi$, where $N_2 = N_1 + m + 1$. Then

$$\begin{aligned}\mathscr{F}(\omega_2 a\,\tau_x\chi)(\xi) &= \omega_2(x)\mathscr{F}(a\,\tau_x\chi_0\,\Phi(x,\,\cdot\,))(\xi)\\ &= (2\pi)^{-m/2}\omega_2(x)\big(\mathscr{F}(a\,\tau_x\chi_0) * \Phi_2(x,\,\cdot\,)\big)(\xi).\end{aligned}$$

Here the convolution of $\Phi_2(x,\xi)$ should be applied on the ξ-variable only, considering the x-variable as fix parameter.

Now let $\omega(x,\xi) = \omega_1(x,\xi)\omega_2(x)$, $h(\xi) = \sup_x |\Phi_2(x,\xi)|$, and set $H(\xi) = \langle\xi\rangle^{N_1}h(\xi)$. Then $\|H\|_{L^1} < \infty$ by Lemma 3.1. Hence for some positive constants C and C', Minkowski's inequality gives

$$\begin{aligned}\|\omega_2 a\|_{M^{p,q}_{(\omega_1)}} &= \Big(\int\Big(\int|\mathscr{F}(\omega_2 a\,\tau_x\chi)(\xi)\omega_1(x,\xi)|^p\,dx\Big)^{q/p}d\xi\Big)^{1/q}\\ &\le C\Big(\int\Big(\int|(\mathscr{F}(a\,\tau_x\chi_0) * \Phi_2(x,\,\cdot\,))(\xi)\omega(x,\xi)|^p\,dx\Big)^{q/p}d\xi\Big)^{1/q}\\ &\le C\Big(\int\Big(\int\Big(\int|\mathscr{F}(a\,\tau_x\chi_0)(\xi-\eta)\omega(x,\xi)|^p\,dx\Big)^{1/p}h(\eta)\,d\eta\Big)^{q}d\xi\Big)^{1/q}\\ &\le C\Big(\int\Big(\int\Big(\int|\mathscr{F}(a\,\tau_x\chi_0)(\xi-\eta)\omega(x,\xi-\eta)|^p\,dx\Big)^{1/p}H(\eta)\,d\eta\Big)^{q}d\xi\Big)^{1/q}\\ &\le C\Big(\int\Big(\int|\mathscr{F}(a\,\tau_x\chi_0)(\xi)\omega(x,\xi)|^p\,dx\Big)^{q/p}d\xi\Big)^{1/q}\int H(\eta)\,d\eta\\ &\le C'\|a\|_{M^{p,q}_{(\omega_1\omega_2)}}\|H\|_{L^1},\end{aligned}$$

where the last step follows from Proposition 1.4 (1).

On the other hand, from the first part of the proof together with the obvious fact that $\omega_2^{-1} \in \mathscr{P}_0$, it follows that

$$\|a\|_{M^{p,q}_{(\omega)}} = \|\omega_2^{-1}(\omega_2 a)\|_{M^{p,q}_{(\omega)}} \le C\|\omega_2 a\|_{M^{p,q}_{(\omega/\omega_2)}} = C\|\omega_2 a\|_{M^{p,q}_{(\omega_1)}}$$

This proves the proposition in this case.

The case $\omega_2(x,\xi) = \omega_2(\xi)$ follows by similar arguments. In fact, assume that Φ is the same as the above, and let $\chi_0 = \langle\,\cdot\,\rangle^{N_2}\widehat{\chi}$. Then

$$\begin{aligned}|\mathscr{F}(\omega_2(D)a\,\tau_x\check{\chi})(\xi)| &= |\mathscr{F}(\omega_2\widehat{a}\,\tau_\xi\widehat{\chi})(-x)|\\ &= \omega_2(\xi)|\mathscr{F}(\widehat{a}\,\tau_\xi\chi_0\Phi(\xi,\,\cdot\,))(-x)|\\ &= (2\pi)^{-m/2}\omega_2(\xi)\Big|\int \mathscr{F}(\widehat{a}\,\tau_\xi\chi_0)(-x-y)\Phi_2(\xi,y)\,dy\Big|,\end{aligned}$$

by Remark 1.6. Now let

$$\omega(x,\xi) = \omega_1(x,\xi)\omega_2(\xi), \quad \check{\psi} = \widehat{\chi}_0 \quad \text{and} \quad H(y) = \langle y\rangle^{N_2}\sup_\xi|\Phi_2(\xi,y)|.$$

Then Remark 1.6 and Minkowski's inequality gives

$$\begin{aligned}&\|\omega_2(D)a\|_{M^{p,q}_{(\omega_1)}}\\ &\le C\Big(\int\Big(\int\Big(\int|\mathscr{F}(a\,\tau_{x+y}\psi)(\xi)\Phi_2(\xi,y)\omega(x,\xi)|\,dy\Big)^p\,dx\Big)^{q/p}\,d\xi\Big)^{1/q}\\ &\le C'\|H\|_{L^1}\Big(\int\Big(\int|\mathscr{F}(a\,\tau_x\psi)(\xi)\omega(x,\xi)|^p\,dx\Big)^{q/p}\,d\xi\Big)^{1/q}\\ &= C'\|H\|_{L^1}\|a\|_{M^{p,q}_{(\omega)}},\end{aligned}$$

for some constants C and C'. This gives the result and the proof is complete. □

Corollary 3.3. *Assume that $s, s_0, t, t_0 \in \mathbf{R}$ and that $p,q \in [1,\infty]$. Then the following are true:*

(1) *the map $a \mapsto \langle\,\cdot\,\rangle^{t_0}a$ is a homeomorphism from $M^{p,q}_{s,t+t_0}(\mathbf{R}^m)$ to $M^{p,q}_{s,t}(\mathbf{R}^m)$;*

(2) *the map $a \mapsto \langle D\rangle^{s_0}a$ is a homeomorphism from $M^{p,q}_{s+s_0,t}(\mathbf{R}^m)$ to $M^{p,q}_{s,t}(\mathbf{R}^m)$.*

Proof. The result follows immediately from Theorem 3.2 by letting $\omega_1(x,\xi) = \sigma_{s,t}(x,\xi)$, and $\omega_2(x,\xi) = \sigma_{t_0}(x)$ or $\omega_2(x,\xi) = \sigma_{s_0}(\xi)$. □

The following result is now an immediate consequence of Corollary 3.3.

Theorem 3.4. *Assume that $s,t \in \mathbf{R}$ and that $p,q \in [1,\infty]$. Then*

$$\begin{aligned}M^{p,q}_{s,t}(\mathbf{R}^m) &= \{\,a \in \mathscr{S}'(\mathbf{R}^m)\,;\,\langle x\rangle^t\langle D\rangle^s a \in M^{p,q}(\mathbf{R}^m)\,\}\\ &= \{\,a \in \mathscr{S}'(\mathbf{R}^m)\,;\,\langle D\rangle^s\langle x\rangle^t a \in M^{p,q}(\mathbf{R}^m)\,\}.\end{aligned}$$

Remark 3.5. It follows from Theorem 3.4 that $M^{p,q}_s$ for $s \ge 0$ may be constructed from the space $M^{p,q}$. In fact, Theorem 3.4 gives

$$\begin{aligned}M^{p,q}_s(\mathbf{R}^m) &= M^{p,q}_{s,0}(\mathbf{R}^m) \cap M^{p,q}_{0,s}(\mathbf{R}^m)\\ &= \{\,a \in \mathscr{S}'(\mathbf{R}^m)\,;\,\langle D\rangle^s a,\ \langle x\rangle^s a \in M^{p,q}(\mathbf{R}^m)\,\}.\end{aligned}$$

Since $L^2 = M^2$ it follows from the above that the well-known identities $M^2_{s,0} = H^2_s$ and $M^2_{0,t} = L^2_t$ when $s,t \in \mathbf{R}$, and $M^2_s = L^2_s \cap H^2_s$ when $s \ge 0$ hold. (Cf. [7] or Proposition 11.3.1 in [10].)

Remark 3.6. Assume that $\omega_0 \in \mathscr{P}_0(\mathbf{R}^m)$, and $\omega, \omega_1, \omega_2 \in \mathscr{P}(\mathbf{R}^{2m})$ such that $\omega_1(x,\xi) = \omega_0(x)$ and $\omega_2(x,\xi) = \omega_0(\xi)$. Assume also that $\kappa \in C^\infty(\mathbf{R}^m)$ such that $\partial^\alpha \kappa/\omega_0 \in L^\infty$ for every α. Then it follows from the proof of Theorem 3.3 that $a \mapsto \kappa \cdot a$ and $a \mapsto \kappa(D)a$ are continuous from $M^{p,q}_{(\omega)}$ to $M^{p,q}_{(\omega/\omega_1)}$, and from $M^{p,q}_{(\omega)}$ to $M^{p,q}_{(\omega/\omega_2)}$ respectively.

4. Embedding Properties for Modulation Spaces

In this section, the results from the previous sections are used in order to generalize the embedding properties between modulation spaces and Sobolev/Besov spaces in [16] or [18].

We start by discussing Sobolev and Besov spaces of general types. For any $p \in [1,\infty]$ and $\omega \in \mathscr{P}_0(\mathbf{R}^m)$, the Sobolev space $H^p_{(\omega)}(\mathbf{R}^m)$ is the Banach space which consists of all $a \in \mathscr{S}'(\mathbf{R}^m)$ such that $\|a\|_{H^p_{(\omega)}} \equiv \|\omega(D)a\|_{L^p}$ is finite. If in addition $\omega = \sigma_s$ for some $s \in \mathbf{R}$, then the standard Sobolev space H^p_s and the norm $\|\cdot\|_{H^p_s}$ are obtained.

Some preparations are needed in order to define Besov spaces of general types. For every $m \geq 1$, assume that

$$\psi_0^m = \psi_0 \in C_0^\infty(\mathbf{R}^m), \quad \psi^m = \psi \in C_0^\infty(\mathbf{R}^m \setminus 0), \quad \psi_k^m = \psi_k = \psi(\,\cdot\,/2^k)$$

for $k \geq 1$, are fixed, non-negative and satisfy $\sum_0^\infty \psi_k = 1$.

For any integers $m, N \geq 1$, let $\Upsilon_N(m)$ be the set of all N-tuples $n = (n_1,\dots,n_N)$ of positive integers such that $n_1 + \dots + n_N = m$, and set $\Upsilon(m) = \cup_{N=1}^m \Upsilon_N(m)$. Assume next that $m, N \geq 1$ and $n = (n_1,\dots,n_N) \in \Upsilon_N(m)$ are fixed. For any $k = (k_1,\dots,k_N) \in \mathbf{N}^N$, set

$$\psi_k^n = \psi_{k_1}^{n_1} \otimes \dots \otimes \psi_{k_N}^{n_N},$$

i. e. $\psi_k^n(\xi) = \psi_{k_1}^{n_1}(\xi_1)\cdots\psi_{k_N}^{n_N}(\xi_N)$ where

$$\xi_j \in \mathbf{R}^{n_j}, \text{ when } j = 1,\dots,N, \quad \text{and } \xi = (\xi_1,\dots,\xi_N) \in \mathbf{R}^m. \tag{4.1}$$

Then $\sum_{k\in\mathbf{N}^N} \psi_k^n = 1$, and $\{\psi_k^n\}_{k\in\mathbf{N}^N}$ is considered as a generalized Calderon-Zygmund decomposition. In the case $N = 1$, then $n = m \in \Upsilon_1(m)$ and usual Calderon-Zygmund decomposition appears. If instead $N = m$, then $n = (1,\dots,1) \in \mathbf{N}^m$, and $\{\psi_k^n\}_{k\in\mathbf{N}^N}$ is a modified decomposition which has been considered in [3].

For any $n \in \Upsilon_N(m)$, $p,q \in [1,\infty]$ and $\omega \in \mathscr{P}_0(\mathbf{R}^m)$, the Besov space $B^{p,q}_{(\omega,n)}(\mathbf{R}^m)$ is the Banach space which consists of all $a \in \mathscr{S}'(\mathbf{R}^m)$ such that

$$\|a\|_{B^{p,q}_{(\omega,n)}} \equiv \Big(\sum_{k\in\mathbf{N}^N} \|\omega(D)\psi_k^n(D)a\|_{L^p}^q \Big)^{1/q}$$

is finite. It follows from the analysis in Chapter 6 in [1] that $B^{p,q}_{(\omega,n)}$ is independent of the choice of the ψ_0 and ψ in the Calderon-Zygmund decomposition above, and that different choices of such functions give rise to equivalent norm.

In the case $N = 1$, i.e. $n = m$, the notations $B^{p,q}_{(\omega)}$ and $\|\cdot\|_{B^{p,q}_{(\omega)}}$ are used instead of $B^{p,q}_{(\omega,n)}$ and $\|\cdot\|_{B^{p,q}_{(\omega,n)}}$.

The following proposition shows that if ω satisfies appropriate extra conditions, then the corresponding Besov spaces can be formulated in alternative ways.

Proposition 4.1. *Assume that $N \geq 1$, $0 < \theta < 1$, and $n \in \Upsilon_N(m)$ are fixed. Also assume that*

$$\omega(\xi) = \omega_0(\langle \xi_1 \rangle, \dots, \langle \xi_N \rangle),$$

where ξ is given by (4.1) and $\omega_0 \in \mathscr{P}_0(\mathbf{R}^N)$ such that the map

$$[\theta, \theta^{-1}]^N \times \mathbf{R}^N \ni (s,t) \mapsto \omega_0(t_1, \dots, t_N)/\omega_0(s_1 t_1, \dots, s_N t_N)$$

is bounded. Then $B^{p,q}_{(\omega,n)}(\mathbf{R}^m)$ consists of all $a \in \mathscr{S}'(\mathbf{R}^m)$ such that

$$|||a|||_{B^{p,q}_{(\omega_0,n)}} \equiv \Big(\sum_{k \in \mathbf{N}^N} \big(\omega_0(2^{k_1}, \dots, 2^{k_N}) \|\psi^n_k(D)a\|_{L^p}\big)^q \Big)^{1/q}$$

is finite. Moreover, $|||\cdot|||_{B^{p,q}_{(\omega_0,n)}}$ and $\|\cdot\|_{B^{p,q}_{(\omega,n)}}$ are equivalent norms.

Proof. From the assumptions it follows that

$$C^{-1} \leq \omega_0(t_1, \dots, t_N)/\omega_0(s_1 t_1, \dots, s_N t_N) \leq C \tag{4.2}$$

holds for some constant C which does not dependent on both $s \in [2^{-1}, 2]^N$ and $t \in \mathbf{R}^N$. Since (4.2) is invariant under the transformation $\omega_0 \mapsto \omega_0^{-1}$, it suffices to prove that

$$\|\omega(D)\psi^n_k(D)a\|_{L^p} \leq C_0 \omega_0(2^{k_1}, \dots, 2^{k_N}) \|\psi^n_k(D)a\|_{L^p} \tag{4.3}$$

for some constant C_0 independent of a and k. For any $m_0 \geq 1$, let $0 \leq \phi^{m_0}_0 \in C^\infty_0(\mathbf{R}^{m_0})$ and $\phi^{m_0} \in C^\infty_0(\mathbf{R}^{m_0} \setminus 0)$ be chosen such that $\phi^{m_0}_0 = 1$ and $\phi^{m_0} = 1$ in the supports of $\psi^{m_0}_0$ and ψ^{m_0} respectively, and set $\phi^{m_0}_{k_0} = \phi^{m_0}(\,\cdot\,/2^{k_0})$ when $k_0 \geq 1$. Let also $\phi^n_k = \phi^{n_1}_{k_1} \otimes \cdots \otimes \phi^{n_N}_{k_N}$ when k and n are the same as the above, and set $\varphi^n_k = \omega_0(2^{k_1}, \dots, 2^{k_N})^{-1} \omega\, \phi^n_k$. Then $\phi^n_k = 1$ in the support of ψ^n_k, and (3.3) and Fourier's inversion formula gives,

$$\begin{aligned} \omega(D)\psi^n_k(D)a &= \omega(D)\phi^n_k(D)(\psi^n_k(D)a) \\ &= (2\pi)^{-m/2} \omega_0(2^{k_1}, \dots, 2^{k_N})\, \widehat{\varphi}^n_k * (\psi^n_k(D)a). \end{aligned} \tag{4.4}$$

From the assumptions, it follows that $\sup_k \|\partial^\alpha \varphi^n_k\|_{L^\infty} < \infty$ for every α, which in combination with the support properties for ϕ^n_k implies that $\{\widehat{\varphi}^n_k\}_{k \in \mathbf{N}^N}$ is a bounded set in L^1. Hence (4.3) follows from (4.4) and Young's inequality. The proof is complete. □

The following result is an immediate consequence of Proposition 4.1. Here and in what follows, $B^{p,q}_s(\mathbf{R}^m)$ for $p, q \in [1, \infty]$ and $s \in \mathbf{R}$ is the standard Besov space which consists of all $a \in \mathscr{S}'(\mathbf{R}^m)$ such that

$$\|a\|_{B^{p,q}_s} \equiv \Big(\sum_{k=0}^{\infty} \big(2^{ks} \|\psi_k(D)a\|_{L^p}\big)^q \Big)^{1/q} < \infty.$$

Corollary 4.2. *Assume that $s, s_0 \in \mathbf{R}$ and $p, q \in [1,\infty]$. Then the following are true:*

(1) $H^p_s(\mathbf{R}^m) = H^p_{(\sigma_s)}(\mathbf{R}^m)$ *and* $B^{p,q}_s(\mathbf{R}^m) = B^{p,q}_{(\sigma_s)}(\mathbf{R}^m)$*;*

(2) *the map $\sigma_{s_0}(D)$ is a homeomorphism from $H^p_{s+s_0}$ to H^p_s, and from $B^{p,q}_{s+s_0}$ to $B^{p,q}_s$.*

By straight-forward verifications, it follows that the analysis concerning the standard Sobolev/Besov spaces (cf. Chapter 5 and Chapter 6 in [1]) carry over to Sobolev and Besov spaces of the type $H^p_{(\omega)}$ and $B^{p,q}_{(\omega,n)}$. In particular, the usual embeddings between Sobolev and Besov spaces (cf. (2.3) in [16] or [18]) takes the form

$$\begin{aligned} &B^{p,q_1}_{(\omega,n)} \hookrightarrow B^{p,q_2}_{(\omega,n)}, \qquad B^{p,\min(p,p')}_{(\omega,n)} \hookrightarrow H^p_{(\omega)} \hookrightarrow B^{p,\max(p,p')}_{(\omega,n)}, \\ &B^{2,2}_{(\omega,n)} = H^2_{(\omega)}, \qquad B^{p,\infty}_{(\omega\sigma_\varepsilon,n)} \hookrightarrow H^p_{(\omega)} \hookrightarrow B^{p,1}_{(\omega\sigma_{-\varepsilon},n)}, \end{aligned} \tag{4.5}$$

when $\omega \in \mathscr{P}(\mathbf{R}^m)$, $\varepsilon > 0$, $p, q, q_1, q_2 \in [1,\infty]$ such that $q_1 \le q_2$, and $n \in \Upsilon(m)$.

Next embeddings between modulation spaces and Besov spaces are discussed. The following result generalizes Proposition 1.6 (3) and (4) in [16] or Proposition 1.7 (3) and (4) in [18].

Proposition 4.3. *Assume that $\omega_1, \omega_2 \in \mathscr{P}(\mathbf{R}^{2m})$ such that $\omega_1(x,\xi) = \omega_1(x)$ and $\omega_2(x,\xi) = \omega_2(\xi)$, and that $p, q, r \in [1,\infty]$. Then the following are true:*

(1) *if $q \le p \le r \le q'$, then*

$$M^{p,q}_{(\omega_1)}(\mathbf{R}^m) \hookrightarrow L^r_{(\omega_1)}(\mathbf{R}^m) \quad \text{and} \quad M^{p,q}_{(\omega_2)}(\mathbf{R}^m) \hookrightarrow H^r_{(\omega_2)}(\mathbf{R}^m); \tag{4.6}$$

(2) *if $q' \le r \le p \le q$, then* (4.6) *holds with reversed inclusions.*

Proof. If $\omega = 1$, then the result agrees with Proposition 1.6 (3) and (4) in [16] and Proposition 1.7 (3) and (4) in [18]. For general ω, the result is a consequence of Theorem 3.2. □

By using Besov spaces instead of Sobolev spaces, Proposition 4.3 is refined in the following result, which also generalizes Theorem 4.1 in [16] or [18].

Theorem 4.4. *Assume that $s \in \mathbf{R}$, $p, q, p_j, q_j \in [1,\infty]$ for $j = 1,2$ satisfy that $p_1 \le p \le p_2$ and $q_1 \le q \le q_2$, and assume that θ_1 and θ_2 are given by* (0.3)*. Then*

$$B^{p_1,q_1}_{s+m\theta_1(p_1,q_1)}(\mathbf{R}^m) \hookrightarrow M^{p,q}_{s,0}(\mathbf{R}^m) \hookrightarrow B^{p_2,q_2}_{s+m\theta_2(p_2,q_2)}(\mathbf{R}^m),$$

and for some constant $C_m > 0$, depending on m only, then

$$C_m^{-1}\|a\|_{B^{p_2,q_2}_{s+m\theta_2(p_2,q_2)}} \le \|a\|_{M^{p,q}_{s,0}} \le C_m\|a\|_{B^{p_1,q_1}_{s+m\theta_1(p_1,q_1)}}, \qquad a \in \mathscr{S}'(\mathbf{R}^m).$$

In particular, (0.4) *holds.*

Proof. If $s = 0$, then the statement coincides with Theorem 4.1 in [16] or [18]. For general s, the result now follows from Theorem 3.3, Corollary 4.2 and (4.5). □

Remark 4.5. In [9], P. Gröbner makes comparisons between Besov spaces and α-modulation spaces. Embeddings between Besov spaces and modulation spaces can then be obtained by choosing certain parameters in appropriate ways. In this context, Theorem 4.4 agrees with Gröbner's result in the case $\min(p,p') \le q \le \max(p,p')$ while Theorem 4.4 strictly improves Gröbner's result in the remaining case $\min(q,q') < p < \max(q,q')$.

Remark 4.6. By similar arguments as in the proofs of Theorem 4.1 in [16] or [18] and Theorem 4.4, the following generalization is obtained. Assume that $p,q,p_j,q_j \in [1,\infty]$, $j=1,2$, satisfy $p_1 \le p \le p_2$ and $q_1 \le q \le q_2$. Assume also that $n \in \Upsilon_j(m)$ is fixed for some $j \ge 1$, and that $\omega \in \mathscr{P}_0(\mathbf{R}^{2m})$ such that $\omega(x,\xi) = \omega(\xi)$, and for $k = 1,2$ set

$$\gamma_k(\xi) = \gamma_{p_k,q_k,k}(\xi) = \omega(\xi)\sigma_{n_1 s_k}(\xi_1)\cdots\sigma_{n_j s_k}(\xi_j),$$

where $s_k = \theta_k(p_k,q_k)$, $\xi_i \in \mathbf{R}^{n_i}$ and $\xi = (\xi_1,\dots,\xi_j) \in \mathbf{R}^m$. Here θ_1 and θ_2 are given by (0.3). Then

$$B^{p_1,q_1}_{(\gamma_1,n)}(\mathbf{R}^m) \hookrightarrow M^{p,q}_{(\omega)}(\mathbf{R}^m) \hookrightarrow B^{p_2,q_2}_{(\gamma_2,n)}(\mathbf{R}^m),$$

and

$$C_m^{-1}\|a\|_{B^{p_2,q_2}_{(\gamma_2,n)}} \le \|a\|_{M^{p,q}_{(\omega)}} \le C_m\|a\|_{B^{p_1,q_1}_{(\gamma_1,n)}}, \quad a \in \mathscr{S}'(\mathbf{R}^m),$$

for some constant $C_m > 0$, depending on m only.

5. Convolution Properties for Modulation Spaces

In this section, the convolution results in [16] or [18] for standard modulation spaces are generalized to weighted modulation spaces. It is proved that the convolution on $\mathscr{S}(\mathbf{R}^m)$ extends to a continuous multiplication from $M^{p_1,q_1}_{(\omega_1)}(\mathbf{R}^m) \times \cdots \times M^{p_N,q_N}_{(\omega_N)}(\mathbf{R}^m)$ to $M^{p_0,q_0}_{(\omega_0)}(\mathbf{R}^m)$, provided $p_j,q_j \in [1,\infty]$ and $\omega_j \in \mathscr{P}(\mathbf{R}^{2m})$ for $0 \le j \le N$ satisfy

$$\frac{1}{p_1}+\frac{1}{p_2}+\cdots+\frac{1}{p_N} = N-1+\frac{1}{p_0}, \qquad \frac{1}{q_1}+\frac{1}{q_2}+\cdots+\frac{1}{q_N} = \frac{1}{q_0}, \tag{5.1}$$

and

$$\omega_0(x_1+\cdots+x_N,\xi) \le C\omega_1(x_1,\xi)\cdots\omega_N(x_N,\xi), \tag{5.2}$$

for some constant C independent of $x_1,\dots,x_N,\xi \in \mathbf{R}^m$.

The first step deals with finding an appropriate definition of the convolution. Assume that

$$\chi_0,\chi_1,\dots,\chi_N \in \mathscr{S}(\mathbf{R}^m), \quad (\chi_0,\chi_1\cdots\chi_N) = (2\pi)^{(N-1)m/2}. \tag{5.3}$$

For every $a_1,\dots,a_N,\varphi \in \mathscr{S}(\mathbf{R}^m)$, it follows from Parseval's formula that

$$\begin{gathered}(a_1 * \cdots * a_N,\varphi) = \iint (f_{1,\xi} * \cdots * f_{N,\xi})(x)\overline{g_\xi(x)}\,dxd\xi,\\ \text{where}\quad f_{j,\xi}(x) = \mathscr{F}(\widehat{a}_j\,\tau_\xi\chi_j)(x) \quad\text{and}\quad g_\xi(x) = \mathscr{F}(\widehat{\varphi}\,\tau_\xi\chi_0)(x).\end{gathered} \tag{5.4}$$

Here and in what follows, the convolution of $f_{j,\xi}(x)$ should be taken only with respect to the x-variable, considering the ξ-variable as parameter.

Lemma 5.1. *Assume that $p_j, q_j \in [1,\infty]$ and $\omega_j \in \mathscr{P}(\mathbf{R}^m)$ for $0 \le j \le N$, satisfy (5.1) and (5.2). Let $a_j \in M^{p_j,q_j}_{(\omega_j)}(\mathbf{R}^m)$ and $f_{j,\xi}$ be as in (5.4) when $1 \le j \le N$. Then $G(x,\xi) = (|f_{1,\xi}| * \cdots * |f_{N,\xi}|)(x)$ is measurable and for some constant C,*

$$\|G\|_{L^{p_0,q_0}_{(\omega_0)}} \le C \prod_{j=1}^{N} \|a_j\|_{M^{p_j,q_j}_{(\omega_j)}} . \tag{5.5}$$

Proof. Since $(x,\xi) \mapsto f_{j,\xi}(x)$ is a continuous function, the result follows by using (5.2) on the left-hand side of (5.5), then by applying Young's inequality with respect to the x-variable, and then Hölder's inequality with respect to the ξ-variable. □

Assume that the hypothesis in Lemma 5.1 is fulfilled. Then (5.4) is taken as the definition of $a_0 = a_1 * \cdots * a_N$ as an element in $\mathscr{S}'(\mathbf{R}^m)$. It follows from Lemma 5.1 that this definition makes sense, and Proposition 1.4 (3) shows that $a_0 \in M^{p_0,q_0}_{(\omega_0)}(\mathbf{R}^m)$. From (5.4) and the proof of Lemma 5.1 it also follows that $a_1 * \cdots * a_N$ is independent of the order of a_j for $j = 1,\dots,N$. If in addition $p_j, q_j < \infty$ for $j = 1,\dots,N$, then Proposition 1.4 (4) shows that the convolution is uniquely defined, independent of $\chi_0,\dots,\chi_N$ above, and satisfies the usual associativity properties.

In the following, similar invariance and associativity is proved for general $p_j, q_j \in [1,\infty]$ for $j = 1,\dots,N$. The first step is to prove that the definition of a_0 is independent of the choice of $\chi_0,\dots,\chi_N$ in (5.3).

Lemma 5.2. *Assume that a_j, $1 \le j \le N$, are the same as in Lemma 5.1. Then $a_1 * \cdots * a_N$ is independent of the choice of $\chi_0, \chi_1, \dots, \chi_N$ in (5.3).*

Proof. The result is proved in the case $N = 2$ and $\chi_0 = \chi_1 = \chi_2 = \chi$. The general case follows by similar arguments and is left to the reader. Assume that $0 \le \psi \in \mathscr{S}(\mathbf{R}^m)$ satisfies $\int \psi^3\, d\xi = (2\pi)^{m/2}$, and set $\chi_\xi = \chi(\cdot - \xi)$ and $\psi_\eta = \psi(\cdot - \eta)$. Then the function $(x,\xi,\eta) \mapsto \mathscr{F}(\widehat{\varphi}\chi_\xi\psi_\eta)(x)$ belongs to $\mathscr{S}(\mathbf{R}^{3m})$, which implies that

$$\begin{aligned} I &= \iiint (\mathscr{F}(\widehat{a}_1\chi_\xi\psi_\eta) * \mathscr{F}(\widehat{a}_2\chi_\xi\psi_\eta))(x)\overline{\mathscr{F}(\widehat{\varphi}\chi_\xi\psi_\eta)(x)}\, dx d\xi d\eta \\ &= \iiiint \mathscr{F}(\widehat{a}_1\chi_\xi\psi_\eta)(x)\mathscr{F}(\widehat{a}_2\chi_\xi\psi_\eta)(y)\overline{\mathscr{F}(\widehat{\varphi}\chi_\xi\psi_\eta)(x+y)}\, dx dy d\xi d\eta \end{aligned}$$

makes sense. Here we have used Lemma 5.1 and its proof in order to conclude that the integrand on the right-hand side belongs to $L^1(\mathbf{R}^{4m})$, which in particular implies that the orders of integration can be interchanged. Hence Parseval's formula applied to the x-variables gives

$$\begin{aligned}I &= \iiiint (\widehat{a}_1\chi_\xi\psi_\eta)(\xi_1)\mathscr{F}(\widehat{a}_2\chi_\xi\psi_\eta)(y)\overline{(\widehat{\varphi}\chi_\xi\psi_\eta)(\xi_1)}e^{i\langle y,\xi_1\rangle}\,dyd\xi d\xi_1 d\eta\\ &= \iiiint (\widehat{a}_1\chi_\xi)(\xi_1)\mathscr{F}(\widehat{a}_2\chi_\xi\psi_\eta)(y)\overline{(\widehat{\varphi}\chi_\xi\psi_\eta^2)(\xi_1)}e^{i\langle y,\xi_1\rangle}\,dyd\xi d\xi_1 d\eta\\ &= \iiiint \mathscr{F}(\widehat{a}_1\chi_\xi)(x)\mathscr{F}(\widehat{a}_2\chi_\xi\psi_\eta)(y)\overline{\mathscr{F}(\widehat{\varphi}\chi_\xi\psi_\eta^2)(x+y)}\,dxdyd\xi d\eta.\end{aligned}$$

By similar arguments it follows that

$$\begin{aligned}I &= \iiiint \mathscr{F}(\widehat{a}_1\chi_\xi)(x)\mathscr{F}(\widehat{a}_2\chi_\xi)(y)\overline{\mathscr{F}(\widehat{\varphi}\chi_\xi\psi_\eta^3)(x+y)}\,dxdyd\xi d\eta\\ &= (2\pi)^{m/2}\iiint \mathscr{F}(\widehat{a}_1\chi_\xi)(x)\mathscr{F}(\widehat{a}_2\chi_\xi)(y)\overline{\mathscr{F}(\widehat{\varphi}\chi_\xi)(x+y)}\,dxdyd\xi.\end{aligned}$$

Here it has been used that $\int \psi_\eta^3(\xi)\,d\eta = (2\pi)^{m/2}$. This gives

$$I = (2\pi)^{m/2}\iint \big(\mathscr{F}(\widehat{a}_1\chi_\xi) * \mathscr{F}(\widehat{a}_2\chi_\xi)\big)(x)\overline{\mathscr{F}(\widehat{\varphi}\chi_\xi)(x)}\,dxd\xi$$

which is equal to $(2\pi)^{m/2}(a_1 * a_2, \varphi)$. In the same way it follows that

$$I = (2\pi)^{m/2}\iint \big(\mathscr{F}(\widehat{a}_1\psi_\eta) * \mathscr{F}(\widehat{a}_2\psi_\eta)\big)(x)\overline{\mathscr{F}(\widehat{\varphi}\psi_\eta)(x)}\,dxd\eta,$$

which proves the announced invariance. The proof is complete. □

Next it is proved that

$$\begin{aligned}(2\pi)^{(N-1)m/2}&\mathscr{F}(\widehat{a}_1\cdots\widehat{a}_N\,\tau_\xi(\chi_1\cdots\chi_N))(x)\\ &\qquad=\big(f_{1,\xi} * \cdots * f_{N,\xi}\big)(x),\end{aligned}\tag{5.6}$$

with equality in $\mathscr{S}'(\mathbf{R}^{2m})$. Here $(2\pi)^{(N-1)m/2}\widehat{a}_1\cdots\widehat{a}_N$ on the left-hand side is the Fourier transform of the distribution a_0 above, while the right-hand side is a measurable function in view of Lemma 5.1.

Lemma 5.3. *Assume that a_j, χ_j and $f_{j,\xi}$, $0 \le j \le N$, are the same as in Lemma 5.1. Then* (5.6) *holds with equality in $\mathscr{S}'(\mathbf{R}^{2m})$.*

Proof. We restrict to the case $N = 2$ and $\chi_0 = \chi_1 = \chi_2 = \chi$. The general case follows by similar arguments and is again left for the reader. Let $\chi_\xi = \chi(\cdot - \xi)$, and set $F_0(x,\xi) = (2\pi)^{m/2}\mathscr{F}(\widehat{a}_1\,\widehat{a}_2\chi_\xi^2)(x)$. If $\Phi \in \mathscr{S}(\mathbf{R}^{2m})$, then Parseval's formula gives

$$(F_0, \Phi) = (2\pi)^{m/2}(\widehat{a}_1\widehat{a}_2, \widehat{\varphi}) = (a_1 * a_2, \varphi),$$

where $\widehat{\varphi}(\xi) = \int \overline{\chi(\xi-\eta)^2\Phi_1(\xi,\eta)}\,d\eta$. Here Φ_1 is the inverse partial Fourier transform of $\Phi(x,\eta)$ with respect to the x-variable. This gives

$$\langle F_0, \Phi\rangle = \iiint \mathscr{F}(\widehat{a}_1\chi_\xi)(x-y)\mathscr{F}(\widehat{a}_2\chi_\xi)(y)\overline{\mathscr{F}(\widehat{\varphi}\chi_\xi)(x)}\,dxdyd\xi$$

$$= \iiiint \mathscr{F}(\widehat{a}_1\chi_\xi)(x-y)\mathscr{F}(\widehat{a}_2\chi_\xi)(y)\overline{\mathscr{F}(\Phi_1(\,\cdot\,,\eta)\overline{\chi_\eta^2}\chi_\xi)(x)}\,dxdyd\xi d\eta.$$

Note that the integrals make sense since their integrands belong to L^1. Hence if $\psi_\xi = \psi(\,\cdot\,-\xi)$ where $\psi = \chi\overline{\chi}^2$, then the same arguments as in the proof of Lemma 5.2 for moving the factors χ_ξ and χ_η, imply that the last integral is equal to

$$\begin{aligned}&\iiiint \mathscr{F}(\widehat{a}_1\chi_\eta)(x-y)\mathscr{F}(\widehat{a}_2\chi_\eta)(y)\overline{\mathscr{F}(\Phi_1(\,\cdot\,,\eta)\psi_\xi)(x)}\,dxdyd\xi d\eta\\ &\quad = (2\pi)^{m/2}\iiint \mathscr{F}(\widehat{a}_1\chi_\eta)(x-y)\mathscr{F}(\widehat{a}_2\chi_\eta)(y)\overline{\Phi(x,\eta)}\,dxdyd\eta.\end{aligned} \tag{5.7}$$

Here it has been used that

$$\int \mathscr{F}(\Phi_1(\,\cdot\,,\eta)\,\psi_\xi)(x)\,d\xi = (2\pi)^{m/2}\mathscr{F}(\Phi_1(\,\cdot\,,\eta))(x) = (2\pi)^{m/2}\Phi(x,\xi),$$

by Fourier's inversion formula, and that $\int \psi_\xi\,d\xi = (2\pi)^{m/2}$. The result now follows since the right-hand side of (5.7) is equal to

$$(2\pi)^{m/2}\iint (\mathscr{F}(\widehat{a}_1\chi_\eta) * \mathscr{F}(\widehat{a}_2\chi_\eta))(x)\overline{\Phi(x,\eta)}\,dxd\eta,$$

which completes the proof. □

The next lemma deals with associativity. Here and in other situations, the map $(a_1,\dots,a_N)\mapsto a_1*\dots*a_N$ is called associative if the corresponding operation $*$ is associative.

Lemma 5.4. *Assume that p_j, q_j and ω_j for $j=0,\dots,N$ are the same as in Lemma 5.1. Then $(a_1,\dots,a_N)\mapsto a_1*\dots*a_N$ is an associative map from $M^{p_1,q_1}_{(\omega_0)}\times\dots\times M^{p_N,q_N}_{(\omega_N)}$ to $M^{p_0,q_0}_{(\omega_0)}$.*

Proof. By induction, it suffices to prove the assertion in the case $N=3$. The result then follows by verifying $a_1*a_2*a_3 = (a_1*a_2)*a_3 = a_1*(a_2*a_3)$.

Let $\chi\in\mathscr{S}$ such that $\int\chi(\xi)^4\,dx = 1$, and set $\chi_\xi = \chi(\,\cdot\,-\xi)$ as before. Then (5.4) and Lemma 5.2 give that

$$\begin{aligned}&(a_1*a_2*a_3,\varphi)\\ &\qquad = (2\pi)^{-m}\iint (\mathscr{F}(\widehat{a}_1\chi_\xi) * \mathscr{F}(\widehat{a}_2\chi_\xi) * \mathscr{F}(\widehat{a}_3\chi_\xi))(x)\overline{\mathscr{F}(\widehat{\varphi}\chi_\xi)(x)}\,dxd\xi.\end{aligned}$$

In the same way it follows that

$$((a_1*a_2)*a_3,\varphi) = (2\pi)^{-m/2}\iint (\mathscr{F}(\widehat{a}_1\widehat{a}_2\chi_\xi^2) * \mathscr{F}(\widehat{a}_3\chi_\xi))(x)\overline{\mathscr{F}(\varphi\chi_\xi)(x)}\,dxd\xi.$$

It now follows from Lemma 5.3 and its proof that $\mathscr{F}(\widehat{a}_1\widehat{a}_2\chi_\xi^2)$ in the last integral may be replaced by $(2\pi)^{-m/2}\mathscr{F}(\widehat{a}_1\chi_\xi) * \mathscr{F}(\widehat{a}_2\chi_\xi)$. This proves that $a_1 * a_2 * a_3 = (a_1 * a_2) * a_3$. By the same arguments it follows that $a_1 * a_2 * a_3 = a_1 * (a_2 * a_3)$. The proof is complete. □

Summing up, the following result is obtained.

Theorem 5.5. *Assume that $p_j, q_j \in [1,\infty]$ when $0 \le j \le N$ satisfy (5.1), and that $\omega_0, \dots, \omega_N \in \mathscr{P}(\mathbf{R}^{2m})$ satisfy (5.2) for some constant C, independent of $x_1, \dots, x_N, \xi \in \mathbf{R}^m$. Then $(a_1, \dots, a_N) \mapsto a_1 * \cdots * a_N$ is a continuous, symmetric and associative map from $M^{p_1,q_1}_{(\omega_1)}(\mathbf{R}^m) \times \cdots \times M^{p_N,q_N}_{(\omega_N)}(\mathbf{R}^m)$ to $M^{p_0,q_0}_{(\omega_0)}(\mathbf{R}^m)$, which is independent of the choice of $\chi_0, \dots, \chi_N$ in (5.3). If $a_j \in M^{p_j,q_j}_{(\omega_j)}(\mathbf{R}^m)$ for $j = 1, \dots, N$, then for some constant $C > 0$,*

$$\|a_1 * \cdots * a_N\|_{M^{p_0,q_0}_{(\omega_0)}} \le C \prod_{j=1}^N \|a_j\|_{M^{p_j,q_j}_{(\omega_j)}}. \tag{5.8}$$

Moreover, $a_1 * \cdots * a_N$ agrees with the usual convolution product when at least $N-1$ factors belong to $\mathscr{S}$.

Remark 5.6. It follows from Theorem 5.5 and its proof that the same conclusion in Theorem 5.5 holds when $M^{p_j,q_j}_{(\omega_j)}$ is replaced by $\mathcal{M}^{p_j,q_j}_{(\omega_j)}$ for every $j = 0, \dots, N$. (See Remark 1.7.)

Remark 5.7. Cordero and Gröchenig have proven independently of the author an estimate similar to (5.8). (See Proposition 2.4 in [4].) However, the arguments in [4] are the same as for (2.7) in [16] or [18], as well as for (5.8) here above. Moreover, similar arguments were already used in [6] in order to establish convolution properties for Wiener-amalgam spaces.

Remark 5.8. There are many different generalizations of Theorem 5.5. For example, when combining Remark 4.6 and Theorem 5.5, the following result is obtained, which also generalizes Theorem 2.6 in [16] or [18].

Assume that $N_0 \ge 0$, and that $p_j, q_j \in [1,\infty]$ when $0 \le j \le N$ satisfy (5.1), and that $q_j \ge \max(p_j, p_j')$ when $N_0+1 \le j \le N$, and let $n_{N_0+1}, \dots, n_N \in \Upsilon(m)$ be fixed. Assume also that $\omega_0, \dots, \omega_N \in \mathscr{P}(\mathbf{R}^{2m})$ satisfy (5.2), and that $\omega_k(x,\xi) = \omega_k(\xi)$ when $N_0 + 1 \le k \le N$. Then $(a_1, \dots, a_N) \mapsto a_1 * \cdots * a_N$ is a continuous, symmetric and associative map from

$$\prod_{j=1}^{N_0} M^{p_j,q_j}_{(\omega_j)}(\mathbf{R}^m) \times \prod_{k=N_0+1}^{N} B^{p_k,q_k}_{(\omega_k,n_k)}(\mathbf{R}^m)$$

to $M^{p_0,q_0}_{(\omega_0)}(\mathbf{R}^m)$. For some constant C, depending on m and N only, the estimate

$$\|a_1 * \cdots * a_N\|_{M^{p_0,q_0}_{(\omega_0)}} \le C\Big(\prod_{j=1}^{N_0} \|a_j\|_{M^{p_j,q_j}_{(\omega_j)}}\Big)\Big(\prod_{k=N_0+1}^{N} \|a_k\|_{B^{p_k,q_k}_{(\omega_k,n_k)}}\Big), \tag{5.9}$$

holds for every $a_j \in M^{p_j,q_j}_{(\omega_j)}(\mathbf{R}^m)$, $1 \le j \le N_0$, and $a_k \in B^{p_k,q_k}_{(\omega_k,n_k)}(\mathbf{R}^m)$, $N_0+1 \le k \le N$.

Moreover, if $n_0 \in \Upsilon(m)$, and in addition $q_0 \le \min(p_0,p_0')$ and $\omega_0(x,\xi) = \omega_0(\xi)$, then the same conclusion holds when $M^{p_0,q_0}_{(\omega_0)}$ is replaced by $B^{p_0,q_0}_{(\omega_0,n_0)}$.

By choosing $\omega_0,\dots,\omega_N$ in Remark 5.8 in appropriate ways, and using (4.5), the following is obtained. The verification is left for the reader.

Corollary 5.9. *Assume that $N_0 \ge 0$, and that $p_j,q_j \in [1,\infty]$ when $0 \le j \le N$ satisfy* (5.1), *and that $q_j \ge \max(p_j,p_j')$ when $N_0+1 \le j \le N$. Assume also that s_j, $0 \le j \le N$, are real numbers such that $s_0 = s_1+\cdots+s_N$. Then $(a_1,\dots,a_N) \mapsto a_1 * \cdots * a_N$ is a continuous map from*

$$M^{p_1,q_1}_{s_1,0}(\mathbf{R}^m) \times \cdots \times M^{p_{N_0},q_{N_0}}_{s_{N_0},0}(\mathbf{R}^m) \times B^{p_{N_0+1},q_{N_0+1}}_{s_{N_0+1}}(\mathbf{R}^m) \times \cdots \times B^{p_N,q_N}_{s_N}(\mathbf{R}^m)$$

to $M^{p_0,q_0}_{s_0,0}(\mathbf{R}^m)$.

If in addition $q_0 \le \min(p_0,p_0')$, then the same conclusion holds when $M^{p_0,q_0}_{s_0,0}$ is replaced by $B^{p_0,q_0}_{s_0}$.

The same conclusion holds when $B^{p_k,q_k}_{s_k}$, $k = N_0+1,\dots,N$, are replaced by $H^{p_k}_{s_k}$.

Remark 5.10. Assume that $p,q \in [1,\infty]$ and $\omega \in \mathscr{P}(\mathbf{R}^{2m})$ such that $\omega(x,\xi) = \omega(\xi)$. Then it follows from Remark 1.5 and Theorem 5.5 that the usual convolution on C_B' extends to a continuous associative and commutative multiplication on $M^{1,\infty}$. Moreover, it follows that $M^{p,q}_{(\omega)}$ is an $M^{1,\infty}$-module under convolution. (See also Corollary 2.12 in [8], Remark 2.11 in [16] or [18].)

Remark 5.11. Assume that P is a polynomial of degree N_0 on $\mathbf{R}^m$, $p_j,q_j \in [1,\infty]$ and $\omega_j \in \mathscr{P}(\mathbf{R}^{2m})$ for $j = 0,\dots,N$ such that (5.1) and (5.2) are fulfilled. Let $a_0 = a_1 * \cdots * a_N$ and $\nu = 1 \otimes \sigma_{N_0}$. Then it follows from Remark 3.6, Theorem 5.5 and (5.4) that $P(D)a_0 = (P(D)a_1) * a_2 * \cdots * a_N \in M^{p_0,q_0}_{(\omega_0/\nu)}$.

It is obvious that Remark 3.6 and Remark 5.11 open up for using modulation space theory when dealing with fundamental solutions and parametrix constructions for linear partial differential equations.

Acknowledgments. I am grateful to Professors P. Boggiatto, H.G. Feichtinger, K.H. Gröchenig and N. Teofanov for fruitful discussions and for valuable advice. I am also grateful for the support and the nice hospitality when I participated in Bimestre Intensivo 2003, and other activities at Universitè di Torino. Especially I thank Professors P. Boggiatto, D. Coriasco and L. Rodino. I also thank Professor B.-W. Schulze for interesting discussions and nice hospitality when I visited Universität Potsdam in November 2002. I am also grateful to Professor M.W. Wong for interesting discussions, and for the nice hospitality when I participated in the Fourth ISAAC Congress at York University in Toronto.

References

[1] J. Bergh and J. Löfström, *Interpolation Spaces: An Introduction*, Springer-Verlag, Berlin Heidelberg NewYork, 1976.

[2] P. Boggiatto, Localization operators with L^p symbols on modulation spaces, preprint.

[3] A. Boulkhemair, L^2 estimates for Weyl quantization, *J. Funct. Anal.* **165** (1999), 173–204.

[4] E. Cordero and K. Gröchenig, Time-frequency analysis of localization operators, *J. Funct. Anal.* **205** (2003), 107–131.

[5] H.G. Feichtinger, Banach spaces of distributions of Wiener's type and interpolation, in Proc. Conf. Oberwolfach, *Functional Analysis and Approximation*, August 1980, Vol 69, Int. Ser. Num. Math., Editors: P. Butzer, B. Sz. Nagy and E. Görlich, Birkhäuser-Verlag, Basel, Boston, Stuttgart, 1981, pp. 153–165.

[6] ———, Banach convolution algebras of Wiener's type, in *Proc. Functions, Series, Operators* in Budapest, Colloquia Math. Soc. J. Bolyai, North Holland Publ. Co., Amsterdam Oxford NewYork, 1980.

[7] ———, *Modulation spaces on locally compact abelian groups*, Technical report, University of Vienna, Vienna, 1983.

[8] H.G. Feichtinger and P. Gröbner, Banach spaces of distributions defined by decomposition methods, I, *Math. Nachr.* **123** (1985), 97–120.

[9] P. Gröbner, *Banachräume Glatter Funktionen und Zerlegungsmethoden*, Thesis, University of Vienna, 1992.

[10] K. Gröchenig, *Foundations of Time-Frequency Analysis*, Birkhäuser, Boston, 2001.

[11] L. Hörmander, *The Analysis of Linear Partial Differential Operators I, III*, Springer-Verlag, Berlin Heidelberg New York Tokyo, 1983, 1985.

[12] K.A. Okoudjou, Embeddings of some classical Banach spaces into modulation spaces, preprint.

[13] J. Sjöstrand, An algebra of pseudo-differential operators, *Math. Res. Lett.* **1** (1994), 185–192.

[14] J. Toft, *Continuity and Positivity Problems in Pseudo-Differential Calculus*, Thesis, University of Lund, 1996.

[15] ———, Subalgebras to a Wiener type algebra of pseudo-differential operators, *Ann. Inst. Fourier* (5) **51** (2001), 1347–1383.

[16] ———, *Modulation spaces and pseudo-differentianl operators*, Research Report 2002:05 (2002), Blekinge Institute of Technology, Karlskrona.

[17] ———, *Weighted modulation spaces and pseudo-differential operators*, Research Report 2003:05 (2003), Blekinge Institute of Technology, Karlskrona.

[18] ———, Continuity properties for modulation spaces with applications in pseudo-differential calculus, I, *J. Funct. Anal.* **207** (2004), 399–429.

[19] M.W. Wong, *Wavelet Transforms and Localization Operators*, Birkhäuser, Basel, 2002.

Joachim Toft
Department of Mathematics and Systems Engineering
Växjö University
Sweden
e-mail: joachim.toft@msi.vxu.se

Operator Theory:
Advances and Applications, Vol. 155, 187–202

Pseudo-Differential Operators, Microlocal Analysis and Image Restoration

Ryuichi Ashino, Steven J. Desjardins, Chris Heil,
Michihiro Nagase and Rémi Vaillancourt

Abstract. Pseudo-differential operators with symbols supported on sectors of dyadic annuli in the Fourier domain are used to perform microlocal analysis of tempered distributions. Microlocal analysis is recalled. The above symbols are made of smooth wavelet frames which are constructed in the Fourier domain by means of modulated smooth tapered functions. The method is used to localize a line of singularities in an image.

Mathematics Subject Classification (2000). Primary 35S05; Secondary 35A27, 65T60.

Keywords. Pseudo-differential operator, microlocal analysis, smooth wavelet frame.

1. Introduction

Pseudo-differential operators whose symbols are modulated smooth tapered functions with support on sectors of dyadic annuli in the Fourier domain are designed to perform microanalysis of tempered distribution on $\mathbb{R}^2$. These tapered functions form a smooth tight frame with frame bound 1, called *Parseval frame.*

Nonsmooth orthonormal multiwavelets have been constructed in [3] for performing microlocal analysis of tempered distributions in $\mathbb{R}^n$. The multiwavelets, whose Fourier transforms consist of characteristic functions of cubes, have perfect localization in the Fourier domain but poor localization in the x domain. To obtain good localization in both the x and Fourier domains in $\mathbb{R}^n$, in [4], the block-wavelets are smoothed by convolution and normalized to produce Parseval frames.

This research was partially supported by the Japanese Ministry of Education, Culture, Sports, Science and Technology, Grant-in-Aid for Scientific Research (B) 14340045(2002–2004), (C) 15540170(2003–2004), NSF Grant DMS-0139261, and the Natural Sciences and Engineering Research Council of Canada.

In [5], [8], nonorthogonal rectangular and polar smooth frame wavelets in $\mathbb{R}^2$ are obtained from a multiresolution analysis. In [5], one also finds a general construction of microlocal frame wavelets supported on a general ring of sets surrounding the origin and whose dyadic dilations cover $\mathbb{R}^2 \setminus \{0\}$.

In [1], Aldroubi, Cabrelli, and Molter have constructed wavelets on irregular grids with arbitrary dilation matrices, and frame atoms for $L^2(\mathbb{R}^d)$.

In the present paper, attention is restricted to microlocal analysis in the two-dimensional case with the goal of localizing the singularities of a function f in the time domain by analyzing the growth of its Fourier transform, $\widehat{f}$, in wedges in the Fourier domain.

In view of numerical applications on domains with pixels at integer points, the continuous Fourier transform $\widehat{f}(\xi)$ of a function $f(x)$ defined over $\mathbb{R}^2$ and the inverse Fourier transform of $\widehat{f}(\xi)$ will be

$$\widehat{f}(\xi) = \int e^{-2\pi i \xi \cdot x} f(x)\, dx, \qquad f(x) = \int e^{2\pi i x \cdot \xi}\, \widehat{f}(\xi)\, d\xi \tag{1}$$

and Plancherel's theorem will be

$$\langle f, g\rangle = \langle \widehat{f}, \widehat{g}\rangle, \tag{2}$$

where

$$\langle f, g\rangle = \int_{\mathbb{R}^2} f(x)\, \overline{g(x)}\, dx, \qquad \langle \widehat{f}, \widehat{g}\rangle = \int_{\mathbb{R}^2} \widehat{f}(\xi)\, \overline{\widehat{g}(\xi)}\, d\xi.$$

Since wedges in $\mathbb{R}^2$ are generalizations of the positive real axis in $\mathbb{R}$, smooth frame wavelets are generalized to $L^2(\mathbb{R}^2)$ by properly tapering the Fourier transforms of the orthonormal multiwavelets given in [3]. Thus, the dyadically scaled Fourier transforms of the frame wavelets $\{\psi^\ell(x)\}$, $\ell = 1, 2, \dots, L$, satisfy the identity

$$G(\xi) := \sum_{\ell=1,2,\dots,12,\, j\in\mathbb{Z}} |\widehat{\psi}^\ell(2^j\xi)|^2 \equiv 1, \qquad \xi \neq 0. \tag{3}$$

Our smooth frame wavelets can be derived from a frame multiresolution analysis by two-scale equations. However, they are not orthogonal to each other, nor to the spaces V_j spanned by the scaled and shifted scaling functions.

The paper proceeds as follows. In Section 2, the concepts of microlocal analysis are briefly reviewed. In Section 3, necessary and sufficient conditions are given to characterize tight frame wavelets. In Section 4, two-dimensional tapered characteristic functions are constructed. Section 5 introduces radial paving of the Fourier domain for microlocal analysis. In Section 6 an image restoration algorithm is described and the location of a line of singularities in an image is found by means of the frame coefficients.

2. Microlocal Analysis

Our smooth tight frame wavelets are designed for investigating the microanalytic content of distributions. Hyperfunctions, which were introduced by Sato [13] and extensively developed by the Kyoto school of mathematics, can be considered intuitively as sums of boundary values of holomorphic functions defined in infinitesimal wedges (see [12] and the references therein).

We write $\mathbb{C}^2 = \mathbb{R}_x^2 + i\mathbb{R}_y^2$ and let Γ be a cone with vertex at the origin in $\mathbb{R}_y^2$. If Δ is an open set in $\mathbb{R}_y^2$ which approaches Γ asymptotically near the origin from the interior of Γ, then the subset $U = \mathbb{R}^2 + i\Delta$ of $\mathbb{C}^2$ is called an *infinitesimal wedge* with opening Γ, and is denoted by $\mathbb{R}^2 + i\Gamma 0$ (see Fig. 1).

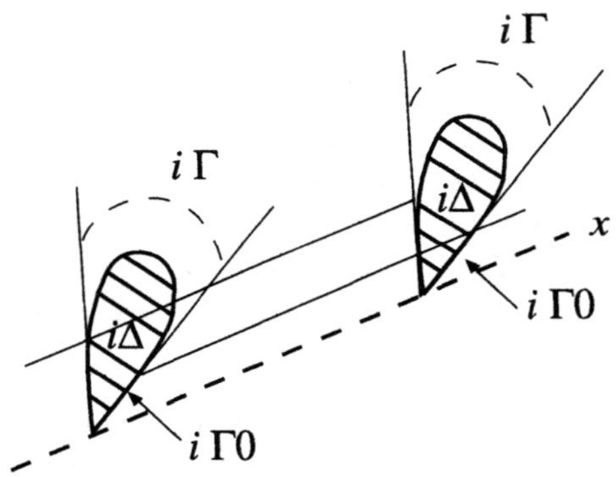

FIGURE 1. An infinitesimal wedge $\mathbb{R}^2 + i\Gamma 0$.

Let $f(x)$ be a "generalized boundary value" of a holomorphic function in an infinitesimal wedge $\mathbb{R}^2 + i\Gamma 0$; that is,

$$f(x) = \lim_{y\to 0\, y\in\Gamma} f(x+iy) := f(x+i\Gamma 0).$$

Definition 1. A distribution $f(x)$ is said to be *analytic* with respect to a direction ξ_0 if it can be represented as a finite sum of limits $f_j(x+i\Gamma_j 0)$ of slowly increasing holomorphic functions $f_j(z)$ in $\mathbb{R}^2 + i\Gamma_j 0$ such that, for every j, we have

$$\Gamma_j \cap \{y \in \mathbb{R}^2 \ ; \ y\cdot\xi_0 < 0\} \neq \emptyset.$$

To characterize the microanalyticity of a slowly increasing distribution $f \in \mathcal{S}'(\mathbb{R}^2)$ by its Fourier transform, $\widehat{f}$, we introduce the *dual cone*, Γ°, of the cone Γ defined by

$$\Gamma^\circ := \{\xi \in \mathbb{R}^2 \ ; \ y\cdot\xi \geq 0 \ \text{ for every } y \in \Gamma\}$$

(see Fig. 2); Γ° is a proper closed convex cone in $\mathbb{R}^2$. The complement of Γ° is denoted by $(\Gamma^\circ)^c$.

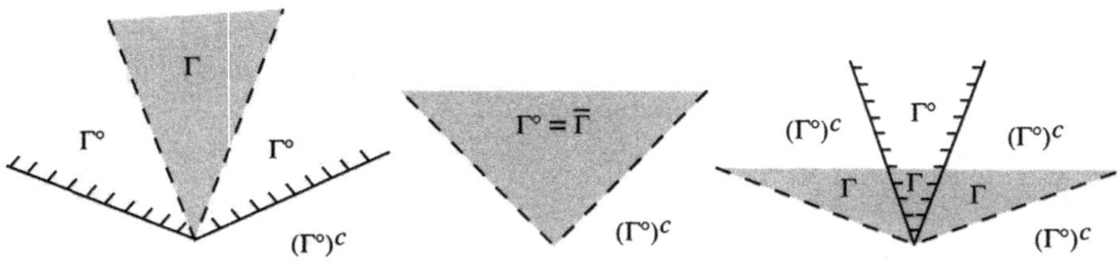

FIGURE 2. Open cone Γ, dual cone Γ°, and complement $(\Gamma^\circ)^c$ of dual cone.

The following lemma is standard (see [12]).

Lemma 1. *Let Γ be an open convex cone. A slowly increasing distribution $f(x) \in \mathcal{S}'(\mathbb{R}^2)$ can be represented as the limit $f(x+i\Gamma 0)$ of a slowly increasing holomorphic function $f(z)$ in the convex hull of the infinitesimal wedge $\mathbb{R}^2 + i\Gamma 0$ if and only if the Fourier transform, $\widehat{f}$, of f is exponentially decreasing in the open cone $(\Gamma^\circ)^c$, the complement of Γ°; that is, $\widehat{f}$ is exponentially decreasing on every closed proper subcone $\Gamma' \subset\subset (\Gamma^\circ)^c$.*

The larger the opening of Γ, the more regular a slowly increasing distribution $f(x + i\Gamma 0)$ will be. The cone Γ with largest opening is the whole space, in which case $f(x+i\Gamma 0)$ is analytic. For the next largest possible openings, Γ are half-spaces.

It is desirable to localize the directional decay of $\widehat{f}(\xi)$, because local nonsmoothness of a function $f(x)$ corresponds to slow decay of $\widehat{f}(\xi)$ along some directions at infinity. Each such direction corresponds to a point on the unit sphere $\mathbb{S}^1$ in ξ-space. Therefore, we shall use the coordinates $(x, \xi) \in \mathbb{R}^2 \times \mathbb{S}^1$ to represent a point $x \in \mathbb{R}^2$ together with a direction $\xi \in \mathbb{S}^1$.

Definition 2. A distribution $f(x) \in \mathcal{D}'(\mathbb{R}^2)$ is said to be *analytic* at $x_0 \in \mathbb{R}^2$ if there exists an open neighborhood $V \subset \mathbb{R}^2$ of x_0 such that the restriction $f|_V$ of f on V is analytic in V. The set of all points $x \in \mathbb{R}^2$ where f is not analytic is called the *singular support* of f.

Definition 3. A distribution $f(x)$ is said to be *microanalytic* or *microlocal analytic* at $(x_0, \xi_0) \in \mathbb{R}^2 \times \mathbb{S}^1$ if there exists a distribution $g(x)$ which is analytic with respect to the direction ξ_0 such that $f(x) - g(x)$ is analytic in a neighborhood of x_0. The set of all points $(x, \xi) \in \mathbb{R}^2 \times \mathbb{S}^1$ where f is not microanalytic is called the *singular spectrum* of f.

3. Necessary and Sufficient Conditions for Tight Frame Wavelets

We refer to [7], [10] for detailed background on frames.

Definition 4. Let I be an index set. A sequence of vectors $\{f_i\}_{i\in I}$ in a Hilbert space H is called a *frame* for H if there exist constants A, $B > 0$ (called *frame* *bounds*) such that

$$\forall\, f \in H, \quad A\,\|f\|^2 \le \sum_{i\in I} |\langle f, f_i\rangle|^2 \le B\,\|f\|^2.$$

If $A = B$, the frame is said to be an *A-tight frame.* A 1-tight frame is called a *Parseval frame.*

It can be shown that $\|f_i\|^2 \le B$ for each $i \in I$. Further, $\{f_i\}_{i\in I}$ is a tight frame with frame bound A if and only if

$$\forall\, f \in H, \quad \sum_{i\in I} \langle f, f_i\rangle\, f_i = Af.$$

A frame is *redundant* if it is possible to remove some element from the frame and still leave a complete set. For the case of tight frames, we can give the following easy characterization of redundancy. Redundancy of frames is explored in more detail in [6].

Lemma 2. *Let $\{f_i\}_{i\in I}$ be an A-tight frame for a Hilbert space H. If $j \in I$ is such that $\|f_j\|^2 < A$, then $\{f_i\}_{i\neq j}$ is still a frame for H, with frame bounds $A' = A - \|f_j\|^2$ and $B' = A$.*

Proof. If $f \in H$, then clearly

$$\sum_{i\neq j} |\langle f, f_i\rangle|^2 \le \sum_{i\in I} |\langle f, f_i\rangle|^2 = A\,\|f\|^2.$$

This establishes the upper frame bound. On the other hand, using the Cauchy-Schwarz inequality, we have

$$\begin{aligned}\sum_{i\neq j} |\langle f, f_i\rangle|^2 &= \sum_{i\in I} |\langle f, f_i\rangle|^2 - |\langle f, f_j\rangle|^2 \\ &\ge A\,\|f\|^2 - \|f\|^2\,\|f_j\|^2 = (A - \|f_j\|^2)\,\|f\|^2,\end{aligned}$$

which establishes the lower frame bound. □

Given $f \in L^2(\mathbb{R}^2)$, let $f_{j,k}(x)$ denote the scaled and shifted function

$$f_{j,k}(x) = 2^j f(2^j x - k), \qquad j \in \mathbb{Z}, \quad k \in \mathbb{Z}^2,$$

with Fourier transform

$$\widehat{f}_{j,k}(\xi) = e^{-2\pi i k\cdot\xi} 2^{-j} \widehat{f}(2^{-j}\xi).$$

Let $\mathbb{L}$ be a finite index set. A system $\{\psi^\ell_{j,k}\}_{\ell\in\mathbb{L}, j\in\mathbb{Z}, k\in\mathbb{Z}^2} \subset L^2(\mathbb{R}^2)$ is called an *A-tight wavelet frame* if

$$f(x) = \frac{1}{A} \sum_{\ell\in\mathbb{L}, j\in\mathbb{Z}, k\in\mathbb{Z}^2} \langle f, \psi^\ell_{j,k}\rangle\, \psi^\ell_{j,k}(x), \qquad \forall f \in L^2(\mathbb{R}^2). \tag{4}$$

A system $\{\psi^\ell_{j,k}\}_{\ell\in\mathbb{L}, j\in\mathbb{Z}, k\in\mathbb{Z}^2} \subset L^2(\mathbb{R}^2)$ is called an *orthonormal wavelet basis* if it is an orthonormal basis for $L^2(\mathbb{R}^2)$. An extension of Lemma 2 shows that this is equivalent to saying that the system $\{\psi^\ell_{j,k}\}_{\ell\in\mathbb{L},j\in\mathbb{Z},k\in\mathbb{Z}^2}$ is a Parseval wavelet frame and $\|\psi^\ell\|_{L^2(\mathbb{R}^2)} = 1$ for $\ell \in \mathbb{L}$.

The following general theorem, which is stated and proved for $\mathbb{R}^n$ in [9, Theorem 1], gives necessary and sufficient conditions to have a Parseval wavelet frame in $\mathbb{R}^2$.

Theorem 1. *If $\{\psi^1, \psi^2, \dots, \psi^L\} \subset L^2(\mathbb{R}^2)$, then*

$$\|f\|^2_{L^2(\mathbb{R}^2)} = \sum_{\ell\in\mathbb{L},j\in\mathbb{Z},k\in\mathbb{Z}^2} \left|\langle f, \psi^\ell_{j,k}\rangle\right|^2 \tag{5}$$

for all $f \in L^2(\mathbb{R}^2)$ if and only if the functions $\{\psi^1, \psi^2, \dots, \psi^L\}$ satisfy the following two equalities:

$$\sum_{\ell\in\mathbb{L},j\in\mathbb{Z}} \left|\widehat{\psi^\ell}(2^j\xi)\right|^2 = 1, \qquad a.e.\ \xi \in \mathbb{R}^2, \tag{6}$$

$$\sum_{\ell\in\mathbb{L},j\in\mathbb{Z}_+} \widehat{\psi^\ell}(2^j\xi)\,\overline{\widehat{\psi^\ell}(2^j(\xi+q))} = 0, \qquad a.e.\ \xi \in \mathbb{R}^2, \quad \forall q \in \mathbb{Z}^2\backslash(2\mathbb{Z})^2, \tag{7}$$

where $\mathbb{Z}_+ := \mathbb{N} \cup \{0\}$ and $q \in \mathbb{Z}^2\backslash(2\mathbb{Z})^2$ means that at least one component q_j is odd.

Corollary 1. *Under the hypotheses of Theorem 1, any function $f \in L^2(\mathbb{R}^2)$ admits the tight wavelet frame expansion*

$$f(x) = \sum_{\ell\in\mathbb{L},j\in\mathbb{Z},k\in\mathbb{Z}^2} \langle f, \psi^\ell_{j,k}\rangle\, \psi^\ell_{j,k}(x). \tag{8}$$

By using the localization property of the frame wavelet in the Fourier domain, one can study the directions of growth of $\widehat{f}(\xi)$ by looking at the size of the frame coefficients

$$\langle f, \psi^\ell_{j,k}\rangle = \langle \widehat{f}, \widehat{\psi}^\ell_{j,k}\rangle. \tag{9}$$

Moreover, by using the localization property of the frame wavelets in x-space, one can localize the singular support of $f(x)$ by varying ℓ, j and k in (9).

An alternate formulation of the problem is by means of a pseudodifferential operator

$$\begin{aligned} Pf(x) &= \int \left(\sum_{j.k.\ell} \psi^\ell_{j,k}(x)\overline{\widehat{\psi}^\ell_{j,k}(\xi)} \right) \widehat{f}(\xi)\, d\xi \\ &= \int e^{2\pi i x\cdot\xi} \left(\sum \psi^\ell_{j,k}(x)\, e^{-2\pi i(k+x)\cdot\xi} 2^{-j} \overline{\widehat{\psi^\ell}(2^{-j}\xi)} \right) \widehat{f}(\xi)\, d\xi \\ &= \int e^{2\pi i x\cdot\xi} p(x,\xi) \widehat{f}(\xi)\, d\xi, \end{aligned} \tag{10}$$

The problem is to find a symbol p such that Pf is strongly localized on the singular support of f and negligible where f is smooth. Pseudo-differential operators with smooth symbols do not extend the singular support of f; that is, the singular support of Pf is contained in the singular support of f. In Section 6, the symbol will involve only the matrix Q_j^ℓ which will be a discretized version of $\widehat{\psi}_j^\ell(\xi) = \widehat{\psi}^\ell(2^{-j}\xi)$, and the values of the shift parameter k in the summation in (8) will be determined indirectly by the singular support of f through the size of $|\langle \widehat{f}, \widehat{\psi}_{j,k}^\ell \rangle|$.

4. Two-Dimensional Tapered Characteristic Functions

The localization property of frame wavelets in the Fourier domain by means of the frame coeffficients (9) depends upon the support of the wavelet functions $\widehat{\psi}_{j,k}^\ell$ on appropriate rectangular or polar pavings of the plane.

In this section, we construct rectangular paving as a steppping stone for polar paving in section 5.

We first define bell functions of one variable. We partition the s axis with points $\{a_j\}$ ($a_j < a_{j+1}$) into intervals, such that the jth interval is $[a_j, a_{j+1}]$ and has length $L_j = a_{j+1} - a_j$. Around each endpoint of an interval, say a_j, we allow a *transition region* $[a_j - \varepsilon_j, a_j + \varepsilon_j]$ of width $2\varepsilon_j$; in this region, the bell function $b_j(s)$ over the interval j rises smoothly from 0 to 1, and the bell function $b_{j-1}(s)$ over the interval $j-1$ decreases smoothly from 1 to 0. The bell function $b_j(s)$ is nonzero for s in the region $(a_j - \varepsilon_j, a_{j+1} + \varepsilon_{j+1})$ and it is 1 for s in $[a_j + \varepsilon_j, a_{j+1} - \varepsilon_{j+1}]$. Bell functions over two adjacent intervals overlap in their transition region.

A bell function, or window, $b_j(s)$ has the following properties:

(i) $0 \leq b_j(s) \leq 1$ for all s and

$$b_j(s) = \begin{cases} 1 \text{ if } a_j + \varepsilon_j \leq s \leq a_{j+1} - \varepsilon_{j+1} \\ 0 \text{ if } s \leq a_j - \varepsilon_j \text{ or } s \geq a_{j+1} + \varepsilon_{j+1} \end{cases}$$

where $\varepsilon_j \geq 0$ and $\varepsilon_j + \varepsilon_{j+1} \leq L_j$;

(ii) $b_j^2(a_j + s) + b_j^2(a_j - s) = 1$ if $|s| \leq \varepsilon_j$;

(iii) $b_j(a_j + s) = b_{j-1}(a_j - s)$ if $|s| \leq \varepsilon_j$.

Condition (i) says that the windows are simply smoothed versions of the rectangular window and b_j can be specified if it is known in the left transition region $LTR = [a_j - \varepsilon_j, a_j + \varepsilon_j]$ and the right transition region $RTR = [a_{j+1} - \varepsilon_{j+1}, a_{j+1} + \varepsilon_{j+1}]$. So we need only consider the window function in the transition regions; we refer to this as a *taper function.* We shall denote the restrictions of b_j on LTR and RTR by

$$t_l(s; \epsilon_j) = b_j(s) \qquad \text{for } s \in [a_j - \varepsilon_j, a_j + \varepsilon_j] \tag{11}$$

and

$$t_r(s; \epsilon_j) = b_j(s) \qquad \text{for } s \in [a_{j+1} - \varepsilon_{j+1}, a_{j+1} + \varepsilon_{j+1}], \tag{12}$$

respectively.

The left part of Fig. 3 shows three tapered characteristic functions: (a) of the interval $[0, 1]$ with transition width 1/2 at both ends, (b) of the interval $[1, 2]$ with transition widths 1/2 at the left end and 1 at the right end, and (c) of the interval $[2,4]$ with transition widths 1 at the left end and 2 at the right end. It is seen in the right part of Fig. 3 that the square root of the sum of the squares of these three functions is equal to 1 over the overlapping tapered parts.

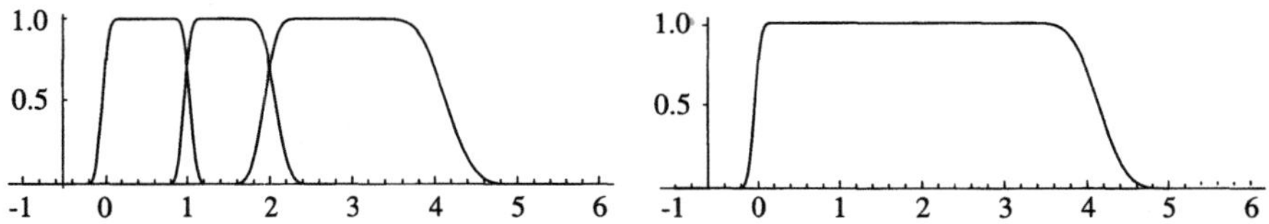

FIGURE 3. Left: three overlapping tapered characteristic functions. Right: square root of sum of squares of the three functions.

Bell functions of two variables are defined as the tensor product of two bell functions of one variable,

$$b(s_1, s_2) = b_1(s_1)b_2(s_2),$$

where b_1 and b_2 have transition regions of appropriate lengths at their left and right ends.

We illustrate two-dimensional paving of the Fourier domain by the following example. The tapered characteristic functions of the 12 squares of side $1/2$ of the ring $\mathcal{R}^{[0]}$, shown in Fig. 4, have transition widths 4ε on the outside edges and 2ε on all the other edges. Similarly, the three squares of side 1 of the second ring $\mathcal{R}^{[1]}$

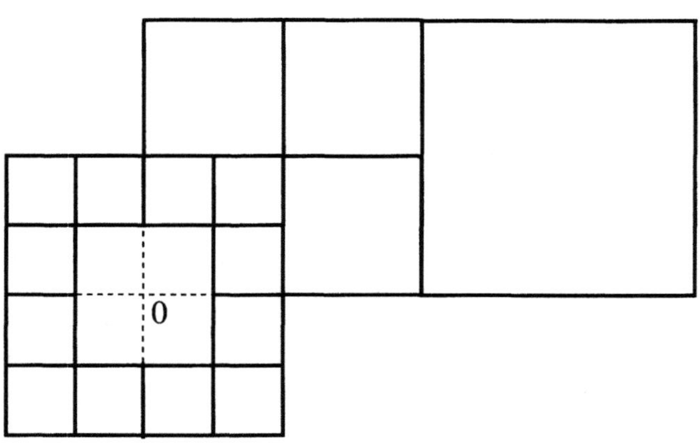

FIGURE 4. Paving of the Fourier domain by rings of 12 dyadic squares.

in the first quadrant, produced by dilation by 2, have transition widths 8ε on the four outside edges and 4ε on the remaining edges.

In general, the ring $\mathcal{R}^{[j]}$, parametrized by $j \in \mathbb{Z}$, is the support of tapered characteristic functions

$$\widehat{\psi}_j^{\ell}(\xi) := \widehat{\psi}^{\ell}(2^j \xi), \qquad \ell = 1, 2, \dots, 12,$$

where the Fourier transform is defined by (1). For fixed j, the only rings that intersect with $\mathcal{R}^{[j]}$ are $\mathcal{R}^{[j-1]}$ and $\mathcal{R}^{[j+1]}$. Given one ring $\mathcal{R}^{[j]}$, the other rings are simply obtained by dilation.

To prove identity (3), one needs only check points where two, three and four tapered wavelet frame functions overlap. Let us check identity (3) at an arbitrary point ξ in the intersection of the four tapered parts of $\widehat{\psi}_1^2$ and $\widehat{\psi}_2^\ell$, $\ell = 1, 2, 3$. Since tapering is of the same width for all four transition regions, it is obtained by the same taper functions (11) and (12), $t_l(s, \varepsilon)$ and $t_r(s, \varepsilon)$, respectively. Proceeding counter-clockwise from the top right corner of the frame $\widehat{\psi}_1^2$, we have

$$\begin{aligned}
t_r(\xi,\varepsilon)^2 t_r(\xi,\varepsilon)^2 &+ t_r(\xi,\varepsilon)^2 t_l(\xi,\varepsilon)^2 + t_l(\xi,\varepsilon)^2 t_l(\xi,\varepsilon)^2 + t_l(\xi,\varepsilon)^2 t_r(\xi,\varepsilon)^2 \\
&= t_r(\xi,\varepsilon)^2[t_r(\xi,\varepsilon)^2 + t_l(\xi,\varepsilon)^2] + t_l(\xi,\varepsilon)^2[t_l(\xi,\varepsilon)^2 + t_r(\xi,\varepsilon)^2] \\
&= t_r(\xi,\varepsilon)^2 + t_l(\xi,\varepsilon)^2 \\
&= 1.
\end{aligned}$$

We generalize to two dimensions the treatment of smooth frames for $H^2(\mathbb{R})$ given in [11, Section 8.4] and show that the functions $\psi_{j,k}^\ell$ form a Parseval wavelet frame for $L^2(\mathbb{R}^2)$.

Theorem 2. *If $0 < \varepsilon \leq 1/4$, the system $\{\psi_{j,k}^\ell\}$, $j \in \mathbb{Z}$ and $k \in \mathbb{Z}^2$, is a Parseval wavelet frame for $L^2(\mathbb{R}^2)$.*

Proof. Identity (3) has just been proved. For $f \in L^2(\mathbb{R}^2)$, a similar argument to the proof of Theorem 2 implies

$$\sum_{k\in\mathbb{Z}^2} |\langle f, \psi_{j,k}^\ell\rangle|^2 = \int |\widehat{f}(\xi)|^2\, |\widehat{\psi}^\ell(2^{-j}\xi)|^2\, d\xi$$

and

$$\begin{aligned}
\sum_{\ell=1}^{12}\sum_{j=-\infty}^{\infty}\sum_{k\in\mathbb{Z}^2} |\langle f, \psi_{j,k}^\ell\rangle|^2 &= \sum_{\ell=1}^{12}\sum_{j=-\infty}^{\infty} \int |\widehat{f}(\xi)|^2\, |\widehat{\psi}^\ell(2^{-j}\xi)|^2\, d\xi \\
&= \int |\widehat{f}(\xi)|^2\, d\xi = \|f\|^2_{L^2(\mathbb{R}^2)}.
\end{aligned}$$

A folding argument [11, p. 419], shows that

$$\|\psi_{0,0}^\ell\|^2_{L^2(\mathbb{R}^2)} = \frac{1}{4}.$$

Thus we have a family of frames in $L^2(\mathbb{R}^2)$ that do not form an orthonormal basis. □

5. Radial Paving of the Fourier Domain for Microlocal Analysis

To have arbitrarily fine angular resolutions in the Fourier domain it is convenient to consider polar pavings of the plane by dyadic sectors of annuli with appropriate angular divisions. A few sectors of annuli of the first and second dyadic rings are shown in Fig. 5.

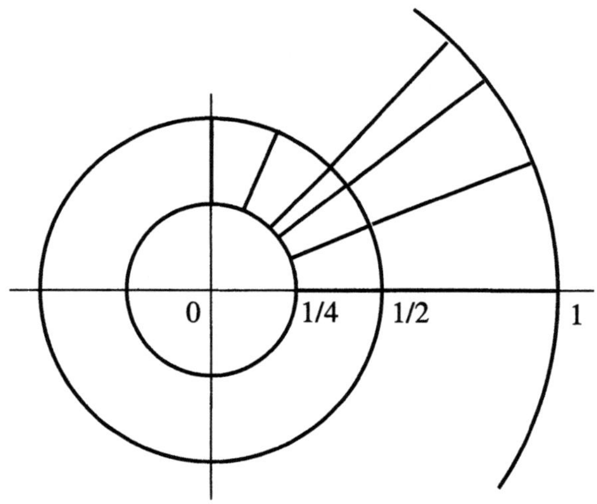

FIGURE 5. Dyadic sectors of annuli with central disk in the Fourier domain. Unequal angular divisions offer directional freedom.

It is to be noticed that the radial interval is of the form $[r, 2r]$. Polar wavelets can be defined by the inverse Fourier transforms of nonoverlapping characteristic functions over sectors of annuli. If the Fourier domain is completely covered by L nonoverlapping wedges, the L family of wavelets $\{\psi_{j,k}^{\ell}\}$, with $\ell = 1, \dots, L$, $j \in \mathbb{Z}$, and $k \in \mathbb{Z}^2$, form an orthonormal basis of $L^2(\mathbb{R}^2)$. This basis can be generated by a frame multiresolution analysis with scaling functions.

To have better localization in x-space, these functions are tapered with width 2ε on the inside arcs, width 4ε on the outside arcs and width εs, $r \leq s \leq 2r$, on the straight radial edges. Provided ε is sufficiently small so that tapering overlaps are restricted to immediately adjacent regions, it will be shown below that identity (3) is satisfied. By taking a ring sufficiently close to the origin, by construction, dyadic polar rings form a Parseval frame.

The description of polar frame wavelets is considerably simplified by using the polar coordinates (r, θ) in the Fourier domain instead of the Cartesian coordinates (ξ_1, ξ_2). The inverse and direct coordinate transformations are

$$\xi_1 = r\cos 2\pi\theta,\ \xi_2 = r\sin 2\pi\theta; \qquad r = \sqrt{\xi_1^2 + \xi_2^2},\ \theta = \frac{1}{2\pi}\arctan\left(\frac{\xi_2}{\xi_1}\right),$$

with the appropriate branches of arctan. In the (r, θ)-plane, the polar frame wavelets are supported in the semi-infinite strip $\{0 \leq r < \infty,\ 0 \leq \theta \leq 1\}$, and are 1-periodic in θ. The strip, shown in Fig. 6, is divided into rectangles with vertical sides at $r = 2^j$, $j \in \mathbb{Z}$ and horizontal sides at

$$0 = \theta_0 < \theta_1 < \cdots < \theta_\ell < \cdots < \theta_L = 1.$$

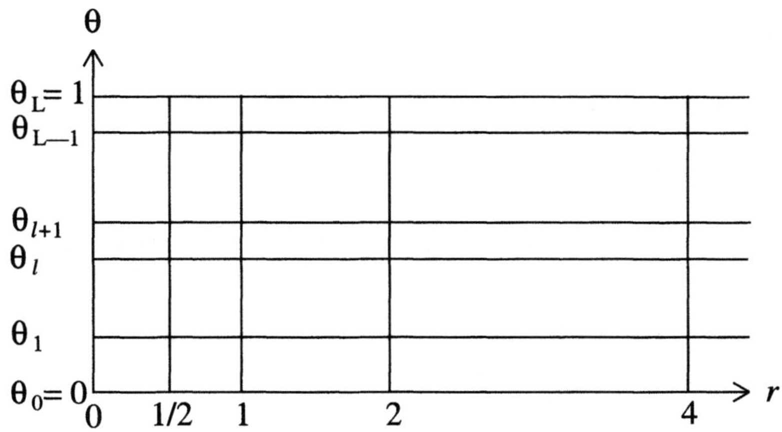

FIGURE 6. Strip in the (r, θ)-plane.

The smooth functions $\widehat{\psi}_j^\ell(r,\theta) = \widehat{\psi}^\ell(2^{-j}r, \theta)$ are easily obtained by tapering the characteristic functions of each rectangle. Tapering is done along vertical sides with transition regions of width $2^{j+1}\varepsilon$ at $r = 2^{j-1}$ and along horizontal sides with transition regions of width $2\varepsilon_\ell$ with $\varepsilon_L = \varepsilon_0$ at $\theta = \theta_L$.

It is then clear that identity (3) is satisfied.

It is shown in [5] that our smooth frame wavelets can be obtained from a frame multiresolution analysis with scaling functions.

6. A Numerical Implementation of the Localization Method

In this section, we describe how to apply the above theory to the restoration of finite images represented by matrices. Since the Fourier transform of a finite region gives rise to oscillations of the cardinal sine type, care must be taken in the restoration process.

The restoration process described in [5] involves the following steps.

(a) A scarred image (A) is to be restored as its original image (F).
(b) Image (A) is Fourier transformed and filtered by multiplication with tapered characteristic functions with support far from the origin and at right angles to the singularities to be localized. This produces image (B).

(c) The frame coefficients (9) in the Fourier domain,

$$\langle \widehat{f}, \widehat{\psi}^{\ell}_{j,k} \rangle = \langle f, \psi^{\ell}_{j,k} \rangle,$$

make image (C).

(d) In view of the Plancherel theorem, image (C) is used in the x domain to obtain the wavelet frame expansion (8) of Corollary 1. This produces a thick image (D) of the singularities in image (A)

(e) The extra width of (D), caused by the side lobes in the support of $\psi^{\ell}_{j,k}$, is narrowed to eliminate oscillations due to the cardinal sine effect when transforming functions with finite support. This is image (E).

(f) A tuned multiple of (E) is subtracted from (A) to restore the original image (F).

In the case of polar frames, two-dimensional bell functions were produced by the Mathematica Wavelet Explorer by tapering characteristic functions over the rectangles with left and right vertical edges at $r = 1/2$ and $r = 1$ with horizontal transition widths 1/8 and 1/4, respectively. The lower and upper transition widths of the rectangle with lower and upper horizontal edges at θ_ℓ and $\theta_{\ell+1}$ are $(\theta_\ell - \theta_{\ell-1})/4$ and $(\theta_{\ell+1} - \theta_\ell)/4$, respectively.

Tapering was done by iterating the sine function twice to get

$$b(s) = \sin\left(\frac{\pi}{2}\sin^2\left(\frac{\pi}{2}\sin^2\left(\frac{\pi}{2}s\right)\right)\right)$$

with the option `Taper -> {Trig[2],epsilon}`. It is easy to see that

$$b(s)^2 + b(1-s)^2 = 1, \quad s \in [0,1],$$

by noting that $\sin((\pi/2)(1-s)) = \cos((\pi/2)s)$ and repeated use of the identity $\cos^2 s = 1 - \sin^2 s$.

Six sets of longer rectangles were produced by horizontal dyadic scaling of the functions of the first set. The support of these functions is a strip with far right end at $r = 287$.

These functions are evaluated in the form of tables by Mathematica in the (ξ_1, ξ_2) plane and exported to MATLAB in the form of matrices:

$$Q_1^1, \quad \ldots, \quad Q_1^L, \quad \ldots, \quad Q_7^1, \quad \ldots, \quad Q_7^L.$$

By construction, the top right 87×87 matrix Q_7^d (d for diagonal), with center line making an angle of $\pi/4$ radians with respect to the ξ_1 axis, has lower left and upper right elements in positions $(87, 201)$ and $(1, 287)$. Therefore, in the sequel we shall work with matrices of dimension 287×287.

A diagonal line is added to the Saturn image making (A) shown in the top left part of Fig. 7.

Let f be the matrix representing image (A). The discrete Fourier transform F of f is filtered by means of the top right frame, Q_7^d, of the fifth ring to recover the singularity and eliminate the smooth part of the image. We write $k = (k_1, k_2)$.

Let the (m,n)-th element of the matrix of exponentials E_k be

$$E_k(m,n) = e^{-2\pi i(m-1,n-1)\cdot(k_1,k_2)/287}. \tag{13}$$

Discretizing the scalar product (9),

$$\langle \widehat{f}, \widehat{\psi}^d_{5,k} \rangle,$$

we obtain the matrix $C = (c_k)$ of frame coefficients

$$c_k = \sum_{m,n=1}^{287} (F.{*}E_k.{*}Q^d_7)(m,n), \qquad k_1 = k_2 = 1, \dots, 287, \tag{14}$$

where $.*$ denotes componentwise matrix multiplication. Note that Q^d_7 is a real matrix. The matrix C represents image (C).

Discretizing the partial sum (8),

$$\sum_{k_1=k_2=1}^{287} \langle \widehat{f}, \widehat{\psi}^d_{5,k} \rangle \, \widehat{\psi}^d_{5,k}(\xi),$$

we obtain the matrix $W = (w_k)$ of the wavelet frame expansion of the discrete inverse Fourier transform of the filtered image (B). The entries of W are

$$w_k = \left(\sum_{k_1=k_2=1}^{287} c_{m_1,m_1} E_{-m_1,-m_1}.{*}Q^d_7 \right)_k, \tag{15}$$

where summation is over the diagonal segment S.

Since the Fourier transform of the smooth component of the image is localized mainly near the origin and the Fourier transform of the singularity consists mainly of details with high frequency, the properly modulated matrix filter Q^d_7 picks up only the singularity.

The frame $\widehat{\psi}^d_{5,k}(\xi)$, with support in the top right corner in the Fourier domain, picks up the singularities across segments parallel to the main diagonal.

In the top left part of Fig. 7, a diagonal line of height 100 has been added to the Saturn image making image (A). The Fourier transform of this image is filtered by a smooth polar frame supported on a sector of an annulus of aperture of 0.02π radians and inner and outer radii 220 and 287, along the secondary diagonal in the upper right corner of a 287×287 matrix in the Fourier domain and shown in the top right part of the figure (image (B)). The image of the frame coefficients of the filtered image is shown in the bottom left part of the figure (image (C)). This image clearly locates the diagonal line singularity added to the Saturn image. The bottom right part of the figure shows the frame expansion of the filtered image in the x domain (image (D)). It is seen that the smooth image has been filtered out.

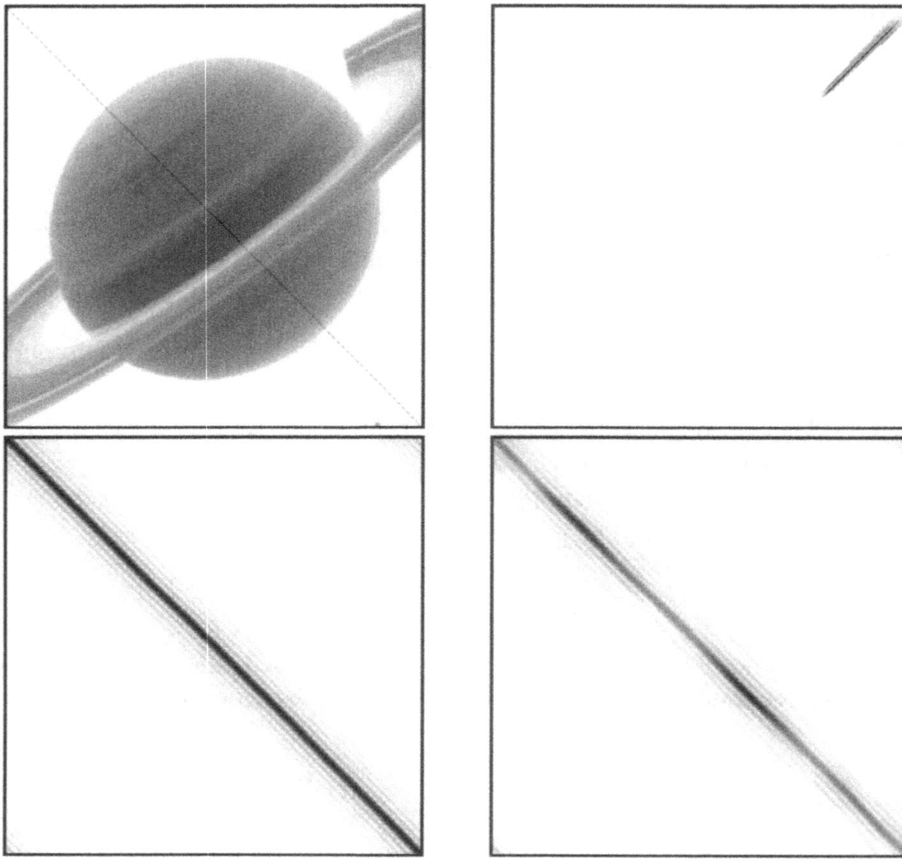

FIGURE 7. Top left: framed negative image of diagonal line added to the Saturn image. Top right: framed negative image of the filtered image in the Fourier domain. Bottom left: framed negative image of the frame coefficients of the filtered image in the Fourier domain. Bottom right: framed negative image of the frame reconstruction of the filtered image in the x domain.

References

[1] A. Aldroubi, C. Cabrelli and U.M. Molter, Wavelets on irregular grids with arbitrary dilation matrices, and frame atoms for $L^2(\mathbb{R}^d)$, preprint.

[2] R. Ashino, C. Heil, M. Nagase and R. Vaillancourt, Microlocal analysis and multiwavelets in *Geometry, Analysis and Applications*, Editor: R.S. Pathak, World Scientific, River Edge, NJ, 2001, 293–302.

[3] R. Ashino, C. Heil, M. Nagase and R. Vaillancourt, Microlocal filtering with multiwavelets, *Comput. Math. Appl.* **41** (2001), 111–133.

[4] R. Ashino, S.J. Desjardins, C. Heil, M. Nagase and R. Vaillancourt, Microlocal analysis, smooth frames and denoising in Fourier space, *J. Asian Information-Science-Life* **1** (2002), 153–160.

[5] R. Ashino, S.J. Desjardins, C. Heil, M. Nagase and R. Vaillancourt, Smooth tight frame wavelets and image microanalysis in the Fourier domain, *Comput. Math. Appl.* **45** (2003), 1551–1579.

[6] R. Balan, P.G. Casazza, C. Heil and Z. Landau, Deficits and excesses of frames, *Adv. Comput. Math.* **18** (2003), 93–116.

[7] I. Daubechies, *Ten Lectures on Wavelets*, Society for Industrial and Applied Mathematics, Philadelphia PA, 1992.

[8] S.J. Desjardins, *Image Analysis in Fourier Space*, Ph.D. Thesis, University of Ottawa, 2002.

[9] M. Frazier, G. Garrigós, K. Wang and G. Weiss, A characterization of functions that generate wavelet and related expansion, *J. Fourier Anal. Appl.* **3** (1997), 883–906.

[10] C. Heil and D.F. Walnut, Continuous and discrete wavelet transforms, *SIAM Rev.* **31** (1989), 628–666.

[11] E. Hernández and G. Weiss, *A First Course on Wavelets*, CRC Press, Boca Raton FL, 1996.

[12] A. Kaneko, Microlocal Analysis, in *Encyclopaedia of Mathematics*, Kluwer Academic Publishers, Dordrecht, 1997.

[13] M. Sato, Theory of hyperfunctions I, *J. Fac. Sci. Univ. Tokyo, Sec. I* **8** (1959), 139–193.

Ryuichi Ashino
Division of Mathematical Sciences
Osaka Kyoiku University
Kashiwara, Osaka 582-8582
Japan
e-mail: `ashino@cc.osaka-kyoiku.ac.jp`

Steven J. Desjardins
Department of Mathematics and Statistics
University of Ottawa
585 King Edward Ave.
Ottawa ON Canada KiN 6N5
e-mail: `desjards@mathstat.uottawa.ca`

Chris Heil
School of Mathematics
Georgia Institute of Technology
Atlanta GA 30332 U.S.A.
e-mail: `heil@math.gatech.edu`

Michihiro Nagase
Department of Mathematics
Graduate School of Science
Osaka University
Toyonaka, Osaka 560-0043
Japan
e-mail: nagase@math.wani.osaka-u.ac.jp

Rémi Vaillancourt
Department of Mathematics and Statistics
University of Ottawa
585 King Edward Ave.
Ottawa ON Canada KiN 6N5
e-mail: remi@uottawa.ca

Operator Theory:
Advances and Applications, Vol. 155, 203–218

Applications of Wavelet Transforms to System Identification

Ryuichi Ashino, Takeshi Mandai and Akira Morimoto

In memory of Professor Michihiro Nagase

Abstract. A review of system identification based on distribution theory is given. By the Schwartz kernel theorem, to every continuous linear system, there corresponds a unique distribution, called *kernel distribution*.

Formulae using wavelet transform to access time-frequency information of the kernel distribution are deduced. An application of the formula to system identification of a health monitoring system is given.

Mathematics Subject Classification (2000). Primary 42C40; Secondary 93B30, 46F12, 65T60.

Keywords. System identification, distribution, wavelet transform, stationary wavelet.

1. Introduction

Many results on system identification based on distribution theory are already known, but, as far as we know, there is no reference that collects those results and states them in the system identification context. Therefore we give a brief system identification theory starting with the Schwartz kernel theorem in §2 based on [1]. In §3, the continuous wavelet transform on $L^2(\mathbb{R}^n)$ is introduced and time-frequency analysis using the continuous wavelet transform of continuous linear systems on $L^2(\mathbb{R}^n)$ is discussed. Formulae using the continuous wavelet transform to access time-frequency information of the kernel distributions are deduced. In §4, the stationary wavelet transform is introduced as a discrete version of the continuous wavelet transform having translation-invariant property, which is called

This research was partially supported by the Japanese Ministry of Education, Culture, Sports, Science and Technology, Grant-in-Aid for Scientific Research (C), 15540170 (2003–2004), and by the Japan Society for the Promotion of Science, Japan-U.S. Cooperative Science Program (2003–2004).

time-invariant for the one dimensional case, $n = 1$. In §5 and §6, using a discrete version of the formula to access time-frequency information of the kernel distribution corresponding to translation-invariant systems, a wavelet method of system identification is proposed. In §7, as an application of our wavelet based system identification method for a health monitoring system for structures, we give a numerical experiment to identify the structural parameters. By using the de-noising property of wavelets, our wavelet based system identification method performs much better than the conventional method when white noise is added to input and output.

2. System Identification

Let us review what is system identification based on [8]. A *system* L illustrated in Figure 1 is an object in which variables of different kinds interact and produce observable signals. The observable signals g that are of interest to us are called *outputs*. The system is also affected by external stimuli. External signals f that can be manipulated by the observer are called *inputs*. Others are called *disturbances* and can be divided into those, denoted by w, that are directly measured and those, denoted by v, that are only observed through their influence on the output.

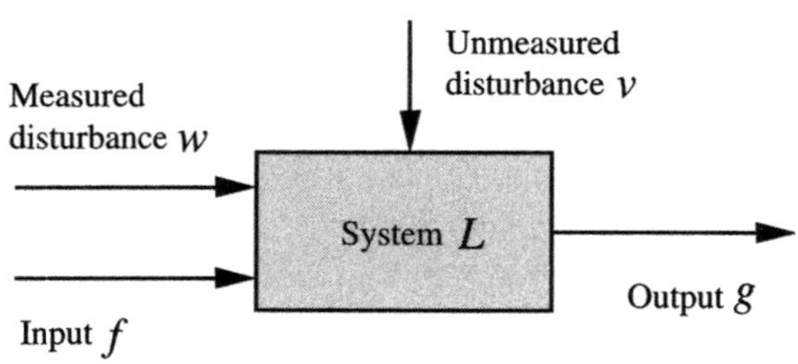

FIGURE 1. System L.

When we interact with a system, we need some concept on how its variables relate to each other. With a broad definition, we shall call such an assumed relationship among observed signals a *model* of the system. A model is called a *mathematical model* or *analytic model* if the model describes the relationships among the system variables in terms of mathematical expressions like difference or differential equations. The construction of a model from data involves three basic entities:

1. *The data.* The input-output data are sometimes recorded during a specifically designed identification experiment, where we may determine which signals to measure and when to measure them and may also choose the input signals. In other cases, we may not have the possibility to affect the experiment, but must use data from the normal operation of the system.

2. *A set of candidate models.* A set of candidate models must be specified by considering various factors, such as formal properties of models, a priori knowledge, engineering intuition and insight.
3. *A rule by which candidate models can be assessed using the data.* Determining the "best" model in the set is guided by the data. This is *the identification method.*

The assessment of model quality is typically based on how the models perform when they attempt to reproduce the measured data. Hence, the system identification procedure are the followings:

1. Collect data.
2. Choose a model set.
3. Pick the "best" model in the model set.
4. Validate the "best" model.

Schwartz Kernel Theorem

Mathematically, a system can be regarded as a mapping which relates inputs to outputs. Hereafter, we will use the terminology "system" rather than "mapping". A system L is said to be *linear* if

$$L[\alpha_1 f_1 + \alpha_2 f_2] = \alpha_1 L[f_1] + \alpha_2 L[f_2],$$

for arbitrary constants α_1, α_2 and arbitrary inputs f_1, f_2. To choose a model set for linear systems, the most general setting should be given by using the Schwartz kernel theorem [16] based on the Schwartz distribution theory [13].

Let us denote by $\mathscr{D}(\mathbb{R}^n)$, the space of compactly supported C^∞ functions with the canonical LF-topology [16]. The space $\mathscr{D}(\mathbb{R}^n)$ is also called the *space of test functions.* A *distribution* T in $\mathbb{R}^n$ is a continuous linear form on $\mathscr{D}(\mathbb{R}^n)$. The set of all the distributions in $\mathbb{R}^n$ is denoted by $\mathscr{D}'(\mathbb{R}^n)$ and the duality of distributions is denoted by $\langle\cdot,\cdot\rangle^*$, that is, the duality between $T \in \mathscr{D}'$ and $\phi \in \mathscr{D}$ is written as

$$T(\phi) = \langle T, \phi\rangle^* = \langle T(x), \phi(x)\rangle^*_x.$$

The space of C^∞ functions in $\mathbb{R}^n$ with the canonical Fréchet-topology is denoted by $\mathscr{E}(\mathbb{R}^n)$ and its topological dual space is denoted by $\mathscr{E}'(\mathbb{R}^n)$, which is the space of compactly supported distributions in $\mathbb{R}^n$. On the other hand, as the inner product of L^2 is denoted by $\langle\cdot,\cdot\rangle$, we have

$$\langle f, g\rangle^* = \langle f, \overline{g}\rangle, \qquad f, g \in L^2.$$

The space of C^∞ functions in $\mathbb{R}^n$ rapidly decreasing at infinity is called the *Schwartz space* and denoted by $\mathscr{S}(\mathbb{R}^n)$. The topological dual space of $\mathscr{S}(\mathbb{R}^n)$ is denoted by $\mathscr{S}'(\mathbb{R}^n)$ and an element of $\mathscr{S}'(\mathbb{R}^n)$ is called a *tempered distribution.*

Theorem 1 (The Schwartz kernel theorem). *Let $L : \mathscr{D}(\mathbb{R}^n) \to \mathscr{D}'(\mathbb{R}^n)$ be a continuous linear system. Then, there is a unique distribution $k \in \mathscr{D}'(\mathbb{R}^{2n})$ such that*

$$\langle L[f], \phi\rangle^* = \langle k(x,y), \phi(x) f(y)\rangle^*_{x,y},$$

for every f, $\phi \in \mathscr{D}(\mathbb{R}^n)$, *that is,*

$$L[f](x) = \langle k(x,y), f(y)\rangle_y^*, \qquad f \in \mathscr{D}(\mathbb{R}^n).$$

The distribution k *is called the kernel distribution of* L.

If $L : \mathscr{S}(\mathbb{R}^n) \to \mathscr{S}'(\mathbb{R}^n)$, *then* $k \in \mathscr{S}'(\mathbb{R}^{2n})$.

Let us compare it to those of a more "classical" nature, such as, for instance, the one occurring in the L^2 theory. It is trivially false that every bounded operator from $L^2(\mathbb{R}^n_x)$ to $L^2(\mathbb{R}^n_y)$ can be represented by a kernel $k(x,y) \in L^2(\mathbb{R}^n_{x,y})$, that is, can be written as

$$f \mapsto \int k(x,y) f(y)\, dy,$$

because the kernel for the identity system must be a Dirac kernel $\delta(x-y)$.

Translation-Invariant Systems

Let us define three fundamental operators for time-frequency analysis. They are unitary operators when they act on $L^2(\mathbb{R}^n)$. Define the *translation operator* T_a by

$$T_a f(x) = f(x-a), \qquad a \in \mathbb{R}^n,$$

the *modulation operator* M_ξ by

$$M_\xi f(x) = e^{ix\xi} f(x), \qquad \xi \in \mathbb{R}^n,$$

and the *dilation operator* D_ρ by

$$D_\rho f(x) = \rho^{-n/2} f(\rho^{-1}x), \qquad \rho \in \mathbb{R}_+ := \{\, x \in \mathbb{R} \,;\, x > 0 \,\}.$$

For simplicity, we consider D_ρ only for $\rho > 0$ although we can consider D_ρ for $\rho \in \mathbb{R}\backslash\{0\}$.

A continuous linear system $L : \mathscr{D}(\mathbb{R}^n) \to \mathscr{D}'(\mathbb{R}^n)$ is said to be *translation-invariant* if

$$\langle L[f], T_{-a}\phi\rangle^* = \langle L[T_a f], \phi\rangle^*,$$

for every f, $\phi \in \mathscr{D}(\mathbb{R}^n)$ and every $a \in \mathbb{R}^n$, or simply,

$$T_a L[f] = L[T_a f],$$

for every $f \in \mathscr{D}(\mathbb{R}^n)$ and every $a \in \mathbb{R}^n$. Proposition 1 stated below follows from the argument given in [13], Chapter 6, §7 (see the part before Theorem 24).

Proposition 1. *Let* $L : \mathscr{D}(\mathbb{R}^n) \to \mathscr{D}'(\mathbb{R}^n)$ *be a continuous linear system and* $k(x,y)$ *be its kernel distribution. The system* L *is translation-invariant if and only if there exists a unique* $h \in \mathscr{D}'(\mathbb{R}^n)$ *such that* $k(x,y) = h(x-y)$, *that is,* $L[f] = h*f$. *As a result, we have* $L\big(\mathscr{D}(\mathbb{R}^n)\big) \subset \mathscr{E}(\mathbb{R}^n)$.

If L *is continuous from* $\mathscr{S}(\mathbb{R}^n)$ *to* $\mathscr{S}'(\mathbb{R}^n)$, *then* $h \in \mathscr{S}'(\mathbb{R}^n)$, *and hence we have* $L\big(\mathscr{S}(\mathbb{R}^n)\big) \subset \mathscr{O}_M(\mathbb{R}^n)$, *where* $\mathscr{O}_M$ *is the space of slowly increasing* C^∞ *functions* ([13]).

Causality

Causality is natural for a physical system in which the variable is time. It means that the response at time t depends only on what has happened before and at t. In particular, a system does not respond before there is an input. Thus causality is a necessary condition for a system to be physically realizable.

Let L be a continuous linear system $\mathscr{D}(\mathbb{R}^n) \to \mathscr{D}'(\mathbb{R}^n)$, and $k \in \mathscr{D}'(\mathbb{R}^{2n})$ be its kernel distribution. A continuous linear system L is said to be *causal* if

$$\operatorname{supp} L[f] \subset \operatorname{supp} f + \overline{\mathbb{R}_+}^n,$$

for every $f \in \mathscr{D}(\mathbb{R}^n)$. Here, $A + B := \{ a + b \, ; \; a \in A, \, b \in B \}$ and $\overline{\mathbb{R}_+} := [0, \infty)$. We simply write $a + B$ for $\{a\} + B$. Then, we can show that the following three conditions are equivalent.

(a) L is causal.
(b) For every $a \in \mathbb{R}^n$, if $f \in \mathscr{D}(\mathbb{R}^n)$ and $\operatorname{supp} f \subset a + \overline{\mathbb{R}_+}^n$, then $\operatorname{supp} L[f] \subset a + \overline{\mathbb{R}_+}^n$.
(c) $\operatorname{supp} k \subset \Delta_+ := \{ (x, y) \in \mathbb{R}^{2n} \; ; \; x_j \geq y_j, \; j = 1, \ldots, n \}$.

Let us define the causality of distributions. A distribution $f \in \mathscr{D}'(\mathbb{R}^n)$ is said to be *causal* if $\operatorname{supp} f \subset \overline{\mathbb{R}_+}^n$. When L is translation-invariant, there exists $h \in \mathscr{D}'(\mathbb{R}^n)$ such that $L[f] = h * f$ for $f \in \mathscr{D}(\mathbb{R}^n)$. Then the following three conditions are equivalent.

(a) L is causal.
(b') If f is causal then $L[f]$ is causal.
(c') h is causal.

3. Continuous Wavelet Transforms

Assume that $\psi \in L^2(\mathbb{R}^n)$ satisfy

$$0 < C_\psi := \int_0^\infty \frac{|\widehat{\psi}(s\xi)|^2}{s} \, ds < \infty, \tag{3.1}$$

where C_ψ is independent of ξ. Note that this condition is satisfied if ψ is a continuous function with compact support, $\psi \neq 0$, ψ is radially symmetric, and $\int_{\mathbb{R}^n} \psi(x) \, dx = 0$. For simplicity we further assume that $\|\psi\|_{L^2(\mathbb{R}^n)} = 1$.

The *continuous wavelet transform* of $f \in L^2(\mathbb{R}^n)$ with respect to ψ is defined by

$$W_\psi f(b, a) : = |a|^{-n/2} \int_{\mathbb{R}^n} f(x) \, \overline{\psi\Big(\frac{x - b}{a}\Big)} \, dx = \langle f, T_b D_a \psi \rangle_{L^2(\mathbb{R}^n)}, \qquad (b, a) \in \mathbb{H}_n := \mathbb{R}^n \times \mathbb{R}_+. \tag{3.2}$$

It is well-known that $W_\psi f \in \mathscr{H}_1 := L^2(\mathbb{H}_n; dbda/a^{n+1})$ and

$$\langle W_\psi f, W_\psi g \rangle_{\mathscr{H}_1} = C_\psi \langle f, g \rangle_{L^2(\mathbb{R}^n)} \quad \text{for every } f, g \in L^2(\mathbb{R}^n). \tag{3.3}$$

(See, for example, [4].)

Wavelet Analysis of Kernel Distributions

For $(b, a), (u, s) \in \mathbb{H}_n$, set

$$K_\psi(b, a; u, s) := C_\psi{}^{-1} \langle L[T_u D_s \psi], T_b D_a \psi \rangle_{L^2(\mathbb{R}^n)}. \tag{3.4}$$

Then, we have the following result, whose proof can be found in [1].

Theorem 2. *Assume that the system L is a Hilbert–Schmidt operator, that is, the kernel distribution k belongs to $L^2(\mathbb{R}^{2n})$. Then, we have*

$$K_\psi(b, a; u, s) = C_\psi{}^{-1} \langle k, T_b D_a \psi \otimes \overline{T_u D_s \psi} \rangle_{L^2(\mathbb{R}^{2n})}, \tag{3.5}$$

where $(f \otimes g)(x, y) := f(x) g(y)$,

$$K_\psi \in \mathscr{H}_2 := L^2\Big((\mathbb{H}_n)^2_{(b,a,u,s)}; db \frac{da}{a^{n+1}} du \frac{ds}{s^{n+1}}\Big),$$

$$\|K_\psi\|_{\mathscr{H}_2} = \|k\|_{L^2(\mathbb{R}^{2n})},$$

and

$$W_\psi(L[f])(b, a) = \int_{\mathbb{H}_n} K_\psi(b, a; u, s)\, W_\psi f(u, s)\, du \frac{ds}{s^{n+1}}, \tag{3.6}$$

for every $f \in L^2(\mathbb{R}^n)$.

We also have an inversion formula for k:

$$k(x, y) = C_\psi^{-1} \int_{(\mathbb{H}_n)^2} K_\psi(b, a; u, s) T_b D_a \psi(x)\, \overline{T_u D_s \psi(y)}\, db \frac{da}{a^{n+1}} du \frac{ds}{s^{n+1}}, \tag{3.7}$$

where the integral is, for example, in the weak sense:

$$\begin{aligned} \langle k, h \rangle_{L^2(\mathbb{R}^{2n})} = & C_\psi^{-1} \int_{(\mathbb{H}_n)^2} K_\psi(b, a; u, s) \\ & \times \langle T_b D_a \psi \otimes \overline{T_u D_s \psi}, h \rangle\, db \frac{da}{a^{n+1}} du \frac{ds}{s^{n+1}}, \end{aligned} \tag{3.8}$$

for every $h \in L^2(\mathbb{R}^{2n})$. Note that $\langle T_b D_a \psi \otimes \overline{T_u D_s \psi}, h \rangle \in \mathscr{H}_2$, just like K_ψ.

The wavelet transforms $W_\psi f$ and $W_\psi(L[f])$ are computable from the observed input f and output $L[f]$. The formula (3.6) enables us to access to information of K_ψ. Then, we can access to time-frequency information of k by a similar way to the ordinary wavelet analysis, because (3.5) means that K_ψ is a kind of wavelet transform of k and (3.7) is a kind of inverse wavelet transform.

When the system L is translation-invariant (of non-Hilbert–Schmidt type), we have the following result, whose discrete version will appear in §6. The proof can be found in [1].

Theorem 3. *Let L be a continuous translation-invariant linear system from $L^2(\mathbb{R}^n)$ to $L^2(\mathbb{R}^n)$. Then, we have*

$$(W_\psi(L[f]))(\cdot, a) = L[W_\psi f(\cdot, a)] \tag{3.9}$$

for every $f \in L^2(\mathbb{R}^n)$ and $a \in \mathbb{R}_+$.

Remark 1. The *short-time Fourier transform*, denoted by $\mathscr{F}_w f$, of $f \in L^2(\mathbb{R}^n)$ with respect to a window function w is defined by

$$\mathscr{F}_w f(\xi, b) := \langle f, M_\xi T_b w \rangle_{L^2(\mathbb{R}^n)}, \qquad (\xi, b) \in \mathbb{R}^{2n},$$

which is the most familiar time-frequency analysis. $\mathscr{F}_w$ is an operator from $L^2(\mathbb{R}^n)$ into $L^2(\mathbb{R}^{2n})$. As for Theorem 2, we have very similar results for $\mathscr{F}_w$ instead of W_ψ. As for Theorem 3, since the modulation M_ξ does not commute with translation T_b, the short-time Fourier transform does not commute generally with translation-invariant systems for fixed $\xi \in \mathbb{R}$. If we modify the definition of the short-time Fourier transform as

$$\widetilde{\mathscr{F}_w} f(b, \xi) := \langle f, T_b M_\xi w \rangle_{L^2(\mathbb{R}^n)}, \qquad (b, \xi) \in \mathbb{R}^{2n},$$

then we have the same result as Theorem 3.

Wavelet Transforms and Causality

A function ψ is called a *wavelet function for causality* if $W_\psi f(b, a)$ is causal with respect to b for every causal function f. If we define the *involution* $\mathcal{I}$ by

$$\mathcal{I}[g(x)] := \overline{g(-x)},$$

then the wavelet transform $W_\psi f(b, a)$ can be represented as

$$W_\psi f(b, a) = (f * D_a \mathcal{I} \psi)(b). \tag{3.10}$$

When $a > 0$, the following two conditions are equivalent.

(i) f is causal $\Longrightarrow$ $W_\psi f(b, a)$ is causal with respect to b.

(ii) $\mathcal{I}[\psi]$ is causal.

We will use a discretized version of the wavelet function for causality constructed as follows. Take a continuous orthonormal wavelet function ψ with compact support such as Daubechies' orthonormal wavelet functions. Then, there exists $b \in \mathbb{R}$ such that $\operatorname{supp} T_b \mathcal{I} \psi \subset \overline{\mathbb{R}_+}$. Since $T_b \mathcal{I} \psi = \mathcal{I} T_{-b} \psi$, the function $\mathcal{I} T_{-b} \psi$ is causal, which means that $T_{-b}\psi$ is a wavelet function for causality.

4. Stationary Wavelet Transforms

Until now, we have dealt with analog systems and the continuous wavelet transform. For numerical experiments, we will deal with discrete systems whose inputs and outputs are sequences. Discrete systems are linear transforms on sequences, especially, any translation-invariant discrete system, which is called a *digital filter*, can be represented as a convolution of the input $\{x_n\}$ with the impulse response $\{\alpha_n\}$:

$$y_n = \sum_\ell \alpha_\ell x_{n-\ell}.$$

There are several discretized versions of the continuous wavelet transform. Discrete wavelet transforms can be classified as either redundant or nonredundant.

The nonredundant transforms such as orthogonal or biorthogonal decompositions perform better in the compression ability because only components relevant to reconstruction are added. But such nonredundant discrete wavelet transforms cannot have the translation-invariance, which is important in statistical signal processing. Translation-invariant redundant discrete wavelet transforms studied in [2], [11], and [12] work well for de-noising. In [11], such a translation-invariant redundant discrete wavelet transform was called *stationary wavelet transform*, and we follow their terminology.

Let us explain celebrated nonredundant discrete wavelet transform produced by Mallat [9], called *fast wavelet transform*. We deal with fast wavelet transform only for an orthonormal wavelet basis. Denote by $h_{0,n}$ and $g_{0,n}$, the reconstruction filter coefficients appearing in the dilation equation and the wavelet equation, respectively.

Let $j = 0$ be the finest level and $j = 1, 2, \cdots$ be coarse levels increasingly. The sequences $s_{j+1,k}$ called the *approximation coefficients* of level $j+1$ and $d_{j+1,k}$ called the *detail coefficients* of level $j+1$ are given by the sequence $s_{j,n}$ of level j as

$$\begin{cases} s_{j+1,k} = \sum_n \overline{h_{0,n-2k}}\, s_{j,n} = \sum_n \overline{h_{0,n}}\, s_{j,n+2k}, \\ d_{j+1,k} = \sum_n \overline{g_{0,n-2k}}\, s_{j,n} = \sum_n \overline{g_{0,n}}\, s_{j,n+2k}. \end{cases} \tag{4.1}$$

By looking carefully at the computation, we may keep only one point out of two, say the even or the odd indices, in each of $\sum_n \overline{h_{0,n-k}}\, s_{j,n}$ and $\sum_n \overline{g_{0,n-k}}\, s_{j,n}$ to get the complete information. Hence, in (4.1), we keep only the even indices $2k$ out of $\sum_n \overline{h_{0,n-k}}\, s_{j,n}$ and $\sum_n \overline{g_{0,n-k}}\, s_{j,n}$, respectively. This is what we call *downsampling*.

Applying the fast wavelet transform iteratively, we have the following decomposition:

$$s_{0,n} \mapsto \{d_{1,n}, d_{2,n}, \ldots, d_{j,n}; s_{j,n}\},$$

which is called a *multilevel* or *multistep fast wavelet transform* of level j. By Parseval's identity, we have

$$\sum_n |s_{0,n}|^2 = \sum_{r=1}^{j} \sum_n |d_{r,n}|^2 + \sum_n |s_{j,n}|^2. \tag{4.2}$$

Roughly speaking, the stationary wavelet transform is a kind of fast wavelet transform without downsampling. The sequences $S_{j+1,k}$ and $D_{j+1,k}$ of level $j+1$ are given by the sequence $S_{j,n}$ of level j as

$$\begin{cases} S_{j+1,k} = \sum_n \overline{h_{j,n-k}}\, S_{j,n} = \sum_n \overline{h_{j,n}}\, S_{j,n+k}, \\ D_{j+1,k} = \sum_n \overline{g_{j,n-k}}\, S_{j,n} = \sum_n \overline{g_{j,n}}\, S_{j,n+k}. \end{cases} \tag{4.3}$$

Here the initial filter coefficients $\overline{h_{0,n}}$ and $\overline{g_{0,n}}$ are the decomposition filter coefficients corresponding to an orthonormal wavelet basis. It must be remarked that filter coefficients depend on their levels and the filter coefficients $h_{j+1,n}$ and $g_{j+1,n}$ of level $j+1$ can be given as upsamplings of the filter coefficients $h_{j,n}$ and $g_{j,n}$ of level j, respectively. That is,

$$h_{j,k} = \begin{cases} h_{0,n}, & k = 2^j n, \\ 0, & \text{otherwise}. \end{cases} \tag{4.4}$$

Substitute $k = 2^{j+1}m$ in (4.3), then (4.4) implies

$$S_{j+1,2^{j+1}m} = \sum_n \overline{h_{j,n-2^{j+1}m}}\, S_{j,n} = \sum_\ell \overline{h_{j,2^j\ell-2^{j+1}m}}\, S_{j,2^j\ell} = \sum_\ell \overline{h_{0,\ell-2m}}\, S_{j,2^j\ell}.$$

Hence, $S_{j+1,2^{j+1}m}$ is the approximation coefficients of the fast wavelet transform decomposition of $S_{j,2^j\ell}$. By the same argument, $D_{j+1,2^{j+1}m}$ is the detail coefficients of the fast wavelet transform decomposition of $S_{j,2^j\ell}$. Thus, we can show that the following subsequences of the stationary wavelet transform of $S_{0,n}$:

$$\{D_{1,2n}, D_{2,2^2n}, \dots, D_{j,2^jn}; S_{j,2^jn}\}, \tag{4.5}$$

is the multistep fast wavelet transform of level j. Namely, $D_{r,2^rn} = d_{r,n}$ and $S_{r,2^rn} = s_{r,n}$. Substituting (4.5) into (4.2), we have

$$\sum_n |S_{0,n}|^2 = \sum_{r=1}^{j} \sum_n |D_{r,2^rn}|^2 + \sum_n |S_{j,2^jn}|^2. \tag{4.6}$$

5. System Identification of Discrete Systems

The most commonly used model structure in system identification is a convolution equation which relates the current output $\{y_n\}$ to a finite number of past outputs $\{y_{n-k}\}$ and inputs $\{x_{n-k}\}$ as

$$y_n = \sum_{\ell=0}^{P} a_\ell x_{n-\ell} + \sum_{m=1}^{Q} b_m y_{n-m}, \qquad P, Q \in \mathbb{N}, \tag{5.1}$$

where $\{a_\ell\,; \ell = 0, 1, \dots, P\}$ and $\{b_m\,; m = 1, 2, \dots, Q\}$ are filter coefficients. We will denote

$$A = [a_0, a_1, \dots, a_P]^T, \qquad B = [b_1, b_2, \dots, b_Q]^T.$$

The structure is thus entirely defined by the two integers P and Q. The simplest model with a disturbance $\{e_n\}$, which is called an *error*, is

$$y_n = \sum_{\ell=0}^{P} a_\ell x_{n-\ell} + \sum_{m=1}^{Q} b_m y_{n-m} + e_n. \tag{5.2}$$

Model (5.2) is often called an *equation error model* or an *autoregressive model with exogenous input*, which is simply called an *ARX model*.

The discrete versions of some linear differential equations are expressed by ARX models. One such example is the following health monitoring system.

Health Monitoring Systems

Structures under dynamic load, such as buildings, bridges, and so on, store cumulative damages on their structural members. The main concern of health monitoring systems is to have an efficient identification method of the structural parameters and to find when those parameters have been changed. Although these damages are generally estimated by continuous observation of several measurements, such as acceleration, velocity and displacement at several observing points, health monitoring systems based on these measurements could be expensive. Therefore, approaches to health monitoring systems utilizing only one measurement are growing in importance. Such health monitoring systems have only one input-output pair of long length. In this case, we subdivide the input-output pair into an enough number of input-output pairs of short length.

6. Identification Method

We use the same identification method for our wavelet method as for the conventional method. The difference is that the wavelet method preprocesses observed input-output pairs and uses the preprocessed input-output pairs to identify, while the conventional method uses observed input-output pairs themselves.

Denote by Y the column vector whose elements are y_n, and by Z the matrix whose elements are

$$Z(n,j) = \begin{cases} x_{n-j+1}, & 1 \leq j \leq P+1, \\ y_{n-j+P+1}, & P+2 \leq j \leq P+Q+1. \end{cases}$$

Then, equation (5.1) can be represented in matrix form as

$$Z \begin{bmatrix} A \\ B \end{bmatrix} = Y. \tag{6.1}$$

Equation (6.1) can be solved for the filter coefficients A and B in several ways. We will solve equation (6.1) by the least square method. Our problem is to find filter coefficients A and B which minimize

$$E = \sum_n |e_n|^2 = \sum_n \left(\sum_{\ell=0}^{P} a_\ell x_{n-\ell} + \sum_{m=1}^{Q} b_m y_{n-m} - y_n \right)^2, \tag{6.2}$$

when enough input-output pairs (x_n, y_n) are given.

Conventional Method

First, let us explain the conventional method of system identification. Put $M = \max\{P, Q\}$. Let k be a starting index. Take N successive elements starting with the index k from the output $\{y_n\}$ and denote the N-dimensional column vector by

$$Y = [y_k, y_{k+1}, y_{k+2}, \cdots, y_{k+N-1}]^T.$$

Construct the $N\times(P+Q+1)$ matrix Z in equation (6.1) by using the input-output pairs

$$(x_n, y_n), \qquad n = k-M, k-M+1, \ldots, k+N-1.$$

Then solve the equation (6.1) by the least square method.

Wavelet Method

Next, we propose our wavelet method of system identification. We use a wavelet function for causality. Then, the stationary wavelet transform (4.3) is represented as

$$S_{j+1,k} = \sum_{n=-\infty}^{k} \overline{h_{j,n-k}} S_{j,n}, \qquad D_{j+1,k} = \sum_{n=-\infty}^{k} \overline{g_{j,n-k}} S_{j,n}. \tag{6.3}$$

Let a pair of input $\{x_n\}$ and output $\{y_n\}$ be given. Put

$$S_{0,n}^{\mathrm{in}} = x_n, \qquad S_{0,n}^{\mathrm{out}} = y_n.$$

Applying the stationary wavelet transform (6.3) for causal systems, calculate inductively the approximation $S_{j,k}^{\mathrm{in}}$ and the detail $D_{j,k}^{\mathrm{in}}$ of level j for the input and the approximation $S_{j,k}^{\mathrm{out}}$ and the detail $D_{j,k}^{\mathrm{out}}$ of level j for the output. Then, as a discrete version of Theorem 3, we have the following Theorem 4, whose proof is omitted.

Theorem 4. *Let a pair of input $\{x_n\}$ and output $\{y_n\}$ be given. Assume that the system to be identified has the form* (5.1). *Then,*

$$S_{j,k}^{\mathrm{out}} = \sum_{\ell=0}^{P} a_\ell S_{j,k-\ell}^{\mathrm{in}} + \sum_{m=1}^{Q} b_m S_{j,k-m}^{\mathrm{out}},$$

$$D_{j,k}^{\mathrm{out}} = \sum_{\ell=0}^{P} a_\ell D_{j,k-\ell}^{\mathrm{in}} + \sum_{m=1}^{Q} b_m D_{j,k-m}^{\mathrm{out}}.$$

Choosing enough approximation pairs

$$\{(j_i, k_i)\}_{i\in I_a}, \qquad I_a = \{1, 2, \ldots, N_a\},$$

and detail pairs

$$\{(j_i, k_i)\}_{i\in I_d}, \qquad I_d = \{1, 2, \ldots, N_d\},$$

we have the following system to solve for A and B:

$$\begin{cases} S_{j_i,k_i}^{\mathrm{out}} = \displaystyle\sum_{\ell=0}^{P} a_\ell S_{j_i,k_i-\ell}^{\mathrm{in}} + \sum_{m=1}^{Q} b_m S_{j_i,k_i-m}^{\mathrm{out}}, & i \in I_a, \\ D_{j_i,k_i}^{\mathrm{out}} = \displaystyle\sum_{\ell=0}^{P} a_\ell D_{j_i,k_i-\ell}^{\mathrm{in}} + \sum_{m=1}^{Q} b_m D_{j_i,k_i-m}^{\mathrm{out}}, & i \in I_d. \end{cases} \tag{6.4}$$

The wavelet method solves (6.4) by the least square method.

7. Numerical Experiment

We will deal with a mathematical model of simplified health monitoring systems. The model changes its structural parameters at a moment, that is, the filter coefficients of the system are not constants but step functions. The system is translation-invariant both before and after the critical moment, but the filter coefficients are different between before and after.

Model for Identification

Let us consider a damped forced linear oscillation model, which is a simple model of vehicle suspension systems, illustrated in Figure 2. The equation of motion of this model is

$$m\ddot{y} + C\dot{y} + Ky = -m\ddot{x}, \tag{7.1}$$

where m is the mass, C is the damping constant, and K is the spring constant. The output y is called the *response*. Let us discretize the above model. The difference equation corresponding to (7.1) can be represented as an ARX model. Since the order of this difference equation is two, it implies that $P = Q = 2$.

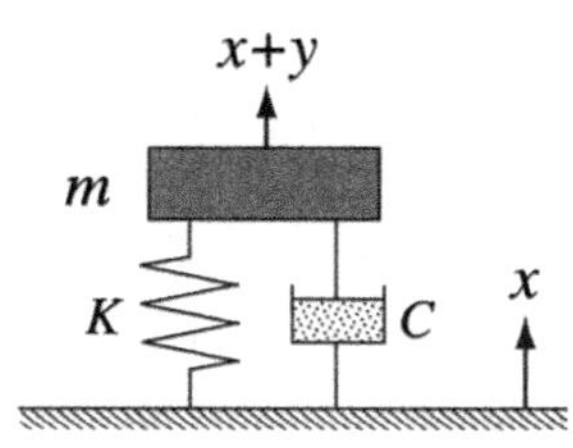

FIGURE 2. Damped forced linear oscillation model.

Let the step size of time be 0.05 seconds. Assume that the structural parameters for the first half of the input, which consists of 512 points corresponding to $t \leq 25.6$, are

$$m = 1, \quad C = 50, \quad K = 500,$$

and those for the second half of the input, which consists of 512 points corresponding to $t > 25.6$, are

$$m = 1, \quad C = 35, \quad K = 350.$$

Then, the ARX model to be identified, whose parameters are listed in Table 7, is given by replacing derivatives with differences in the equation of motion.

	Parameters	
Time	$t \leq 25.6$	$t > 25.6$
Filter a_0	-0.22222	-0.26667
Filter a_1	0.44444	0.53333
Filter a_2	-0.22222	-0.26667
Filter b_1	0.33333	0.60000
Filter b_2	0.11111	-0.06667

TABLE 1. The parameters of the ARX model.

Input-Output Pair

Let the ideal input x be

$$x = -5\sin\left(\frac{2\pi t}{3}\right).$$

We add white noise with values in $[-2.5, 2.5]$ on the input x and denote the noised input by

$$\widetilde{x} = -5\sin\left(\frac{2\pi t}{3}\right) + \text{WhiteNoise}(-2.5, 2.5),$$

where WhiteNoise$(-2.5, 2.5)$ means a white noise with values in $[-2.5, 2.5]$. Then the output y is given by solving the equation of motion with the noised input $\widetilde{x}$. We add white noise with values in $[-0.15, 0.15]$ on the output y and denote the noised output by

$$\widetilde{y} = y + \text{WhiteNoise}(-0.15, 0.15).$$

These input $\widetilde{x}$ and output $\widetilde{y}$ are illustrated in Figure 3.

The aim of the following numerical experiment is to compare the conventional method with the wavelet method. The outline and our conclusion of the numerical experiment are as follows.

Outline of the Numerical Experiment

Assume that $P = Q = 2$. Then, the lengths of the filter coefficients A and B are $P+1 = 3$ and $Q = 2$, respectively. Identify the filter coefficients A and B at three positions: before, around, and after the critical moment, for input-output pairs with white noise.

Conclusion of the Numerical Experiment

Both the conventional and the wavelet methods can identify the filter coefficients A at two positions: before and after the critical moment, for input-output pairs. Only the wavelet method can identify the filter coefficients B at two positions: before and after the critical moment, for input-output pairs.

We will use MATLAB's colon operator. The expression $J : K$ is the same as the row vector $[J, J+1, \ldots, K]$, where $J, K \in \mathbb{Z}$ and $J \leq K$. For a real number X, fix(X) rounds X to the nearest integer towards zero. The expression $J : D : K$ is the same as $[J, J+D, \ldots, J+MD]$, where $M = \text{fix}((K-J)/D)$. For a row vector

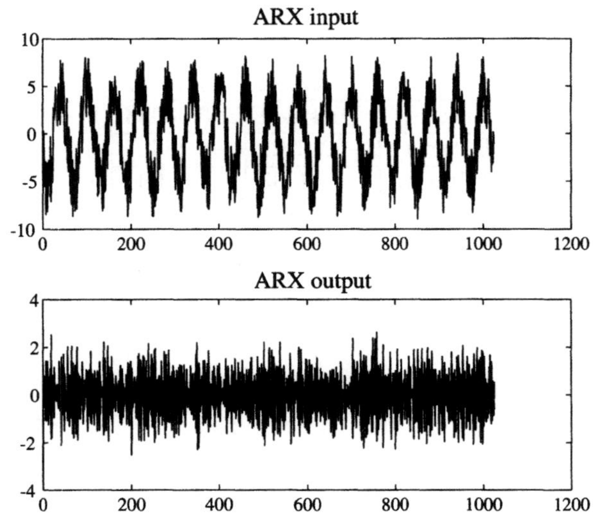

FIGURE 3. The input $\widetilde{x}$ and the output $\widetilde{y}$ of the ARX model.

or a sequence x_n, the expression $x_n(J:K)$ is the same as $[x_J, x_{J+1}, \ldots, x_K]$, the expression $x_n(J:D:K)$ is the same as $[x_J, x_{J+D}, \ldots, x_{J+MD}]$, and so on.

Conventional method. For the input-output pairs $(\widetilde{x}_n, \widetilde{y}_n)$ with white noise, denote by

$$\widetilde{x}_n^{(k)} = \widetilde{x}_n(k-2:k+255), \qquad k = 54, 384, 704,$$

the subsequences of $\widetilde{x}_n$ with length 258 starting with indices $k = 54$, 384, 704 and by

$$\widetilde{y}_n^{(k)} = \widetilde{y}_n(k-2:k+255), \qquad k = 54, 384, 704,$$

the subsequences of $\widetilde{y}_n$ with length 258 starting with indices $k = 54$, 384, 704. Apply the identification method explained before to input-output pairs $(\widetilde{x}_n^{(k)}, \widetilde{y}_n^{(k)})$ for $k = 54$, 384, 704. Under the assumption that $P = Q = 2$, solve equation (6.1) with respect to A and B. As the result is given in Table 7, A can be identified but B cannot.

Wavelet method. We take stationary wavelet transforms of the input and output, up to 3 levels, by using Daubechies' $N = 3$ orthonormal wavelet function for causality constructed in §3.

The detail pairs and the approximation pairs in the left-hand side of equation (6.4) are chosen as

$$\begin{aligned} &D_{2,n}^{\text{out}}, \qquad n = k:2:k+255, \quad 128 \text{ points}, \\ &D_{3,n}^{\text{out}}, \qquad n = k:4:k+255, \quad 64 \text{ points}, \\ &S_{3,n}^{\text{out}}, \qquad n = k:4:k+255, \quad 64 \text{ points}, \end{aligned}$$

	Conventional method		
Starting index	$k = 54$	$k = 384$	$k = 704$
Filter a_0	−0.224	−0.246	−0.270
Filter a_1	0.426	0.408	0.488
Filter a_2	−0.202	−0.156	−0.214
Filter b_1	0.254	0.176	0.437
Filter b_2	0.088	−0.032	−0.079
ℓ^1 norm of the error of A	0.040		0.101
ℓ^1 norm of the error of B	0.102		0.175

TABLE 2. Identification by the conventional method.

where k is the starting index. It must be noticed that the level 1 detail coefficients are not used.

Equation (6.4) is solved by the least square method, for three starting indices $k = 54$, $k = 384$, $k = 704$. As the result is given in Table 7, the filters A and B can be identified.

	Wavelet method		
Starting index	$k = 54$	$k = 384$	$k = 704$
Filter a_0	−0.220	−0.218	−0.262
Filter a_1	0.444	0.428	0.514
Filter a_2	−0.224	−0.204	−0.252
Filter b_1	0.338	0.429	0.568
Filter b_2	0.116	−0.063	−0.107
ℓ^1 norm of the error of A	0.004		0.039
ℓ^1 norm of the error of B	0.010		0.072

TABLE 3. Identification by the wavelet method.

References

[1] R. Ashino, T. Mandai and A. Morimoto, System identification based on distribution theory and wavelet transform, *Applic. Anal.*, to appear.

[2] R. Coifman and D. Donoho, Translation invariant de-noising, in *Lecture Notes in Statistics* **103**, Springer-Verlag, New York, NY, 1995, 125–150,

[3] I. Daubechies, *Ten Lectures on Wavelets*, SIAM, Philadelphia, PA, 1992.

[4] K. Gröchenig, *Foundations of Time-Frequency Analysis*, Birkhäuser, Boston, 2001.

[5] L. Hörmander, Estimates for translation invariant operators in L^p spaces, *Acta Math.* **104** (1960), 93–140.

[6] L. Hörmander, *The Analysis of Linear Partial Differential Operators I*, Second Edition, Springer-Verlag, Berlin, 1990.

[7] P.D. Lax, *Functional Analysis*, John Wiley & Sons, New York, NY, 2002.

[8] L. Ljung, *System Identification: Theory for the User*, Prentice Hall PTR, Upper Saddle River, NJ, 1987.

[9] S. Mallat, A theory for multiresolution signal decomposition: The wavelet representation, *IEEE Pattern Anal. and Machine Intell.* **11** (1989), 674–693.

[10] S. Mallat, *A Wavelet Tour of Signal Processing*, Second Edition, Academic Press, San Diego, CA, 1999.

[11] G. Nason and B. Silverman, The stationary wavelet transform and some statistical applications, in *Lecture Notes in Statistics* **103**, Springer-Verlag, New York, NY, 1995, 281–299.

[12] J. Pesquet, H. Krim and H. Carfatan, Time-invariant orthonormal wavelet representations, *IEEE Trans. Sign. Proc.* **44** (1996), 1964–1970.

[13] L. Schwartz, *Théorie des Distributions*, Nouvelle Édition, Hermann, Paris, 1966.

[14] E.M. Stein, *Singular Integrals and Differentiability Properties of Functions*, Princeton University Press, Princeton, NJ, 1970.

[15] E.M. Stein and G. Weiss, *Introduction to Fourier analysis on Euclidean spaces*, Princeton University Press, Princeton, NJ, 1971.

[16] F. Treves, *Topological Vector Spaces, Distributions and Kernels*, Academic Press, New York, NY, 1967.

Ryuichi Ashino
Mathematical Sciences
Osaka Kyoiku University
Kashiwara, Osaka 582-8582
Japan
e-mail: ashino@cc.osaka-kyoiku.ac.jp

Takeshi Mandai
Research Center for Physics and Mathematics
Osaka Electro-Communication University
Neyagawa, Osaka 572-8530
Japan
e-mail: mandai@isc.osakac.ac.jp

Akira Morimoto
Information Science
Osaka Kyoiku University
Kashiwara, Osaka 582-8582
Japan
e-mail: morimoto@cc.osaka-kyoiku.ac.jp

Operator Theory:
Advances and Applications, Vol. 155, 219–233

Two-Dimensional Wavelet Bases for Partial Differential Operators and Applications

M.A. Hajji and S. Melkonian and Rémi Vaillancourt

Abstract. Partial differential operators are represented in non-standard form in separable two-dimensional orthonormal wavelet bases. A formal pseudodifferential approach is suggested for numerical applications. Applications are to the fifth-order quasilinear thin-film equation on an inclined plane.

Mathematics Subject Classification (2000). Primary 65T60; Secondary 35Q51.

Keywords. Periodic wavelet, differential operator, wavelet basis, non-standard representation, thin film equation.

1. Introduction

Wavelets have found applications in the representation of differential operators. The representation of operators in compactly supported, one-dimensional wavelet bases, as well as the construction of their corresponding matrices, has been considered by Beylkin [3] and Beylkin, Coifman and Rokhlin [4]. The foregoing results have been employed by Beylkin and Keiser [5] in the numerical solution of evolution equations in time and one space dimension.

In this paper, we give the non-standard representation of linear partial differential operators in two-dimensional compactly supported wavelet bases. Periodized Daubechies' wavelets [6, 7] of period 1 are used in the construction of the matrices representing linear differential operators of the form $L = g(\partial_x, \partial_y)$, where g is a real-valued function coming from partial differential equations of interest [5, 8, 9]. A formal pseudo-differential approach is suggested for numerical applications.

This work was partially supported by the Natural Sciences and Engineering Research Council of Canada.

Applications include the numerical realization of horseshoe-shaped solitons in the solution of a fifth-order partial differential equation governing the flow of a viscous fluid film on an inclined plane [10, 8].

2. Separable Two-Dimensional Multiresolution Analysis

Consider an orthogonal one-dimensional multiresolution analysis (MRA),

$$\cdots \subset V_j \subset V_{j+1} \subset \cdots, \qquad j \in \mathbb{Z},$$

with scaling function $\varphi(x)$ and wavelet function $\psi(x)$ supported on $[0, L-1]$ and satisfying the two-scale equations:

$$\varphi(x) = \sqrt{2} \sum_{k=0}^{L-1} h_k \varphi(2x-k), \qquad \psi(x) = \sqrt{2} \sum_{k=0}^{L-1} g_k \varphi(2x-k).$$

The scaled and shifted φ and ψ are

$$\varphi_{j,k}(x) = 2^{j/2}\varphi(2^j x - k), \qquad \psi_{j,k}(x) = 2^{j/2}\psi(2^j x - k).$$

The spaces

$$\cdots \subset \boldsymbol{V}_j = V_j \otimes V_j \subset \boldsymbol{V}_{j+1} \subset \cdots, \qquad j \in \mathbb{Z},$$

form a separable two-dimensional MRA with scaling function $\Phi(x,y) = \varphi(x)\varphi(y)$. The orthogonal complement of $\boldsymbol{V}_j$ in $\boldsymbol{V}_{j+1}$ is

$$\boldsymbol{W}_j = \boldsymbol{W}_j^h \oplus \boldsymbol{W}_j^v \oplus \boldsymbol{W}_j^d,$$

where

$$\boldsymbol{W}_j^h = V_j \otimes W_j, \qquad \boldsymbol{W}_j^v = W_j \otimes V_j, \qquad \boldsymbol{W}_j^d = W_j \otimes W_j,$$

and $\boldsymbol{W}_0$ is defined by the horizontal, vertical and diagonal wavelet functions:

$$\Psi^h(x,y) = \varphi(x)\psi(y), \quad \Psi^v(x,y) = \psi(x)\varphi(y), \quad \Psi^d(x,y) = \psi(x)\psi(y).$$

The scaled and shifted version of Φ, Ψ^h, Ψ^v and Ψ^d are

$$\Phi^j_{k_1,k_2}(x,y) = \varphi_{j,k_1}(x)\varphi_{j,k_2}(y), \qquad \Psi^{j,h}_{k_1,k_2}(x,y) = \varphi_{j,k_1}(x)\psi_{j,k_2}(y),$$
$$\Psi^{j,v}_{k_1,k_2}(x,y) = \psi_{j,k_1}(x)\varphi_{j,k_2}(y), \qquad \Psi^{j,d}_{k_1,k_2}(x,y) = \psi_{j,k_1}(x)\psi_{j,k_2}(y).$$

In the following, the superscript λ will take values in the index set $\{h, v, d\}$. For each $j \in \mathbb{Z}$, bases for $\boldsymbol{V}_j$ and $\boldsymbol{W}_j$ are:

$$\{\Phi^j_{k_1,k_2},\ k_1,k_2 \in \mathbb{Z}\}, \qquad \{\Psi^{j,\lambda}_{k_1,k_2},\ k_1,k_2 \in \mathbb{Z},\ \lambda = h,v,d\}.$$

A separable wavelet basis of $L^2(\mathbb{R}^2)$ is $\{\Psi^{j,\lambda}_{k_1,k_2},\ j,k_1,k_2 \in \mathbb{Z},\ \lambda = h,v,d\}$.

Let $\langle\cdot,\cdot\rangle$ denote the inner product in $L^2(\mathbb{R}^2)$. A function $f \in L^2(\mathbb{R}^2)$ is approximated by its orthogonal projection on the space $\boldsymbol{V}_n$:

$$f_n = P_n f = \sum_{k_1,k_2} s^n_{k_1,k_2} \Phi^n_{k_1,k_2}, \quad \text{where} \quad s^n_{k_1,k_2} = \langle f, \Phi^n_{k_1,k_2}\rangle,$$

and represented in the wavelet basis by the series

$$f = \sum_{j\in\mathbb{Z}} \sum_{\lambda=h,v,d} \sum_{k_1,k_2\in\mathbb{Z}} d^{j,\lambda}_{k_1,k_2} \Psi^{j,\lambda}_{k_1,k_2}, \quad \text{where} \quad d^{j,\lambda}_{k_1,k_2} = \langle f, \Psi^{j,\lambda}_{k_1,k_2} \rangle.$$

Choosing $\boldsymbol{V}_n$ and $\boldsymbol{V}_{n-J}$ as the finest and coarsest subspaces, respectively, and using $\boldsymbol{V}_j = \boldsymbol{V}_{j-1} \oplus \boldsymbol{W}_{j-1}$, we have

$$\boldsymbol{V}_n = \boldsymbol{V}_{n-J} \oplus \boldsymbol{W}_{n-J} \oplus \boldsymbol{W}_{n-J+1} \oplus \cdots \oplus \boldsymbol{W}_{n-1}.$$

Then $f_n \in \boldsymbol{V}_n$ can be expanded as

$$f_n = P_{n-J} f + \sum_{j=n-J}^{n-1} Q_j f, \quad \text{where} \quad Q_j f = \sum_{k_1,k_2} \sum_{\lambda=h,v,d} d^{j,\lambda}_{k_1,k_2} \Psi^{j,\lambda}_{k_1,k_2}.$$

Here, $P_{n-J} f$ is the coarsest approximation on $\boldsymbol{V}_{n-J}$ and the projections $Q_j = Q_j^h + Q_j^v + Q_j^d$ on $\boldsymbol{W}_j = \boldsymbol{W}_j^h \oplus \boldsymbol{W}_j^v \oplus \boldsymbol{W}_j^d$ give the detail functions $Q_j f$ on $\boldsymbol{W}_j$ which build up the finest approximation, f_n, of f on $\boldsymbol{V}_n$. Compression is achieved by setting $d^{j,\lambda}_{k_1,k_2} = 0$ below a threshold.

The two-dimensional fast wavelet transform (FWT) decomposition is obtained by lowpass and highpass filterings with downsampling:

$$s^j_{k_1,k_2} = \sum_{m_1,m_2} H_{m_1,m_2} s^{j+1}_{m_1+2k_1,m_2+2k_2},$$

$$d^{j,\lambda}_{k_1,k_2} = \sum_{m_1,m_2} G^\lambda_{m_1,m_2} s^{j+1}_{m_1+2k_1,m_2+2k_2},$$

respectively, with real lowpass filter

$$H_{m_1,m_2} = h_{m_1} h_{m_2}$$

and real highpass filters

$$G^h_{m_1,m_2} = h_{m_1} g_{m_2}, \quad G^v_{m_1,m_2} = g_{m_1} h_{m_2}, \quad G^d_{m_1,m_2} = g_{m_1} g_{m_2}.$$

The two-dimensional inverse fast wavelet transform (IFWT) reconstruction of s^{j+1} from s^j and $d^{j,\lambda}$ is obtained by upsampling and filtering:

$$s^{j+1}_{2k_1,2k_2} = \sum_{m_1,m_2} H_{2m_1,2m_2} s^j_{k_1-m_1,k_2-m_2} + \sum_{\lambda=h,v,d} \sum_{m_1,m_2} G^\lambda_{2m_1,2m_2} d^{\lambda,j}_{k_1-m_1,k_2-m_2}, \tag{1}$$

$$s^{j+1}_{2k_1+1,2k_2+1} = \sum_{m_1,m_2} H_{2m_1+1,2m_2+1} s^j_{k_1-m_1,k_2-m_2} + \sum_{\lambda=h,v,d} \sum_{m_1,m_2} G^\lambda_{2m_1+1,2m_2+1} d^{\lambda,j}_{k_1-m_1,k_2-m_2}, \tag{2}$$

$$s^{j+1}_{2k_1,2k_2+1} = \sum_{m_1,m_2} H_{2m_1,2m_2+1} s^j_{k_1-m_1,k_2-m_2} + \sum_{\lambda=h,v,d} \sum_{m_1,m_2} G^\lambda_{2m_1,2m_2+1} d^{\lambda,j}_{k_1-m_1,k_2-m_2}, \tag{3}$$

$$s^{j+1}_{2k_1+1,2k_2} = \sum_{m_1,m_2} H_{2m_1+1,2m_2} s^j_{k_1-m_1,k_2-m_2} + \sum_{\lambda=h,v,d} \sum_{m_1,m_2} G^\lambda_{2m_1+1,2m_2} d^{\lambda,j}_{k_1-m_1,k_2-m_2}. \tag{4}$$

For periodic wavelets, the matrices s^j and $d^{j,\lambda}$ are $(2^j, 2^j)$-periodic. The $[0,1]$-periodic scaling function and wavelets are defined by the formulae

$$\sum_{\ell_1 \in \mathbb{Z}} \sum_{\ell_2 \in \mathbb{Z}} \Phi^j_{k_1,k_2}(x-\ell_1, y-\ell_2)$$

and

$$\sum_{\ell_1 \in \mathbb{Z}} \sum_{\ell_2 \in \mathbb{Z}} \Psi^{j,\lambda}_{k_1,k_2}(x-\ell_1, y-\ell_2), \quad \lambda = h, v, d,$$

respectively.

3. Nonstandard Representation of Differential Operators

Given an operator T, we approximate T by $T_n = P_n T P_n$ and with $\boldsymbol{V}_n$ and $\boldsymbol{V}_{n-J}$ as above, we expand $T_n = P_n T P_n$ in the telescoping sum

$$T_n = P_n T P_n = \sum_{j=0}^{J-1} [P_{n-j} T P_{n-j} - P_{n-j-1} T P_{n-j-1}] + P_{n-J} T P_{n-J},$$

and use $P_{j+1} = P_j + Q_j$ to obtain the NS-representation of T,

$$T_n = \sum_{j=n-J}^{n-1} [A_j + B_j + C_j] + T_{n-J},$$

where $T_{n-J} = P_{n-J} T P_{n-J} : \boldsymbol{V}_{n-J} \to \boldsymbol{V}_{n-J}$ and

$$\begin{aligned}
A_j &= \sum_{\lambda,\lambda'=h,v,d} A_j^{\lambda,\lambda'} = \sum_{\lambda,\lambda'=h,v,d} Q_j^\lambda T Q_j^{\lambda'} &&: \boldsymbol{W}_j^{\lambda'} \to \boldsymbol{W}_j^\lambda, \\
B_j &= \sum_{\lambda=h,v,d} B_j^\lambda = \sum_{\lambda=h,v,d} Q_j^\lambda T P_j &&: \boldsymbol{V}_j \to \boldsymbol{W}_j, \\
C_j &= \sum_{\lambda=h,v,d} C_j^\lambda = \sum_{\lambda=h,v,d} P_j T Q_j^\lambda &&: \boldsymbol{W}_j \to \boldsymbol{V}_j.
\end{aligned}$$

The NS-representation decouples the scales by acting on $\boldsymbol{V}_j$ and $\boldsymbol{W}_j$ at each level.

Then for $f \in L^2(\mathbb{R}^2)$, we have the approximation

$$Tf \approx \sum_{j=n-J}^{n-1} \left[\sum_{\lambda,\lambda'=h,v,d} A_j^{\lambda,\lambda'} f + \sum_{\lambda=h,v,d} (B_j^\lambda + C_j^\lambda) f \right] + T_{n-J} f.$$

The steps in the application of an operator T on periodic functions $f \in L^2([0,1]^2)$ of period 1 are as follows.

- Approximate f in $\boldsymbol{V}_n$ by $P_n f$ to get the $2^n \times 2^n$ coordinate matrix s^n.
- Decompose s^n down the MRA spaces using two-dimensional FWT to get $d^{j,h}$, $d^{j,v}$, $d^{j,d}$ and s^j for $j = n-1, n-2, \ldots, n-J$.
- For $n - J \leq j \leq n-1$, construct the matrices $A^{j,\lambda,\lambda'}$, $B^{j,\lambda}$, $C^{j,\lambda}$ representing the operators $A_j^{\lambda,\lambda'}$, B_j^λ, C_j^λ, respectively.
- The approximation $Tf \simeq T_n f$ is then given by

$$T_n f = \sum_{j=n-J}^{n-1} \left[\sum_{\lambda=h,v,d} \left(\sum_{k_3,k_4=0}^{2^j-1} \tilde{d}_{k_3,k_4}^{j,\lambda} \Psi_{k_3,k_4}^{j,\lambda} \right) + \sum_{k_3,k_4=0}^{2^j-1} \tilde{s}^j \Phi_{k_3,k_4}^j \right],$$

where

$$\begin{aligned}
\tilde{d}^{j,h} &= \sum_{\lambda=h,v,d} A^{j,h,\lambda} \odot d^{j,\lambda} + B^{j,h} \odot s^j, \\
\tilde{d}^{j,v} &= \sum_{\lambda=h,v,d} A^{j,v,\lambda} \odot d^{j,\lambda} + B^{j,v} \odot s^j, \\
\tilde{d}^{j,d} &= \sum_{\lambda=h,v,d} A^{j,d,\lambda} \odot d^{j,\lambda} + B^{j,d} \odot s^j, \\
\tilde{s}^j &= \sum_{\lambda=h,v,d} C^{j,\lambda} \odot d^{j,\lambda}, \quad \text{for} \quad n - J + 1 \leq j \leq n-1, \\
\tilde{s}^{n-J} &= \sum_{\lambda=h,v,d} C^{n-J,\lambda} \odot d^{n-J,\lambda} + T^{n-J} \odot s^{n-J}.
\end{aligned}$$

The operation $\odot$ is defined as follows.

If P is a $k \times k$ block matrix with $k \times k$ block matrices $P^{i,j}$ and D is a $k \times k$ matrix, then

$$(P \odot D)_{i,j} = \sum_{m,n} P_{m,n}^{i,j} D_{m,n}.$$

The function $\tilde{f} = T_n f$ can be reconstructed to the finest space $\boldsymbol{V}_n$ as

$$\tilde{f} = \sum_{k_1=0}^{2^n-1} \sum_{k_2=0}^{2^n-1} s_{k_1,k_2}^n \Phi_{k_1,k_2}^n.$$

The coefficients s^n can be reconstructed from the coefficients $\tilde{d}^{j,\lambda}$ and $\tilde{s}^j$ of the NS-representation of f for $n - J \leq j \leq n-1$ by the following **pseudo-inverse fast wavelet transform**.

For $j = n-J, n-J+1, n-J+2, \dots, n-2$, $\hat{s}^{j+1}$ is constructed from $\tilde{s}^j$ and $\tilde{d}^{j,\lambda}$ by means of (1)–(4) and

$$s^{j+1} = \hat{s}^{j+1} + \tilde{s}^{j+1}.$$

Finally, for $j = n-1$, s^n is constructed from s^{n-1} and $\tilde{d}^{n-1,\lambda}$ again by (1)–(4).

4. Matrix Construction of the NS-Representation

The operator T_n acting on a function $f \in L^2([0,1]^2)$,

$$\begin{aligned} T_n f := P_n T P_n f &= P_n T\Big(\sum_{k_1,k_2} \langle f, \Phi^n_{k_1,k_2}\rangle \Phi^n_{k_1,k_2}\Big) \\ &= \sum_{k_3,k_4} \underbrace{\Big(\sum_{k_1,k_2} \underbrace{\langle T\Phi^n_{k_1,k_2}, \Phi^n_{k_3,k_4}\rangle}_{T^{n,k_3,k_4}_{k_1,k_2}} \underbrace{\langle f, \Phi^n_{k_1,k_2}\rangle}_{s^n_{k_1,k_2}}\Big)}_{\tilde{s}^n_{k_3,k_4}} \Phi^n_{k_3,k_4}, \end{aligned}$$

is represented by the 4-dimensional structure:

$$T^{n,k_3,k_4}_{k_1,k_2} = \langle T\Phi^n_{k_1,k_2}, \Phi^n_{k_3,k_4}\rangle.$$

For each k_3, k_4, the $2^n \times 2^n$ matrix

$$T^{n,k_3,k_4}_{k_1,k_2} = \langle T\Phi^n_{k_1,k_2}, \Phi^n_{k_3,k_4}\rangle, \quad 0 \le k_1, k_2 \le 2^n - 1,$$

is a block in the structure representation T^n of $T_n : \boldsymbol{V}_n \to \boldsymbol{V}_n$,

$$T^n = \begin{bmatrix} T^{n,0,0} & T^{n,0,1} & \dots & T^{n,0,2^n-1} \\ T^{n,1,0} & T^{n,1,1} & \dots & T^{n,1,2^n-1} \\ \vdots & \vdots & \dots & \vdots \\ T^{n,2^n-1,0} & T^{n,2^n-1,1} & \dots & T^{n,2^n-1,2^n-1} \end{bmatrix}.$$

Due to the structure of the MRA, the operators $A_j^{\lambda,\lambda'}$, B_j^λ, C_j^λ and T_j are represented by the structures $A^{j,\lambda,\lambda'}$, $B^{j,\lambda}$, $C^{j,\lambda}$ and T^j, respectively,

$$A^{j,\lambda,\lambda',k_3,k_4}_{k_1,k_2} = \langle T\Psi^{j,\lambda'}_{k_1,k_2}, \Psi^{j,\lambda}_{k_3,k_4}\rangle, \quad B^{j,\lambda,k_3,k_4}_{k_1,k_2} = \langle T\Phi^{j}_{k_1,k_2} \Psi^{j,\lambda}_{k_3,k_4}\rangle,$$
$$C^{j,\lambda,k_3,k_4}_{k_1,k_2} = \langle T\Psi^{j,\lambda}_{k_1,k_2}, \Phi^{j}_{k_3,k_4}\rangle, \quad T^{j,k_3,k_4}_{k_1,k_2} = \langle T\Phi^{j}_{k_1,k_2} \Phi^{j}_{k_3,k_4}\rangle.$$

The entries of the lower-scale structures are obtained recursively from T^n by the formulae:

$$A^{j,\lambda,\lambda',k_3,k_4}_{k_1,k_2} = \sum_{m_1,m_2,m_3,m_4=0}^{L-1} G^{\lambda'}_{m_1,m_2} G^{\lambda}_{m_3,m_4} T^{j+1,m_3+2k_3,m_4+2k_4}_{m_1+2k_1,m_2+2k_2},$$

$$B^{j,\lambda,k_3,k_4}_{k_1,k_2} = \sum_{m_1,m_2,m_3,m_4=0}^{L-1} H_{m_1,m_2} G^{\lambda}_{m_3,m_4} T^{j+1,m_3+2k_3,m_4+2k_4}_{m_1+2k_1,m_2+2k_2},$$

$$C^{j,\lambda,k_3,k_4}_{k_1,k_2} = \sum_{m_1,m_2,m_3,m_4=0}^{L-1} G^{\lambda}_{m_1,m_2} H_{m_3,m_4} T^{j+1,m_3+2k_3,m_4+2k_4}_{m_1+2k_1,m_2+2k_2},$$

$$T^{j,\lambda,k_3,k_4}_{k_1,k_2} = \sum_{m_1,m_2,m_3,m_4=0}^{L-1} H_{m_1,m_2} H_{m_3,m_4} T^{j+1,m_3+2k_3,m_4+2k_4}_{m_1+2k_1,m_2+2k_2}.$$

For periodic wavelets, these matrices are considered as $(2^j, 2^j)$-periodic.

5. The Operator $T = \partial_x$

The operator $T = \partial_x$ has the following representation:

$$T^{n,k_3,k_4}_{k_1,k_2} = \langle \partial_x(\Phi^n_{k_1,k_2}), \Phi^n_{k_3,k_4} \rangle = 2^n r_{k_3-k_1} \delta_{k_2,k_4},$$

where

$$r_l = \int_{-\infty}^{\infty} \phi(x-l)\phi'(x)\, dx, \qquad l \in \mathbb{Z}. \tag{5}$$

Thus, T^n is completely determined by r_l. A similar representation holds for $T = \partial_y$.

The following proposition is proved in [3].

Proposition 1. *If the integrals* (5) *exist, then the coefficients r_l satisfy the following system of linear equations:*

$$r_l = 2r_{2l} + \sum_{k=1}^{L/2} a_{2k-1}(r_{2l-2k+1} + r_{2l+2k-1}) \tag{6}$$

with

$$\begin{cases} \displaystyle\sum_{l=-(L-2)}^{L-2} l r_l = -1, \\ r_{-l} = -r_l, \\ r_l = 0 \quad \text{for} \quad l \notin [-L+2, L-2], \end{cases} \tag{7}$$

and

$$a_n = 2 \sum_{i=0}^{L-1-n} h_i h_{i+n}, \qquad n = 1, 2, \dots L-1.$$

Moreover, if the number of vanishing moments of the wavelet $\psi(x)$ is at least two, then (6) *and* (7) *have a unique solution with $r_i \neq 0$ for $-L+2 \leq l \leq L-2$ and $r_l = -r_l$.*

With a minor change in notation, the block matrix T_x^n ($= T^n$ for $T = \partial_x$) becomes

$$T_x^n = \begin{bmatrix} \boxed{X^{1,1}} & X^{1,2} & \cdots & X^{1,2^n-1} & X^{1,2^n} \\ X^{2,1} & X^{2,2} & \cdots & X^{2,2^n-1} & X^{2,2^n} \\ \vdots & \vdots & \cdots & \vdots & \vdots \\ X^{2^n-1,1} & X^{2^n-1,2} & \cdots & X^{2^n-1,2^n-1} & X^{2^n-1,2^n} \\ X^{2^n,1} & X^{2^n,2} & \cdots & X^{2^n,2^n-1} & X^{2^n,2^n} \end{bmatrix},$$

with

$$X_{k_1,k_2}^{k_3,k_4} = T_{k_1,k_2}^{n,k_3,k_4} = 2^n r_{k_3-k_1}\delta_{k_2,k_4}.$$

Each X^{k_3,k_4} contains only one nonzero column, the $k_4{}^{\text{th}}$ column. The nonzero column of $X^{1,1}$ is given by

$$c = 2^n\left[0, r_{-1}, r_{-2}, \ldots, r_{-(L-2)}, 0, \ldots, 0, r_{L-2}, \ldots, r_1\right]^T,$$

The matrices X^{k_3,k_4} are obtained by means of the 'Forward Shift Rows Wrap aRound' and 'Forward Shift Columns Wrap aRound' formulas

$$X^{k_3,k_4+1} = \text{FSRWR}(X^{k_3,k_4}), \qquad X^{k_3+1,k_4} = \text{FSCWR}(X^{k_3,k_4}),$$

where, for $A = (\vec{c}_1\ \vec{c}_2\ \cdots\ \vec{c}_{n-1}\ \vec{c}_n)$,

$$\text{FSRWR}(A) = [\vec{c}_n\ \vec{c}_1\ \cdots\ \vec{c}_{n-1}],$$

and for

$$A = \begin{bmatrix} \vec{r}_1 \\ \vec{r}_2 \\ \vdots \\ \vec{r}_n \end{bmatrix}, \qquad \text{FSCWR}(A) = \begin{bmatrix} \vec{r}_n \\ \vec{r}_1 \\ \vdots \\ \vec{r}_{n-1} \end{bmatrix},$$

We note that only the top-left matrix, $X^{1,1}$, is calculated and stored as the other blocks, $X^{i,j}$, can be dynamically generated.

To solve for r_l we write the system for r_l as a matrix-vector equation

$$A\vec{r} = \vec{r}, \qquad \vec{r} = \left[r_1, r_2, \ldots, r_{L-2}\right]^T,$$

where we use $r_{-l} = -r_l$ and $r_0 = 0$. Then $\vec{r}$ is the normalized eigenvector of A corresponding to the eigenvalue $\lambda = 1$. The normalization is

$$\sum_{l=-(L-2)}^{L-2} l r_l = -1.$$

6. The Operator $T = g(\partial_x, \partial_y)$

To construct the NS-forms of analytic functions of the differential operator (∂_x, ∂_y) one can introduce two approaches for computing the NS-forms of operator functions [5]:

- Compute the function of the projection of the operator on $\boldsymbol{V}_n$ by

$$T_n = g(P_n\partial_x P_n, P_n\partial_y P_n). \tag{8}$$

- Compute the projection of the operator function on $\boldsymbol{V}_n$ by

$$T_n = P_n g(\partial_x, \partial_y) P_n. \tag{9}$$

The difference between these two approaches depends on how well the absolute value square of the Fourier transform of the scaling function, $|\widehat{\varphi}(\xi)\widehat{\varphi}(\eta)|^2$, acts as a cutoff function and the choice may depend on the applications in hand. In this paper, representation (8) is used for the numerical application. In Subsection 6.2, representation (9) will give rise to a formal pseudodifferential formulation.

6.1. Function of Projected Operator

The matrix T_g^n of $g(\partial_x, \partial_y)$ on $\boldsymbol{V}_n$ is obtained by applying the function g to the matrices T_x^n and T_y^n representing ∂_x and ∂_y, respectively.

We use this representation to calculate T_{k_1,k_2}^{n,k_3,k_4}. With $T_n = g(T_x^n, T_y^n)$, we rearrange the matrices T_x^n and T_y^n as the $2^{2n} \times 2^{2n}$ matrices X and Y, so that

$$T_x^n \odot s^n \sim X\,\vec{s}, \qquad T_y^n \odot s^n \sim Y\,\vec{s}.$$

where $\vec{s}$ is a 2^{2n}-long column vector rearrangement of the matrix s^n. Because of the structure of T_x^n and T_y^n, X and Y are block circulant with circulant blocks (bccb). Therefore,

$$X = (F_{2^n}^* \otimes F_{2^n}^*)\Lambda_x(F_{2^n} \otimes F_{2^n}), \quad Y = (F_{2^n}^* \otimes F_{2^n}^*)\Lambda_y(F_{2^n} \otimes F_{2^n}),$$

where F_{2^n} and $F_{2^n}^*$ are the Fourier matrix and its adjoint and Λ_x, Λ_y are $2^{2n} \times 2^{2n}$ diagonal matrices containing the eigenvalues of X and Y. Then the matrix T_g^n (in bccb form) is

$$T_g^n = g(X,Y) = (F_{2^n}^* \otimes F_{2^n}^*)g(\Lambda_x, \Lambda_y)(F_{2^n} \otimes F_{2^n}).$$

A long manipulation gives the explicit formula for the top left block $T_g^{n,1,1}$ of T_g^n in terms of the top left blocks $T_x^{n,1,1}$ and $T_y^{n,1,1}$ of T_x^n and T_y^n:

$$T_g^{n,1,1} = \frac{1}{2^n} F_{2^n}\, g(2^n F_{2^n}^* T_x^{n,1,1} F_{2^n}^*, 2^n F_{2^n}^* T_y^{n,1,1} F_{2^n}^*)\, F_{2^n}.$$

By construction, $T_g^{n,1,1}$ is real. If $g(\partial_x, \partial_y) = \partial_x$, i.e., $T = \partial_x$, then

$$T^{n,1,1} = \frac{1}{2^n} F_{2^n} \left(2^n F_{2^n}^* T_x^{n,1,1} F_{2^n}^* \right) F_{2^n} = I_n\, T_x^{n,1,1}\, I_n = T_x^{n,1,1}.$$

Similarly, if $g(\partial_x, \partial_y) = \partial_y$, i.e., $T = \partial_y$, then $T^{n,1,1} = T_y^{n,1,1}$.

6.2. A Formal Pseudo-Differential Operator Approach to $T = g(\partial_x, \partial_y)$
Projecting $g(\partial_x, \partial_y)$ on $\boldsymbol{V}_n$, we have

$$T_n = P_n g(\partial_x, \partial_y) P_n.$$

Thus

$$T^{n,k_3,k_4}_{k_1,k_2} = \int\int \Phi^n_{k_3,k_4} g(\partial_x, \partial_y) \Phi^n_{k_1,k_2}(x,y)\,dx\,dy.$$

Using the Fourier transform, we have the formal pseudo-differential representation of the operator $g(\partial_x, \partial_y)$:

$$g(\partial_x, \partial_y)\Phi^n_{k_1,k_2} = \frac{1}{2\pi}\int\int 2^n g(-i2^n\xi, -i2^n\eta)\widehat{\varphi}(\xi)\widehat{\varphi}(\eta) \times e^{-i2^n x\xi}\, e^{-i2^n y\eta}\, e^{ik_1\xi} e^{ik_2\eta}\, d\xi\, d\eta.$$

Now, using $\Phi^n_{k_3,k_4}(x,y) = 2^n\varphi(2^n x - k_3)\varphi(2^n y - k_4)$, and after making the change of variables

$$x \to 2^{-n}(x + k_3), \qquad y \to 2^{-n}(y + k_4),$$

we obtain

$$T^{n,k_3,k_4}_{k_1,k_2} = \int\int g(-i2^n\xi, -i2^n\eta)|\widehat{\varphi}(\xi)|^2|\widehat{\varphi}(\eta)|^2\, e^{-i(k_3-k_1)\xi}\, e^{-i(k_4-k_2)\eta}\, d\xi\, d\eta.$$

Therefore,

$$T^{n,k_3,k_4}_{k_1,k_2} = T^n_{l_1,l_2}, \qquad l_1 = k_3 - k_1, \quad l_2 = k_4 - k_2.$$

To evaluate this expression, we write

$$T^{n,k_3,k_3}_{k_1,k_2} = T^n_{l_1,l_2} = \int_0^{2\pi}\int_0^{2\pi} h(\xi,\eta)\, e^{-il_1\xi}\, e^{-il_2\eta}\, d\xi\, d\eta,$$

where

$$h(\xi,\eta) = \sum_l \sum_m g(-i2^n(\xi + 2\pi l), -i2^n(\eta + 2\pi m))|\widehat{\varphi}(\xi + 2\pi l)|^2|\widehat{\varphi}(\eta + 2\pi m)|^2.$$

Now, for a given accuracy ε, $|\widehat{\varphi}(\xi)|^2$ acts as a cutoff function in the Fourier domain, i.e., $|\widehat{\varphi}(\xi)|^2 < \varepsilon$ for $|\xi| > \xi_{\text{cutoff}}$. Then we approximate $h(\xi,\eta)$ by the following sum:

$$\begin{aligned} h(\xi,\eta) &\approx \tilde{h}(\xi,\eta) \\ &= \sum_{l=-L}^{L}\sum_{m=-M}^{M}\left[g(-i2^n(\xi + 2\pi l), -i2^n(\eta + 2\pi m))|\widehat{\varphi}(\xi + 2\pi l)|^2|\widehat{\varphi}(\eta + 2\pi m)|^2\right], \end{aligned}$$

from which $T^n_{l_1,l_2}$ is finally approximated by the integral

$$T^n_{l_1,l_2} \approx \int_0^{2\pi}\int_0^{2\pi} \tilde{h}(\xi,\eta) e^{-il_1\xi} e^{-il_2\eta}\, d\xi\, d\eta.$$

This integral can be further approximated using the two-dimensional FFT.

It is noted in [5] that this approach can be used for functions $g(u)$ that have rapidly converging Taylor series expansions. In this case, for a given accuracy, the remainder term of sufficiently high order will be small. If f is a polynomial of degree p, the remainder term of order larger than p will be zero. For detail on the use of the dynamic range of the variable u, the reader is referred to [5]. An interesting case would be to consider smooth wavelets or smooth frame wavelets with compact support in the Fourier domain and which are also well localized in the (x, y) domain [1, 2].

More theoretical questions, like admissible classes of symbols, are left for future research.

7. Application to PDEs

We consider the initial-value problem:

$$u_t = \mathcal{L}u + \mathcal{N}f(u), \quad u(t_0, x, y) = u_0(x, y), \quad x, y \in [0, 1]^2,$$

with the periodic boundary conditions:

$$u(t, 0, y) = u(t, 1, y), \qquad u(t, x, 0) = u(t, x, 1).$$

Here $\mathcal{L}$ and $\mathcal{N}$ are constant-coefficient differential operators (independent of t) and f is typically nonlinear. For example, for $u_t = u_{xx} + uu_y$ we have $\mathcal{L} = \partial_x^2$, $\mathcal{N} = \partial_y$ and $f(u) = \frac{1}{2}u^2$.

The PDE is transformed into the integral equation

$$u(t) = e^{(t-t_0)\mathcal{L}}u(t_0) + \int_{t_0}^{t} e^{(t-\tau)\mathcal{L}}\mathcal{N}f(u)\,d\tau, \quad \text{where} \quad e^{t\mathcal{L}} = \sum_{n=0}^{\infty}\frac{(t\mathcal{L})^n}{n!}.$$

For example, for $u_t = u_{xx} + uu_y$, we obtain

$$u(t) = e^{(t-t_0)\partial_x^2}u(t_0) + \int_{t_0}^{t} e^{(t-\tau)\partial_x^2}u(\tau)u_y(\tau)\,d\tau.$$

A quadrature reduces the integral to an implicit difference equation.

In the case $\mathcal{N}f(u) = u(\tau)v(\tau)$ with time step size $\Delta t = h$, $t_1 = t_0 + h$, we have the $O(h^2)$ approximation $\hat{I}$,

$$\int_{t_0}^{t_1} e^{(t_1-\tau)\mathcal{L}}u(\tau)v(\tau)\,d\tau = (e^{h\mathcal{L}} - I)\mathcal{L}^{-1}\left(\sum_{i,j=0}^{1} c_{ij}u_iv_j\right) + O(h^2)$$

$$:= \hat{I}(t_1) + O(h^2),$$

where $u_i = u(t_i)$, $v_i = v(t_i)$ and c_{ij} are coefficients to be determined by comparing $\hat{I}$ with, say, the approximation

$$\bar{I}(t_1) = \sum_{i,j=0}^{1} f_{i,j}u_iv_j, \quad \text{where} \quad f_{i,j} = \int_{t_0}^{t_1} e^{(t_1-\tau)\mathcal{L}}L_i(\tau)L_j(\tau)\,d\tau.$$

Here $L_0(\tau) = 1 - \tau$ and $L_1(\tau) = \tau$ are the Lagrange linear interpolation basis at 0 and 1. Finally, the coefficients c_{ij} are determined such that $|\hat{I} - \bar{I}| = O(h^3)$.

The application **algorithm** is as follows.

- Discretize the PDE to arrive at an implicit difference equation with operator coefficients

$$u(t+\Delta t) = E(u(t)) + N(u(t), u(t+\Delta t)).$$

- Construct the matrices of the various operators appearing in the difference equation.
- Project u_0 onto the finest space $\boldsymbol{V}_n$ to get the coefficients s^n.
- Project s^n down on $\boldsymbol{V}_j$ to get $d^{j,\lambda}$, and s^j, for $j = n-1, n-2, \ldots, n-J$.
- Solve the implicit difference equation by applying the operator to get the solution at the next time step given by $\tilde{d}^{j,\lambda}$, $\tilde{s}^j, n-J \le j \le n-1$.
- Use the pseudo-inverse FWT to reconstruct s^n.

This algorithm is applied to the fifth-order **thin film equation**:

$$u_t = -4\varepsilon u u_x + \mathcal{L}u,$$

where the operator $\mathcal{L}$ is

$$\mathcal{L} = -\partial_x^2 + B\partial_y^2 - C\partial_x^3 - D\partial_x\partial_y^2 - E(\partial_x^2+\partial_y^2)^2 - F\partial_x^5 - G\partial_x^3\partial_y^2 - H\partial_x\partial_y^4$$

with coefficients

$$A = \mu\left(\frac{8}{15}R - \frac{2}{3}\cot\alpha\right), \qquad B = -\frac{2}{3}\mu\cot\alpha,$$

$$C = \mu^2\left(2 + \frac{32}{63}R^2 - \frac{40}{63}R\cot\alpha\right), \qquad D = \mu^2\left(2 - \frac{40}{63}R\cot\alpha\right),$$

$$E = \frac{2}{3}\mu^3 W, \qquad F = \frac{40}{63}\mu^4 RW\csc\alpha,$$

$$G = \frac{401}{315}\mu^4 RW\csc\alpha, \qquad H = \frac{40}{63}\mu^4 RW\csc\alpha.$$

Here R is a measure of viscosity (small $R \leftrightarrow$ large viscosity), W is the surface tension, α is the plane inclination, and μ is the ratio of film thickness to streamwise lengthscale.

We have experimented with several values of the parameters of the problem with the radially symmetric initial disturbance:

$$u_0(x,y) = 0.1\, e^{-60\cos^2(\pi x) - 60\cos^2(\pi y)}$$

shown in the top part of Fig. 1.

In the interesting case: $R = 10$, $W = 100$ and $\alpha = \pi/4$, we observed the formation of **horseshoe-type** waves downstream at $t = 2.5$ shown in the bottom part of Fig. 1 and at $t = 3.5$ and $t = 5.5$ shown in Fig. 2.

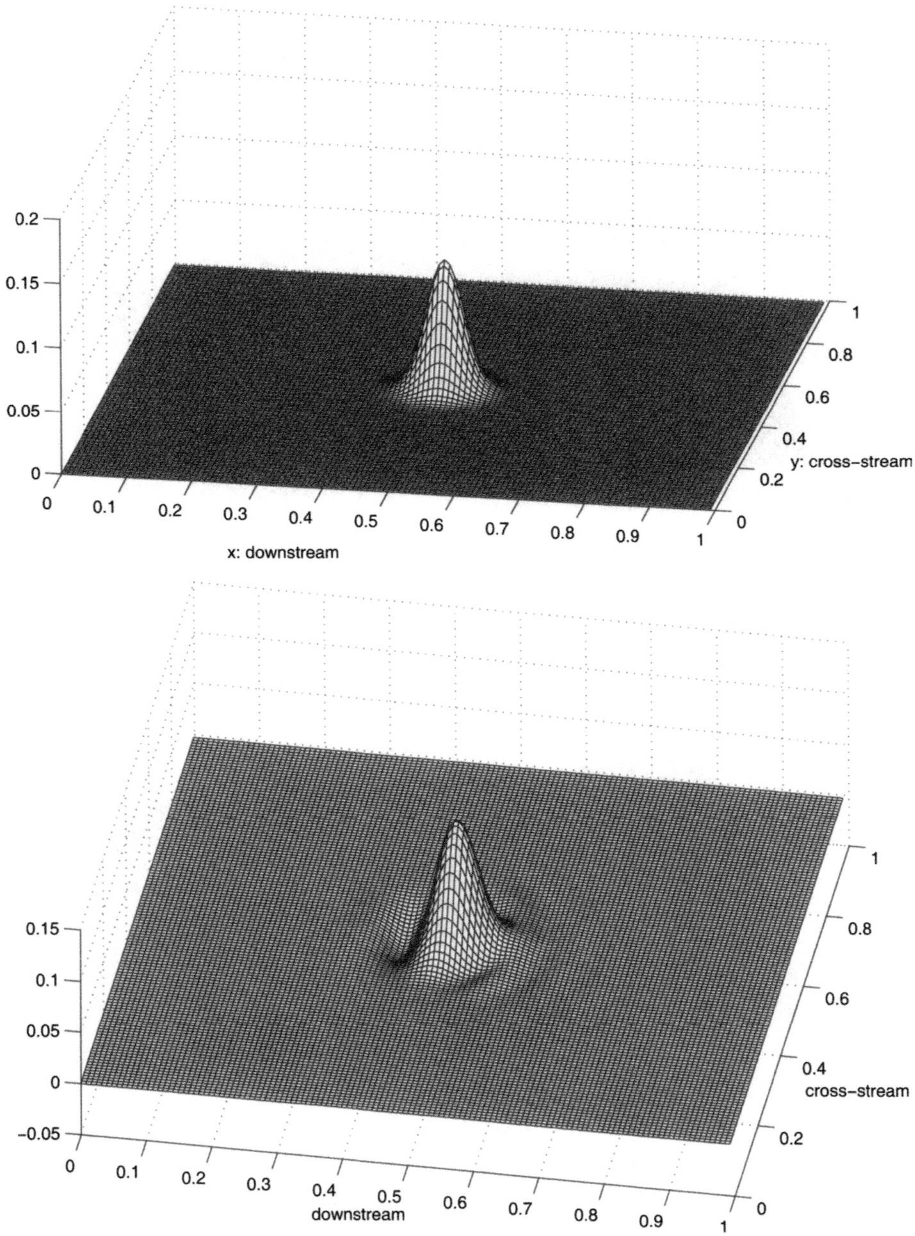

FIGURE 1. Initial disturbance (top) and disturbance at $t = 2.5$ (bottom).

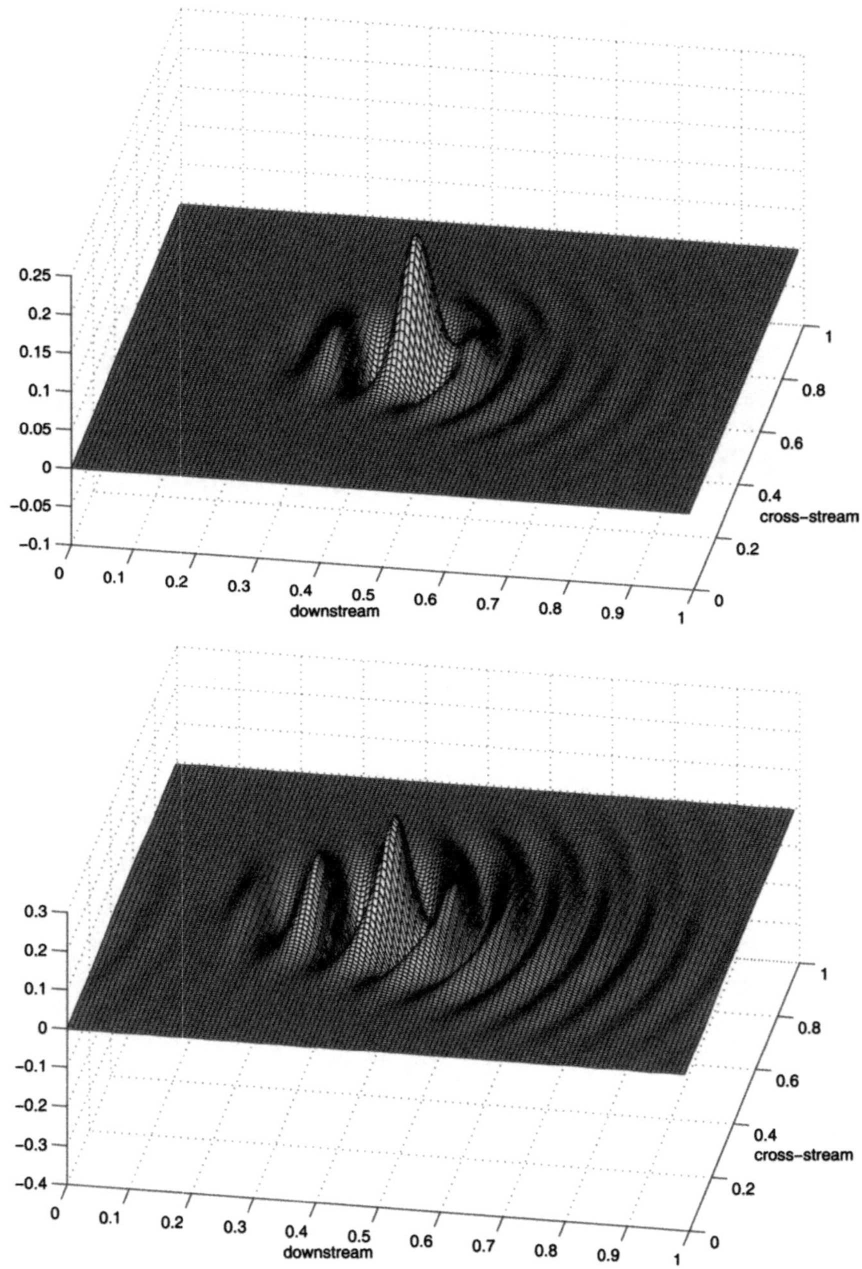

FIGURE 2. Disturbance at $t = 3.5$ (top) and at $t = 5.5$ (bottom).

8. Conclusion and Future Work

In this paper, partial differential operators were represented in non-standard form in periodic two-dimensional separable orthonormal wavelet bases. Applications were to the thin film equation on an inclined plane. Future work can be pursued in representing partial differential operators and pseudo-differential operators in nonseparable and nonperiodic wavelet bases.

Acknowledgement. The authors thank the referee for many suggestions and corrections which greatly improved this work.

References

[1] R. Ashino, S.J. Desjardins, C. Heil, M. Nagase and R. Vaillancourt, Smooth tight frame wavelets and image microanalysis in the Fourier domain, *Computers Math. Applic.* **45** (2003), 1551–1579.

[2] R. Ashino, S.J. Desjardins, C. Heil, M. Nagase and R. Vaillancourt, Pseudo-differential operators, microlocal analysis and image restoration, in *Advances in Pseudo-Differential Operators*, Editors: R. Ashino, P. Boggiatto and M.W. Wong, Birkhäuser, Basel, 2004, 187–202.

[3] G. Beylkin, On the representation of operators in bases of compactly supported wavelets, *SIAM J. Numer. Anal.* **29** (1992), 1716–1740.

[4] G. Beylkin, R. Coifman and V. Rokhlin, Fast wavelet transforms and numerical algorithms, *Comm. Pure Appl. Math.* **44** (1991), 141–183.

[5] G. Beylkin and J.M. Keiser, On the adaptive numerical solution of nonlinear partial differential equations in wavelet bases, *J. Computational Phys.* **132** (1997), 233–259.

[6] I. Daubechies, Orthonormal bases of compactly supported wavelets, *Comm. Pure Appl. Math.* **41** (1988), 909–996.

[7] I. Daubechies, *Ten Lectures on Wavelets*, Society for Industrial and Applied Mathematics, Philadelphia PA, 1992.

[8] M.A. Hajji, *Representation of Differential Operators in 2-Dimensional Periodic Wavelet Basis and Applications to PDEs*, Ph.D. Thesis, Carleton University, 2003.

[9] M.A. Hajji, S. Melkonian and R. Vaillancourt, Representation of differential operators in wavelet basis, *Computers Math. Applic.*, to appear.

[10] V.I. Petviashvili and O.Yu. Tsvelodub, Horseshoe-shaped solitons on a flowing viscous film of fluid, *Dokl. Akad. Nauk SSSR* **138** (1978), 1321–1323.

M.A. Hajji and Rémi Vaillancourt
Department of Mathematics and Statistics, University of Ottawa, 585 King Edward Ave., Ottawa ON Canada KiN 6N5
e-mail: `hemmi@science.uottawa.ca`
`remi@uottawa.ca`

S. Melkonian
School of Mathematics and Statistics, Carleton University, 1125 Colonel By Drive, Ottawa ON Canada K1S 5B6
e-mail: `melkonia@math.carleton.ca`

Operator Theory: Advances and Applications

Your Specialized Publisher in Mathematics

Birkhäuser

For orders originating from all over the world except USA/Canada/Latin America:

Birkhäuser Verlag AG
c/o Springer GmbH & Co
Haberstrasse 7
D-69126 Heidelberg
Fax: +49 / 6221 / 345 4 229
e-mail: birkhauser@springer.de
http://www.birkhauser.ch

For orders originating in the USA/Canada/Latin America:

Birkhäuser
333 Meadowland Parkway
USA-Secaucus
NJ 07094-2491
Fax: +1 201 348 4505
e-mail: orders@birkhauser.com

Edited by
Gohberg, I., School of Mathematical Sciences, Tel Aviv University, Ramat Aviv, Israel

This series is devoted to the publication of current research in operator theory, with particular emphasis on applications to classical analysis and the theory of integral equations, as well as to numerical analysis, mathematical physics and mathematical methods in electrical engineering.

OT 153: Gaspar, D. / Gohberg, I. / Timotin, D. / Vasilescu, F.H. / Zsido, L. Recent Advances in Operator Theory, Operator Algebras, and their Applications. Proceedings of the XIXth International Conference on Operator Theory, Timisoara (Rumania), 2002 (2004).
ISBN 3-7643-7127-7

OT 152: Eidelman, S.D. / Ivasyshen, S.D. / Kochubei, A.N. Analytic Methods in the Theory of Differential and Pseudo-differential Equations of Parabolic Type (2004)
ISBN 3-7643-7115-3

OT 151: Gil, J.B. / Krainer, T. / Witt, I. (Eds.) Aspects of Boundary Problems in Analysis and Geometry (2004)
ISBN 3-7643-7069-6

OT 150: Rabinovich, V. / Roch, S. / Silbermann, B. Limit Operators and their Applications in Operator Theory (2004)
ISBN 3-7643-7081-5

OT 149: Ball, J.A. / Helton, J.W. / Klaus, M. / Rodman, L. (Eds.) Current Trends in Operator Theory and its Applications (2004)
ISBN 3-7643-7067-X

OT 148: Ashyralyev, A. / Sobolevskii, P.E. New Difference Schemes for Partial Differential Equations (2004)
ISBN 3-7643-7054-8

OT 147: Gohberg, I. / dos Santos, A.F. / Speck, F.-O. / Teixeira, F.S. / Wendland, W. (Eds.) Operator Theoretical Methods and Applications to Mathematical Physics. The Erhard Meister Memorial Volume (2004)
ISBN 3-7643-6634-6

OT 146: Kopachevsky, N.D. / Krein, S.G. Operator Approach to Linear Problems of Hydrodynamics. Volume 2: Nonself-adjoint Problems for Viscous Fluids (2003)
ISBN 3-7643-2190-3

OT 145: Albeverio, S. / Demuth, M. / Schrohe, E. / Schulze, B.-W. (Eds.) Nonlinear Hyperbolic Equations, Spectral Theory, and Wavelet Transformations (2003)
ISBN 3-7643-2168-7

OT 144: Belitskii, G. / Tkachenko, V. One-dimensional Functional Equations (2003)
ISBN 3-7643-0084-1

OT 143: Alpay, D. (Ed.) Reproducing Kernel Spaces and Applications (2003)
ISBN 3-7643-0068-X

OT 142: Böttcher, A. / dos Santos, A.F. / Kaashoek, M.A. / Brites Lebre, A. / Speck, F.-O. (Eds.) Singular Integral Operators, Factorization and Applications. International Workshop on Operator Theory and Applications, IWOTA 2000 – Portugal (2003)
ISBN 3-7643-6947-7

GPSR Compliance

The European Union's (EU) General Product Safety Regulation (GPSR) is a set of rules that requires consumer products to be safe and our obligations to ensure this.

If you have any concerns about our products, you can contact us on ProductSafety@springernature.com

In case Publisher is established outside the EU, the EU authorized representative is:

Springer Nature Customer Service Center GmbH
Europaplatz 3
69115 Heidelberg, Germany

Batch number: 10577884

Printed by Printforce, the Netherlands